ELEMENTS OF ALGEBRA

PRENTICE-HALL INTERNATIONAL, INC., *London*
PRENTICE-HALL OF AUSTRALIA, PTY. LTD., *Sydney*
PRENTICE-HALL OF CANADA, LTD., *Toronto*
PRENTICE-HALL OF INDIA PRIVATE LTD., *New Delhi*
PRENTICE-HALL OF JAPAN, INC., *Tokyo*

ELEMENTS OF ALGEBRA

FRANCIS J. MUELLER

PRENTICE-HALL, INC.,
Englewood Cliffs, N. J.

to BUDDY

Current printing (last digit):

10 9 8 7 6 5 4 3 2 1

13–262410–9

Sept 15 78

Library of Congress Catalog Card Number: 79–85665
Printed in the United States of America

PREFACE

This book is intended for the college student whose grasp of algebra is currently weak or marginal. Even those students who come into college without any previous exposure to the subject should find entry into *Elements of Algebra* an easy matter.

From my experience with students at this level of achievement, I have learned the importance of a patient and deliberate pace, of taking little or nothing for granted, of offering an abundance of concrete and illustrative examples, of synthesizing in clear and unequivocal terms "what to do," and of supporting the whole fabric of instruction with an abundance of exercise opportunities and extensive end-of-chapter reviews. All of these are reflected in the presentation of *Elements of Algebra*.

Because students in this category have such diverse backgrounds and abilities, a flexible textbook is almost mandatory. To provide that flexibility, the contents of the book have been organized into ten basic chapters—which contain the staples of an introductory algebra course—and six supplementary units. The latter may be used to extend and amplify the fundamentals of the ten chapters. These units have been written in such a way as to be independent of one another, and may be studied in any order or combination, or even before the final chapter is reached. For example, some instructors may wish to have their students start with the unit on sets (Supplementary Unit C).

Another prominent feature of the book is its statement of operational procedures in essentially linear-program form. These compact statements are not intended to be memorized blindly; rather they are intended to

v

serve as models of the ultimate systematic way in which most persons competent in the subject carry out the basic operations of algebraic computation.

Students for whom this book is appropriate often find difficulty with verbal problems. Consequently, a full chapter has been devoted to that topic. The treatment is extensive and painstaking, the strategy being to keep the student involved long enough for him to develop a useful degree of competence. Problem solving, like much of the mechanics of mathematics, is a skill that can be acquired through practice.

Consistent attention has been given to keeping the wording of the text clear and simple. The style is direct, almost conversational. That and the heavy emphasis upon programmed procedures and illustrative examples should, for many, make the book practically self-teaching.

Substantial portions of *Elements of Algebra* have been drawn from my earlier *Intermediate Algebra*. The repeated acceptance of the latter since its publication in 1960 attests to the successfulness of the approach described above for great numbers of students. In addition the present text in preparation had the benefit of several incisive critiques by highly competent university and community college mathematicians. To them, and to my helpful editors, I am deeply grateful for the excellence of their contributions.

Francis J. Mueller

CONTENTS

SUPPLEMENTARY UNITS

I

FROM ARITHMETIC
TO ALGEBRA

Arithmetic and algebra have common historic roots. Algebra, in many respects, is an extension of arithmetic and, in the past, has been referred to as "generalized arithmetic." But by modern standards, such a characterization is not completely accurate. Algebra has become much more than generalized arithmetic, even though both disciplines continue to share the same operations, the same laws of order, grouping, and distribution, and often the same symbolism. Among the more obvious differences are algebra's extensive use of variables (letters), equations, inequalities, and systems of numbers beyond those of ordinary arithmetic. Even so, one effective approach to the study of algebra is to begin with a discussion of some of the features which arithmetic and algebra have in common. This we do in the present chapter.

1. A COMMON HERITAGE

Mathematics is essentially abstract and much too complicated for a person to carry completely in his mind. Therefore great dependence is placed upon symbolism. In ordinary language it is usually sufficiently clear from the context to tell whether one is referring to an object or its symbol. In mathematics this has not always been so. One important instance is the confusion between a number and its various symbols, or **numerals**. For example, a common characteristic of each of the several collections of objects pictured in Fig. 1.1 is

2. NUMBER AND NUMERAL

"fiveness," which we normally symbolize with 5. (The ancient Romans used V, and the modern Japanese use 五, for example.) Thus the *number* of objects pictured in each case is five, but the *s*-like mark, 5, which we usually put on paper to symbolize that particular characteristic is a *numeral*.

Figure 1.1

Note that the number itself is completely unaffected by the way we choose to symbolize it. The number of stars in the Big Dipper, for example, is fixed at seven, though that number may be expressed by an endless variety of symbols or numerals: 7, VII, 14/2, $9 - 2$, $2 \times 3\frac{1}{2}$, $\dfrac{10 + 11}{3}$, to mention a few.

When it comes to computation—and this holds for the computations of arithmetic, algebra, and higher mathematics beyond—some of the mystery that often plagues the learner will disappear if he remembers that the marks he uses in paper-and-pencil computations are symbols, and not the numbers themselves. Essentially, computational procedures are but proven ways to process numerals; computational shortcuts are liberties we take with numerals, not numbers. In any basic mathematics course, as is the one we are undertaking here, much attention is devoted to such procedures—their details, their implications, and their applications.

3. BASIC PROPERTIES

When we have a collection or **set** of numbers, and a procedure for assigning another number to pairs of numbers in the set, we have what is called a **binary operation**. One of the more familiar binary operations of arithmetic is addition; and perhaps the most basic set of numbers in arithmetic are the **natural numbers**, i.e., zero† and the counting numbers: 0, 1, 2, 3, 4, 5, ... Under the binary operation of addition, 5, for example, is the natural number assigned to the pair of natural numbers 2 and 3. The number

---

†Traditionally, 0 has not been considered one of the natural numbers, but many present-day mathematicians now include it.

5 is also assigned under the binary operation of addition to the pair 1 and 4. Similarly, 15 is assigned under addition to the pair 6 and 9, as well as to the pairs 7 and 8, 10 and 5, 12 and 3, and so on.

Another basic binary operation in arithmetic is multiplication. Under multiplication, 6 is assigned to the pair 2 and 3; 4 to the pair 1 and 4, as well as to 2 and 2; 48 is assigned to the pair 6 and 8 and to other pairs such as 12 and 4, and 3 and 16.

Computational procedures, called **algorithms**, are used to determine which number is assigned to specific pairs of numbers under a given operation. You learned those algorithms in elementary-school arithmetic. In the study of algebra, we move beyond those particular number facts to identify general and fundamentally important properties that numbers exhibit under certain operations. Among them are the following:

I. Closure Property

If the number assigned to any pair of numbers of a set under a binary operation is also a member of the set, then the numbers of the set are said to be *closed* under that operation.

We may illustrate that the set of natural numbers is closed under addition by choosing any pair of natural numbers and noting that their sum is invariably a *unique* (i.e., one and only one) natural number, e.g., $6 + 5 = 11$, $0 + 12 = 12$, $9 + 735 = 744$. Similarly, for multiplication: $8 \times 5 = 40$, $37 \times 0 = 0$, $7 \times 105 = 735$, etc. By way of contrast, note that the set of *odd* natural numbers, 1, 3, 5, 7, 9, . . . , is not closed under addition. That is, the sum of any pair of odd numbers is an even number, e.g., $3 + 5 = 8$, and even numbers are not in the set of odd numbers. On the other hand, the set of odd natural numbers is closed under multiplication, e.g., $3 \times 5 = 15$, and 3, 5, and 15 are all members of the set of odd numbers. In general, the product of two odd numbers is always an odd number.

II. Commutative Property

If the number assigned to any pair of numbers of a set under a binary operation is the same number, regardless of the order of the pair, then the numbers of the set are said to be *commutative* under that operation.

Thus if a and b represent numbers of a set for which the commutative property applies under the binary operation of addition, then $a + b = b + a$. Similarly under the binary operation of multiplication, $a \times b = b \times a$. As an illustration of the commutative property, note that 4 and 6 have the same sum, 10, whether we add 4 to 6, or 6 to 4; and the pair 4 and 6 has the same product, 24, whether we multiply 6 by 4, or 4 by 6. (However, the same cannot be said for the binary operation of subtraction: $6 - 4 \neq 4 - 6$; nor for the binary operation of division: $6 \div 4 \neq 4 \div 6$.)

III. Associative Property

If a, b, c represent any three numbers of a set, and if $(a + b) + c = a + (b + c)$, then the numbers of the set are said to be *associative* under addition; if $(a \times b) \times c = a \times (b \times c)$, then the numbers of the set are said to be *associative* under multiplication.

The associative property is particularly important because it allows us to expand the operations of addition and multiplication from pairs of numbers to more than pairs of numbers, and thus to obtain the same final result no matter how the numbers are grouped as the operation proceeds. Associativity is clearly evident among the natural numbers, for example:

$$6 + 2 + 5 = (6 + 2) + 5 = 8 + 5 = 13$$
$$= 6 + (2 + 5) = 6 + 7 = 13$$
$$4 \times 3 \times 5 = (4 \times 3) \times 5 = 12 \times 5 = 60$$
$$= 4 \times (3 \times 5) = 4 \times 15 = 60$$

IV. Distributive Property

If a, b, c represent any three numbers of a set, and if $a \times (b + c) = (a \times b) + (a \times c)$, and $(b + c) \times a = (b \times a) + (c \times a)$, then the numbers of the set are said to be distributive for multiplication over addition.

The distributive property ties together the two basic binary operations of addition and multiplication and is of considerable consequence in mathematics. That the natural numbers are distributive for multiplication over addition can be readily illustrated:

$$6 \times (3 + 2) = (6 \times 3) + (6 \times 2)$$
$$6 \times \quad 5 \quad = \quad 18 \quad + \quad 12$$
$$30 \quad = \quad 30$$

EXERCISE 1-1

1. Is the set of even natural numbers closed under the binary operations of addition and multiplication?

2. Is the set of natural numbers that have remainders of 1 when divided by 3 closed under the operations of addition and multiplication?

Each of the following pairs of number expressions reflects one of the basic properties described in Section 3. Verify by performing the computations; carry out those within parentheses first.

Commutative—addition:

3. 36,854 + 42,797; 42,797 + 36,854

4. 63,795 + 42,783 + 19,036; 19,036 + 42,783 + 63,795

5. 4632 + 5748 + 615; 615 + 4632 + 5748

Associative—addition:

6. (36,325 + 47,956) + 37,284; 36,325 + (47,956 + 37,284)

7. (463,295 + 527,634 + 876,251) + 476,387; (463,295 + 527,634) + (876,251 + 476,387)

8. (4632 + 8743) + 8702; 4632 + (8743 + 8702)

Commutative—multiplication:

9. 27 × 368; 368 × 27

10. 507 × 40,632; 40,632 × 507

11. 832 × 607; 607 × 832

Associative—multiplication:

12. 47 × (68 × 327); (47 × 68) × 327

13. 395 × (463 × 407); (395 × 463) × 407

14. (51 × 362) × 3; 51 × (362 × 3)

Distributive:

15. 97 × (632 + 517); (97 × 632) + (97 × 517)

16. 285 × (8763 + 402); (285 × 8763) + (285 × 402)

17. 35 × (463 + 27); (35 × 463) + (35 × 27)

Identify which of the basic properties is reflected in each of the following:

18. 8 + 7 = 7 + 8

19. 5 × 4 = 4 × 5

20. (4 + 2) + 8 = 4 + (2 + 8)

21. (3 × 6) × 5 = 3 × (6 × 5)

22. (6 + 3 + 2 + 5) + 1 = (6 + 3) + (2 + 5 + 1)

23. 9 × (1 + 17) = (9 × 1) + (9 × 17)

24. (63 × 42 × 31) × 0 = (63 × 42) × (31 × 0)

25. (18 + 6) × 4 = (18 × 4) + (6 × 4)

26. * + $ = $ + *

27. (# × ?) × * = # × (? × *)

28. # × & = & × #

29. ($ + ?) + % = $ + (? + %)

30. $ × (# + *) = ($ × #) + ($ × *)

4. NONNEGATIVE RATIONALS

Within the set of natural numbers there is closure under the binary operations of addition and multiplication. That is, the sum and product of any pair of natural numbers are invariably natural numbers. However, the set of natural numbers is not closed under the binary operation of division. Even though some pairs of natural numbers have a natural number for a quotient (e.g., $8 ÷ 4 = 2$), one exception is all that is needed to establish the *lack* of closure. In this case, there are many such exceptions, one of which is $6 ÷ 5$; there is no natural number that is the quotient of $6 ÷ 5$. To say this in another way, there is no natural number that will multiply with the divisor, 5, to yield the dividend, 6, as a product—which is the fundamental test for a valid quotient.

To find a suitable quotient for every pair of natural numbers (except when the divisor is 0), we need access to a larger set of numbers, the **nonnegative rationals**. This set of numbers consists of all the previous natural numbers as well as the quotients of all the pairs of natural numbers, except when the divisor is zero. Examples of members of this set are such familiar numbers as $\frac{1}{2}$, the quotient of $1 ÷ 2$; $\frac{3}{4}$, the quotient of $3 ÷ 4$; $\frac{11}{3}$, the quotient of $11 ÷ 3$.

Zero divisors are excluded because the resulting quotient would either be contradictory or indeterminant. An example of the first instance might be $7 ÷ 0$. If we represent the quotient as q, then $7 ÷ 0 = q$ implies that $q × 0 = 7$. But every number, when multiplied with 0, has 0 for the product; so 7 as the product for $q × 0$ is impossible. A similar contradiction occurs for every other nonzero dividend and zero divisor.

When both the dividend and the divisor are zero, we have the case of $0 ÷ 0$. This time let us call the would-be quotient k. If $0 ÷ 0 = k$, then $k × 0 = 0$, which is true for every number in the set that k might represent: $7 × 0 = 0$, $12 × 0 = 0$, $\frac{1}{2} × 0 = 0$, $179 × 0 = 0$, etc. This implies that $0 ÷ 0 = 7, 0 ÷ 0 = 12, 0 ÷ 0 = \frac{1}{2}, 0 ÷ 0 = 179$, etc. The lack of a unique quotient when the divisor and dividend are zero makes $0 ÷ 0$ indeterminant. For these reasons, then, division by zero is excluded.

Returning to the complete set of nonnegative rationals, we may make the following two generalizations:

1. Each number in the set may be expressed as a fraction in which the numerator and denominator are expressions for natural numbers.

2. Under the binary operations of addition and multiplication, the basic

properties of closure, commutativity, associativity, and distributivity, discussed in the previous section, apply as well to the set of nonnegative rationals.

Illustrations are incorporated in Exercise 1-2.

EXERCISE 1-2

Each of the following pairs of number expressions reflects one of the basic properties of Section 3. Verify by performing the computations; carry out those within parentheses first.

Commutative—addition:

1. $\frac{5}{12} + \frac{3}{8}$; $\frac{3}{8} + \frac{5}{12}$

2. $\frac{7}{10} + \frac{4}{3}$; $\frac{4}{3} + \frac{7}{10}$

3. $\frac{7}{5} + \frac{9}{7} + \frac{3}{8}$; $\frac{3}{8} + \frac{9}{7} + \frac{7}{5}$

4. $\frac{8}{13} + \frac{7}{6} + \frac{5}{3}$; $\frac{7}{6} + \frac{5}{3} + \frac{8}{13}$

5. $\frac{3}{7} + \frac{2}{9} + \frac{1}{21} + \frac{2}{3}$; $\frac{2}{3} + \frac{1}{21} + \frac{2}{9} + \frac{3}{7}$

6. $\frac{4}{9} + \frac{5}{8} + 3 + \frac{9}{24}$; $3 + \frac{5}{8} + \frac{9}{24} + \frac{4}{9}$

Associative—addition:

7. $\left(\frac{3}{5} + \frac{2}{7}\right) + \frac{3}{11}$; $\frac{3}{5} + \left(\frac{2}{7} + \frac{3}{11}\right)$

8. $\left(\frac{9}{8} + \frac{3}{25}\right) + \frac{7}{10}$; $\frac{9}{8} + \left(\frac{3}{25} + \frac{7}{10}\right)$

9. $\left(\frac{9}{35} + \frac{3}{28} + \frac{7}{45}\right) + \frac{5}{36}$; $\left(\frac{9}{35} + \frac{3}{28}\right) + \left(\frac{7}{45} + \frac{5}{36}\right)$

10. $\left(\frac{8}{15} + \frac{14}{25}\right) + \left(\frac{7}{5} + \frac{8}{9}\right)$; $\left(\frac{8}{15} + \frac{14}{25} + \frac{7}{5}\right) + \frac{8}{9}$

11. $\left(\frac{9}{35} + \frac{3}{28} + \frac{7}{45}\right) + \left(\frac{5}{36} + \frac{1}{5}\right)$; $\left(\frac{9}{35} + \frac{3}{28}\right) + \left(\frac{7}{45} + \frac{5}{36} + \frac{1}{5}\right)$

12. $\left(\frac{8}{15} + \frac{7}{5} + \frac{8}{9}\right) + \left(\frac{14}{25} + \frac{9}{35} + \frac{10}{7}\right)$; $\left(\frac{8}{15} + \frac{7}{5}\right) + \left(\frac{8}{9} + \frac{14}{25}\right) + \left(\frac{9}{35} + \frac{10}{7}\right)$

Commutative—multiplication:

13. $\frac{3}{8} \times \frac{4}{17}$; $\frac{4}{17} \times \frac{3}{8}$

14. $\frac{9}{8} \times \frac{2}{3}$; $\frac{2}{3} \times \frac{9}{8}$

15. $\frac{12}{35} \times \frac{3}{4} \times \frac{7}{10}$; $\frac{7}{10} \times \frac{3}{4} \times \frac{12}{35}$

16. $\frac{5}{7} \times \frac{3}{2} \times \frac{13}{14}$; $\frac{3}{2} \times \frac{5}{7} \times \frac{13}{14}$

17. $\frac{13}{4} \times \frac{12}{2} \times \frac{19}{13} \times \frac{97}{19}$; $\frac{97}{19} \times \frac{19}{13} \times \frac{12}{2} \times \frac{13}{4}$

18. $\frac{17}{3} \times \frac{20}{7} \times \frac{4}{15} \times \frac{5}{2}$; $\frac{4}{15} \times \frac{17}{3} \times \frac{5}{2} \times \frac{20}{7}$

Associative—multiplication:

19. $\left(\frac{3}{2} \times \frac{5}{7}\right) \times \frac{11}{8}$; $\frac{3}{2} \times \left(\frac{5}{7} \times \frac{11}{8}\right)$

20. $\frac{8}{3} \times \left(\frac{12}{5} \times \frac{7}{5}\right)$; $\left(\frac{8}{3} \times \frac{12}{5}\right) \times \frac{7}{5}$

21. $\left(\frac{16}{5} \times \frac{9}{2} \times \frac{41}{12}\right) \times \frac{3}{8}$; $\frac{16}{5} \times \left(\frac{9}{2} \times \frac{41}{12} \times \frac{3}{8}\right)$

22. $\left(\frac{8}{25} \times \frac{2}{9}\right) \times \left(\frac{3}{16} \times \frac{7}{3}\right)$; $\left(\frac{8}{25} \times \frac{2}{9} \times \frac{3}{16}\right) \times \frac{7}{3}$

23. $\left(\frac{3}{4} \times \frac{6}{7} \times \frac{21}{12}\right) \times \left(\frac{42}{8} \times \frac{3}{10}\right)$; $\left(\frac{3}{4} \times \frac{6}{7}\right) \times \left(\frac{21}{12} \times \frac{42}{8} \times \frac{3}{10}\right)$

24. $(\frac{3}{8} \times \frac{9}{10} \times \frac{12}{6}) \times (\frac{7}{3} \times \frac{5}{21} \times \frac{3}{7})$; $(\frac{3}{8} \times \frac{9}{10}) \times (\frac{12}{6} \times \frac{7}{3}) \times (\frac{5}{21} \times \frac{3}{7})$

Distributive:

25. $\frac{2}{11} \times (\frac{3}{5} + \frac{2}{15})$; $(\frac{2}{11} \times \frac{3}{5}) + (\frac{2}{11} \times \frac{2}{15})$

26. $\frac{2}{3} \times (\frac{3}{7} + \frac{1}{5})$; $(\frac{2}{3} \times \frac{3}{7}) + (\frac{2}{3} \times \frac{1}{5})$

27. $(\frac{7}{8} + \frac{2}{3}) \times \frac{3}{7}$; $(\frac{7}{8} \times \frac{3}{7}) + (\frac{2}{3} \times \frac{3}{7})$

28. $\frac{10}{7} \times (\frac{3}{4} + \frac{2}{5} + \frac{7}{8})$; $(\frac{10}{7} \times \frac{3}{4}) + (\frac{10}{7} \times \frac{2}{5}) + (\frac{10}{7} \times \frac{7}{8})$

29. $\frac{8}{9} \times (\frac{7}{8} + \frac{3}{4} + \frac{5}{6} + \frac{11}{12})$; $(\frac{8}{9} \times \frac{7}{8}) + (\frac{8}{9} \times \frac{3}{4}) + (\frac{8}{9} \times \frac{5}{6}) + (\frac{8}{9} \times \frac{11}{12})$

30. $(\frac{3}{4} \times \frac{2}{3}) \times (\frac{4}{7} + \frac{3}{4})$; $(\frac{3}{4} \times \frac{2}{3} \times \frac{4}{7}) + (\frac{3}{4} \times \frac{2}{3} \times \frac{3}{4})$

5. NEGATIVE NUMBERS

Within the set of nonnegative rational numbers, there is closure under the binary operations of addition, multiplication, and division, except division by zero. However, the set is not closed under the binary operation of subtraction. (The test of a valid difference in a subtraction is that the sum of the difference and the subtrahend should equal the minuend. For example, if $8 - 5 = 3$, then $3 + 5 = 8$.) Lack of closure for the nonnegative rationals under subtraction can be demonstrated by citing just one example, say $7 - 10$: there is no nonnegative rational number which, when added to 10, yields a sum of 7.

To express a difference for every pair of numbers in the set of nonnegative rationals, access to a larger set of numbers is needed. This would be the set of **rationals**, which consists of all the nonnegative rationals *plus* an "opposite" number for each of the nonnegative rationals. We define each "opposite" number (technically, *additive inverse*) as being related to its partner nonnegative rational in such a way that the sum of the two numbers is 0. Thus, for example:

$$8 \text{ (nonnegative rational)} + \text{"opposite 8"} = 0$$

$$37 \text{ (nonnegative rational)} + \text{"opposite 37"} = 0$$

$$\tfrac{1}{2} \text{ (nonnegative rational)} + \text{"opposite } \tfrac{1}{2}\text{"} = 0$$

The numbers of the previous set—the nonnegative rationals—are referred to among the rationals as **positive numbers**, while their "opposites" are referred to as **negative numbers**. The one exception is the nonnegative rational number 0, which is considered to be neither positive nor negative. Numerals for negative numbers are generally expressed with a minus sign, e.g., -7, -3, $-\frac{1}{4}$, whereas numerals for positive numbers are expressed either with a plus sign, e.g., $+7$, $+3$, $+\frac{1}{4}$, or with no sign, e.g., 7, 3, $\frac{1}{4}$.

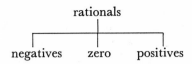

A part of a set of numbers is called a **subset**. For example, the set of numbers, 3, 4, and 5, is a subset of the set of natural numbers, 0, 1, 2, 3, 4, 5, 6, 7, ... The set of positive numbers is also a subset of the rationals. (More on sets and subsets may be found in Supplementary Unit C, pages 289–301.)

In addition to subdividing the set of rational numbers into the nonoverlapping subsets of the positives, negatives, and zero, it is also possible to separate the rationals into two other nonoverlapping subsets: the integers and the nonintegers. The subset of *integers* consists of all of the natural numbers and their respective additive inverses: . . . , −4, −3, −2, −1, 0, +1, +2, +3, . . . ; the remaining numbers in the set of rationals make up the nonintegers.

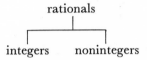

The two generalizations stated in Section 4 with respect to the nonnegative rationals (i.e., zero and the positive rationals) have counterparts among the full set of rationals:

1. Every rational number may be expressed as a fraction in which the numerator and denominator are expressions for integers.
2. Under the binary operations of addition and multiplication, the basic properties of closure, commutativity, associativity, and distributivity, as discussed in Section 3, apply as well to the set of rational numbers. (Examples, however, are best deferred until after the basic computational procedures for the rationals have been established later in the chapter.)

A very useful model for illustrating the set of rational numbers is a **number line**, partially represented in Fig. 1.2. A point of central position, called the **origin**, is identified on the line, and the number 0 is associated with it. By convention, the positive integers, in increasing sequence to the right, and the negative integers, in decreasing sequence to the left, are associated with points spaced at equal intervals in both directions from the

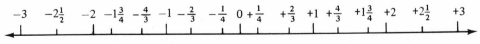

Figure 1.2

origin. Rational numbers that are not integers are associated with points appropriately spaced between those associated with the integers.

The fact that the rational numbers can be ordered in this way is another important property of the set. Each rational number, when compared to any other rational number, is either greater than or less than that number. In terms of the number line, it can be said that each number is *greater than* (symbolically, $>$) every number whose location is to its left, and *less than* (symbolically, $<$) every number whose location is to its right. As can be verified from Fig. 1.2:

$$+3 > +1, \quad +1 < +2, \quad -2\tfrac{1}{2} < 0, \quad 2 > -3, \quad -\tfrac{1}{2} > -1$$

6. NEGATIVE OF A NUMBER

There is an important difference between a negative number and the negative *of* a number. Each rational number except zero has a distinct "opposite" number in the set which, in terms of the number line, is located at an equal distance from 0, but in the opposite direction. For example:

The "opposite" of $+2$ is -2.

The "opposite" of -3 is $+3$.

The "opposite" of $-\tfrac{1}{4}$ is $+\tfrac{1}{4}$.

The "opposite" number for a given number is said to be the **negative of the given number**. Thus, in the foregoing examples, the *negative of* $+2$ is -2 (and the negative of -2 is $+2$); the *negative of* -3 is $+3$; the *negative of* $-\tfrac{1}{4}$ is $+\tfrac{1}{4}$. Note that the negative *of* a number is not necessarily a negative number. The negative of a positive number is always a negative number, and the negative of a negative number is always a positive number. The negative of 0 may be thought to be 0, a unique situation.

EXERCISE 1-3

Compute answers only to those exercises that can be expressed by a nonnegative rational number (i.e., positives and zero).

1. 6374 − 3785	**2.** 14,632 − 9758
3. 46,235 − 46,523	**4.** 830,006 − 794,309
5. 632.56 − 478.3	**6.** 300.24 − 6.037
7. 0.0327 − 0.0063	**8.** 0.000426 − 0.00126
9. 0.000147 − 0.0000389	**10.** 3.0042 − 0.6379

11. $\frac{3}{8} - \frac{1}{4}$

12. $\frac{3}{5} - \frac{1}{4}$

13. $\frac{13}{80} - \frac{1}{16}$

14. $\frac{13}{18} - \frac{20}{27}$

15. $\frac{3}{11} - \frac{4}{19}$

16. $\frac{19}{35} - \frac{17}{42}$

17. $\frac{11}{28} - \frac{5}{12}$

18. $3\frac{3}{4} - 1\frac{5}{8}$

19. $17\frac{3}{4} - 12\frac{1}{3}$

20. $400 - 168\frac{3}{7}$

21. $300\frac{1}{5} - 299\frac{7}{5}$

22. $\frac{863}{7} - \frac{1109}{9}$

Write the negative *of each of the following:*

23. 8

24. -6

25. -5

26. 4

27. $\frac{2}{3}$

28. $-\frac{3}{2}$

29. 1.8

30. $\frac{7}{5}$

31. -2.6

32. 10.03

33. $-4\frac{2}{5}$

34. 0.16

Locate the following on a number line.

35. $-\frac{2}{3}, -0.4, +\frac{5}{6}, -\frac{7}{8}, +1.3, +\frac{1}{2}$

36. $-0.9, -\frac{5}{6}, +\frac{3}{4}, -\frac{9}{10}, +3.4, -2\frac{1}{2}$

7. OPERATIONS ON RATIONAL NUMBERS: ADDITION

Because the rational numbers include both the numbers of arithmetic and the negatives, computational procedures for the rationals vary somewhat from those learned in arithmetic. These new procedures can be expressed more uniformly by means of the concept of absolute value. The **absolute value of a number** may be thought of as the "distance" or "displacement," without regard to direction, that a number lies from 0 on the number line. Note in Fig. 1.2 that the numbers -3 and $+3$ are both located at a distance of 3 units from 0. Using the standard symbolism for absolute value, $|\ \ |$, we can state:

$$|+3| = |-3| = 3$$

which is read, "the absolute value of $+3$ equals the absolute value of -3 equals 3."

Similarly:

$$|-1\tfrac{1}{2}| = |+1\tfrac{1}{2}| = 1\tfrac{1}{2}$$

$$|+0.06| = |-0.06| = 0.06$$

Thus, a number and its "opposite" or negative have the same absolute value. Because it is always nonnegative, the absolute value of a number can be readily established if we ignore the sign of its numeral and consider the numeral as one of the unsigned numerals of elementary arithmetic. Computations involving absolute values are identical to those performed on the nonnegative numbers of arithmetic.

[*Note*: In this book, statements of basic procedure are given in linear flow or step-by-step form. In keeping with the language of modern technology, these statements are referred to as *programs*.]

1.1

To add two or more rational numbers that agree in sign (i.e., all terms positive or all negative):

Step 1. Compute the sum of the absolute values of the rational numbers.

Step 2. Prefix the numeral for the sum of Step 1 with the common sign of the rational numbers.

EXAMPLES

1. Find the sum of -2 and -4 according to Program **1.1**.

Step 1.　　The absolute value of -2 is 2.
　　　　　　The absolute value of -4 is 4.
　　　　　　Sum of the absolute values: $2 + 4 = 6$.

Step 2.　　Common sign of the rational numbers: $-$ (minus).
　　　　　　Sum of -2 and -4 is -6.

2. Add $+3$ and $+4$.

Step 1.　　$|+3| + |+4| = 3 + 4 = 7$

Step 2.　　$(+3) + (+4) = +7$

3. Add: $(-5) + (-2) + (-4)$.

Step 1.　　$|-5| + |-2| + |-4| = 5 + 2 + 4 = 11$

Step 2.　　$(-5) + (-2) + (-4) = -11$

4. Add: $(+6) + (+2) + (+5.7)$.

Step 1.　　$|+6| + |+2| + |+5.7| = 6 + 2 + 5.7 = 13.7$

Step 2.　　$(+6) + (+2) + (+5.7) = +13.7$

5. Add: $(-\frac{2}{3}) + (-\frac{1}{2}) + (-\frac{1}{4})$.

Step 1.　　$|-\frac{2}{3}| + |-\frac{1}{2}| + |-\frac{1}{4}| = \frac{2}{3} + \frac{1}{2} + \frac{1}{4} = \frac{8}{12} + \frac{6}{12} + \frac{3}{12} = \frac{17}{12} = 1\frac{5}{12}$

Step 2.　　$(-\frac{2}{3}) + (-\frac{1}{2}) + (-\frac{1}{4}) = -1\frac{5}{12}$

6. Add: $(-4.2) + (-3.1) + (-0.6)$.

Step 1.　　$4.2 + 3.1 + 0.6 = 7.9$

Step 2.　　$(-4.2) + (-3.1) + (-0.6) = -7.9$

When a positive and a negative number are to be added (e.g., $+8$ and -5), the computation could be carried out in this way:

$$(+8) \qquad + (-5) = ?$$
$$\overline{(+3) + (+5)} + (-5) = ?$$
$$(+3) + \underbrace{\qquad\quad} \qquad\qquad \text{(associative—addition)}$$
$$(+3) + \qquad (0) \qquad = +3$$

Similarly, for the sum of $+9$ and -14:

$$(+9) + \underbrace{(-14)} = ?$$
$$\underbrace{(+9) + (-9)} + (-5) = ?$$
$$(0) + (-5) = -5$$

An equivalent and more direct procedure is stated in the following program.

1.2

To add two rational numbers that differ in sign:

Step 1. Subtract the smaller absolute value of the two rational numbers from the larger.

Step 2. Prefix the numeral for the difference of Step 1 with the sign of the rational number having the larger absolute value.

EXAMPLES

1. Find the sum of $+2$ and -6 according to Program **1.2**.

Step 1.   The absolute value of $+2$ is 2.
The absolute value of -6 is 6, and $6 > 2$.
Difference of absolute values: $6 - 2 = 4$.

Step 2.   The sign of the rational number having the greater absolute value is $-$; so $(+2) + (-6) = -4$.

2. Add: $(+6) + (-5)$.

Step 1. $|+6| = 6$ (greater absolute value)
$|-5| = 5$
$6 - 5 = 1$

Step 2. $(+6) + (-5) = +1$

3. Add: $(-\frac{3}{5}) + (\frac{2}{3})$.

Step 1. $|-\frac{3}{5}| = \frac{3}{5} = \frac{9}{15}$
$|+\frac{2}{3}| = \frac{2}{3} = \frac{10}{15}$ (greater absolute value)
$\frac{10}{15} - \frac{9}{15} = \frac{1}{15}$

Step 2. $(-\frac{3}{5}) + (+\frac{2}{3}) = +\frac{1}{15}$

4. Add: $(+0.362) + (-1.427)$.

Step 1. $|+0.362| = 0.362$
$|-1.427| = 1.427$ (greater absolute value)
$1.427 - 0.362 = 1.065$

Step 2. $(+0.362) + (-1.427) = -1.065$

When there are more than two addends that do not agree in sign, a different procedure is required. By the commutative property, we may reorder the addends; by the associative property, we may add all the positive numbers together and all the negative numbers together; and then by Program **1.2**, we may add the resulting positive and negative sums. To illustrate this combination:

$$(+4) + (-3) + (-2) + (+5) + (-6)$$

$$(+4) + (+5) + (-3) + (-2) + (-6) \quad \text{(commutative—addition)}$$

$$\underbrace{}_{(+9)} \quad + \quad \underbrace{}_{(-11)} \quad \text{(associative—addition)}$$

$$\underbrace{}_{(-2)} \quad \text{(Program } \mathbf{1.2}\text{)}$$

1.3

To add three or more rational numbers that differ in sign:

Step 1. Find the sum of the positive numbers.

Step 2. Find the sum of the negative numbers.

Step 3. Add the sums of Steps 1 and 2 using Program **1.2** for the sum of all the rational numbers.

EXAMPLES

1. Add: $(+3) + (-2) + (+5) + (+6) + (-14)$.

Step 1. $(+3) + (+5) + (+6) = +14$ (by **1.1**)

Step 2. $(-2) + (-14) = -16$ (by **1.1**)

Step 3. $(+14) + (-16) = -2$ (by **1.2**)

Thus, $(+3) + (-2) + (+5) + (+6) + (-14) = -2$.

2. Add: $(+\frac{1}{2}) + (-\frac{1}{3}) + (+4) + (-\frac{1}{6}) + (-1)$.

Step 1. $(+\frac{1}{2}) + (+4) = +4\frac{1}{2}$

Step 2. $(-\frac{1}{3}) + (-\frac{1}{6}) + (-1) = -1\frac{3}{6} = -1\frac{1}{2}$

Step 3. $(+4\frac{1}{2}) + (-1\frac{1}{2}) = +3$ (sum)

3. Add: $(+0.2) + (-0.3) + (-0.4) + (-1.3) + (+1.6)$.

Step 1	*Step 2*	*Step 3*
$+0.2$	-0.3	$+1.8$
$+1.6$	-0.4	-2.0
$(+)\ \overline{}$	-1.3	$(+)\ \overline{}$
$+1.8$	$(+)\ \overline{}$	-0.2 (sum)
	-2.0	

EXERCISE 1-4

Add:

1. $\begin{array}{r} +13 \\ +\ 6 \\ \hline \end{array}$ **2.** $\begin{array}{r} -86 \\ -13 \\ \hline \end{array}$ **3.** $\begin{array}{r} +362 \\ +168 \\ \hline \end{array}$

4. -86.3
-17.5

5. $+163.28$
$+7.437$

6. $+\frac{7}{8}$
$+\frac{2}{3}$

7. $-\frac{17}{3}$
$-\frac{14}{5}$

8. $+63$
-15

9. $+128$
-67

10. -384
$+126$

11. $+17.63$
-8.25

12. $+376$
-954

13. -1836
$+2748$

14. $+63.24$
-187.57

15. $+\frac{3}{4}$
$-\frac{1}{2}$

16. $+\frac{8}{5}$
$-\frac{3}{10}$

17. $+\frac{87}{8}$
$+\frac{42}{5}$

18. $-\frac{127}{81}$
$+\frac{63}{40}$

19. $-16\frac{3}{4}$
$+12\frac{1}{5}$

20. $+76\frac{1}{8}$
$+82\frac{3}{5}$

21. -42.63
$+1.782$

22. $+0.0362$
-4.071

23. -14
-6
-8

24. -42
$+36$
-51

25. -86
$+12$
-48
$+63$

26. -372
$+426$
$+837$
-792
-99

27. $+1.8$
-1.4
$+2.7$
-1.8

28. $+32.6$
-1.43
$+26.2$

29. $+1.23$
-12.3
$+123$

30. $-\frac{2}{3}$
$+\frac{3}{4}$
$-\frac{1}{6}$

31. $-\frac{3}{4}$
$-\frac{5}{8}$
$+\frac{2}{7}$

32. $-6\frac{2}{3}$
$+5\frac{7}{8}$
$-1\frac{3}{2}$
$+6\frac{1}{6}$

8. OPERATIONS ON RATIONAL NUMBERS: SUBTRACTION

The binary operation of subtraction is the inverse or reverse operation of addition. In a sense, addition and subtraction "undo" one another. For example, $4 + 3 = 7$, while $7 - 4 = 3$ and $7 - 3 = 4$. The concept of the negative of a number, explained in Section 6, is important in computing the difference for two rational numbers. (Recall that the negative of a number is its "opposite"; e.g., the negative of $+7$ is -7, and the negative of -3 is $+3$.) By definition, the difference, $a - b$, for two rational numbers, a and b, may be expressed as

$$a - b = a + (-b)$$

In other words, the result of subtracting rational number b (subtrahend) from rational number a (minuend) is the same as that of adding the negative of the subtrahend to the minuend. For instance:

$$8 - 3 = 8 + (-3) = 5 \qquad \text{(by Program 1.2)}$$
$$6 - 9 = 6 + (-9) = -3 \quad \text{(by Program 1.2)}$$

The operation of subtraction on rational numbers, then, may be expressed in a single-step program as follows.

1.4

To subtract one rational number from another add the negative of the subtrahend (the number to be subtracted) to the minuend. [*Note*: An alternative statement, at the numeral level: Reverse the sign of the subtrahend, and add.]

EXAMPLES

1. Subtract: $(+8) - (+5)$.

The negative of $+5$ is -5:
$$(+8) - (+5) = (+8) + (-5) = +3$$

2. Subtract: $(-3) - (-4)$.

The negative of -4 is $+4$:
$$(-3) - (-4) = (-3) + (+4) = +1$$

3. Subtract: $(-6) - (+3)$.

The negative of $+3$ is -3:
$$(-6) - (+3) = (-6) + (-3) = -9$$

4. Subtract: $(-\tfrac{3}{5}) - (+\tfrac{1}{5})$.
$$(-\tfrac{3}{5}) - (+\tfrac{1}{5}) = (-\tfrac{3}{5}) + (-\tfrac{1}{5}) = -\tfrac{4}{5}$$

5. Subtract: $(+1.63) - (-4.25)$.
$$(+1.63) - (-4.25) = (+1.63) + (+4.25) = +5.88$$

EXERCISE 1-5

Subtract; the lower term is the subtrahend.

1. $+14$ $+\ 8$	**2.** $+43$ $+28$	**3.** -18 $-\ 6$
4. -27 -13	**5.** $+1.8$ $+1.4$	**6.** $+81.7$ $+42.2$
7. -32.64 -83.46	**8.** -54.827 -29.378	**9.** $+632$ -817

10. $+7739$
$\quad-4832$

11. $-43,206$
$\quad+17,302$

12. $-83,914$
$\quad+92,145$

13. $+\frac{3}{5}$
$\quad-\frac{1}{5}$

14. $+\frac{2}{7}$
$\quad-\frac{5}{14}$

15. $-\frac{13}{21}$
$\quad-\frac{11}{18}$

16. $-\frac{27}{35}$
$\quad-\frac{48}{49}$

17. $-\frac{82}{9}$
$\quad-\frac{17}{18}$

18. $-\frac{31}{40}$
$\quad+\frac{99}{8}$

19. $+36\frac{2}{5}$
$\quad-16\frac{1}{3}$

20. $+15\frac{2}{3}$
$\quad-71\frac{5}{11}$

21. -12.03
$\quad-\ 1.675$

22. -12.38
$\quad-52.417$

23. $+20\frac{2}{7}$
$\quad-19\frac{9}{7}$

24. $-26\frac{5}{4}$
$\quad+15\frac{3}{2}$

25. -400.032
$\quad-\ 61.247$

26. $+72.8391$
$\quad+28.4973$

9. OPERATIONS ON RATIONAL NUMBERS: MULTIPLICATION

The product of two rational numbers, a and b, is defined to be $+\cdot(|a| \times |b|)$ if the factors a and b are both positive or both negative; the product is $-(|a| \times |b|)$ if one factor is positive and the other is negative. For example:

(i) $(+4) \times (+3) = +(|+4| \times |+3|) = +12$
(ii) $(+4) \times (-3) = -(|+4| \times |-3|) = -12$
(iii) $(-4) \times (+3) = -(|-4| \times |+3|) = -12$
(iv) $(-4) \times (-3) = +(|-4| \times |-3|) = +12$

The first two products, (i) and (ii), can be verified by repeated addition:

(i) $(+4) \times (+3) = (+3) + (+3) + (+3) + (+3) = +12$
(ii) $(+4) \times (-3) = (-3) + (-3) + (-3) + (-3) = -12$

The third product, (iii), can be verified in a similar way, after the commutative property is applied:

(iii) $(-4) \times (+3) = (+3) \times (-4) = (-4) + (-4) + (-4) = -12$

Verifying the fourth product,

(iv) $(-4) \times (-3) = +12$

is somewhat more involved. One way is to recognize that 0 is the product of every rational number and 0; so

$$-4 \times 0 = 0$$

Next, express the factor 0 as the sum, $[(+3) + (-3)]$:

$$(-4) \times [(+3) + (-3)] = 0$$

Then apply the distributive property:

$$(-4) \times [(+3) + (-3)] = \underbrace{[(-4) \times (+3)]} + \underbrace{[(-4) + (-3)]} = 0$$
$$\qquad\qquad\qquad -12 \qquad + \qquad ? \qquad = 0$$

Since $(-4) \times (+3)$ has been shown to be -12, and since the only number that will add to -12 to yield a sum of 0 is the additive inverse of -12, or $+12$, there is no choice but to declare $+12$ as the product of $(-4) \times (-3)$.

When there are more than two factors involved in a product of rational numbers, repeated use of the associative property under multiplication makes it possible to carry out the multiplication by pairs of factors. For example:

$$\underbrace{(-3) \times (+2)} \times (-5) \times (-4) \times (+7)$$
$$\underbrace{(-6)} \quad \times (-5) \times (-4) \times (+7)$$
$$\underbrace{(+30)} \qquad \times (-4) \times (+7)$$
$$\underbrace{(-120)} \qquad\quad \times (+7)$$
$$(-840)$$

Analysis of such examples affirms the generalization that, when the number of negative factors in a product is odd, the resulting product is a negative number; when the number of negative factors is even, the resulting product is a positive number. In the example just given, there are three negative factors, viz., -3, -5, -4, and the resulting product is a negative number.

1.5

To multiply two or more rational numbers:

Step 1. Find the product of the absolute values of the rational numbers.

Step 2. Prefix the numeral for the product of Step 1 with a plus sign if the number of negative factors in the product is even (i.e., none, two, four, etc.), or with a minus sign if the number of negative factors in the product is odd (i.e., one, three, five, etc.).

EXAMPLES

1. Multiply: $(-6) \times (-3) \times (+10)$.

Step 1. $|-6| \times |-3| \times |+10| = 6 \times 3 \times 10 = 180$

Step 2.   The number of negative factors (two) is even; therefore, the product is positive:

$$(-6) \times (-3) \times (+10) = +180$$

2. Multiply: $(-3) \times (-2) \times (-1) \times (-7) \times (-3)$.

Step 1. $3 \times 2 \times 1 \times 7 \times 3 = 126$

Step 2.   The number of negative factors is odd (five); therefore the product is negative:

$$(-3) \times (-2) \times (-1) \times (-7) \times (-3) = -126$$

3. Multiply: $(-\frac{1}{2}) \times (+\frac{2}{5}) \times (-\frac{2}{3}) \times (+\frac{5}{7})$.

Step 1. $\dfrac{1}{\cancel{2}} \times \dfrac{\cancel{2}}{\cancel{5}} \times \dfrac{2}{3} \times \dfrac{\cancel{5}}{7} = \dfrac{2}{21}$

Step 2.   The number of negative factors is even (two); so the product is positive:

$$(-\tfrac{1}{2}) \times (+\tfrac{2}{5}) \times (-\tfrac{2}{3}) \times (+\tfrac{5}{7}) = +\tfrac{2}{21}$$

4. Multiply: $(-1.2) \times (-3.4) \times (-0.5)$.

Step 1. $1.2 \times 3.4 \times 0.5 = 2.04$

Step 2.   The number of negative factors is odd· (three); so the product is negative:

$$(-1.2) \times (-3.4) \times (-0.5) = -2.04$$

EXERCISE 1-6

Compute the following products:

1. $(-2) \times (+3)$

2. $(+6) \times (-8)$

3. $(-3) \times (-7)$

4. $(+5) \times (-32)$

5. $(-\tfrac{2}{3}) \times (+6)$

6. $(-27) \times (-\tfrac{1}{3})$

7. $(-\tfrac{3}{5}) \times (-\tfrac{5}{8})$

8. $(-1.62) \times (+0.03)$

9. $(+0.0042) \times (+0.007)$

10. $(+6.027) \times (-0.03)$

11. $(-8) \times (-3) \times (+6)$

12. $(-2) \times (-5) \times (-7)$

13. $(+\tfrac{2}{3}) \times (-\tfrac{3}{5}) \times (+\tfrac{9}{11})$

14. $(+0.02) \times (-0.06) \times (-0.7)$

15. $(-2) \times (-1) \times (+3) \times (-4)$

16. $(-8.1) \times (-\tfrac{2}{3}) \times (+0.3)$

17. $(+\tfrac{2}{7}) \times (-\tfrac{1}{5}) \times (-\tfrac{7}{8}) \times (+\tfrac{3}{2})$

18. $(-0.003) \times (-0.02) \times (-0.1) \times (-\tfrac{1}{5})$

19. $(-2) \times (+1) \times (+3) \times (-6) \times (-4) \times (-1) \times (-3) \times (+2)$

20. $(-4) \times (-12) \times (+2) \times (-6) \times (-\tfrac{1}{2}) \times (-4) \times (-2) \times (+10)$

21. $(-6) \times (-2 + 5)$

22. $(-3) \times (-8 - \tfrac{1}{2})$

23. $(-7) \times (-3 + \tfrac{1}{4} - \tfrac{7}{8})$

24. $(-3.1) \times (-8.6 + 3.7 - 0.5)$

10. OPERATIONS ON RATIONAL NUMBERS: DIVISION

Division is the inverse operation of multiplication, that is, if $14 \div 7 = 2$, then $2 \times 7 = 14$. In general, if $a \div b = c$, then $c \times b = a$. Because of this relationship between multiplication and division, the distribution of signs in both operations is parallel. To illustrate:

$$(+8) \div (+4) = (+2) \quad \text{because} \quad (\mathbf{+2}) \times (+4) = (+8)$$
$$(+8) \div (-4) = (-2) \quad \text{because} \quad (\mathbf{-2}) \times (-4) = (+8)$$
$$(-8) \div (+4) = (-2) \quad \text{because} \quad (\mathbf{-2}) \times (+4) = (-8)$$
$$(-8) \div (-4) = (+2) \quad \text{because} \quad (\mathbf{+2}) \times (-4) = (-8)$$

In general, if a and b represent rational numbers, $b \neq 0$, then $a \div b = +(|a| \div |b|)$ when a and b are both positive or both negative, and $-(|a| \div |b|)$ when one of the two numbers is positive and the other is negative. When $a = 0$, then, of course, $a \div b = 0$.

1.6

To divide one nonzero rational number by another:

Step 1. Find the quotient of the absolute values of the two numbers.

Step 2. Prefix the numeral for the quotient of Step 1 with a plus sign if the two rational numbers are both positive or both negative, and with a minus sign if one of the rational numbers is positive and the other is negative.

EXAMPLES

1. Divide: $(-36) \div (-9)$.

Step 1. $|-36| \div |-9| = 36 \div 9 = 4$

Step 2. (-36) and (-9) are both negative; therefore, the quotient is positive:

$$(-36) \div (-9) = +4$$

2. Divide: $(-42) \div (+7)$.

Step 1. $|-42| \div |+7| = 42 \div 7 = 6$

Step 2. (-42) and $(+7)$ are negative and positive, respectively; therefore, the quotient is negative:

$$(-42) \div (+7) = -6$$

3. Divide: $(+\tfrac{3}{5}) \div (-\tfrac{3}{2})$.

Step 1. $|+\tfrac{3}{5}| \div |-\tfrac{3}{2}| = \tfrac{3}{5} \div \tfrac{3}{2} = \tfrac{3}{5} \times \tfrac{2}{3} = \tfrac{2}{5}$

Step 2. $(+\tfrac{3}{5})$ and $(-\tfrac{3}{2})$ differ in sign; so the quotient is negative:

$$(+\tfrac{3}{5}) \div (-\tfrac{3}{2}) = -\tfrac{2}{5}$$

4. Divide: $(-0.27) \div (-0.9)$.

Step 1. $0.27 \div 0.9 = 0.3$

Step 2.   Signs alike, quotient positive:

$$(-0.27) \div (-0.9) = +0.3$$

11. SIGNS FOR FRACTIONS

The numeral known as a **fraction** may be used to express the quotient of two numbers (i.e., $n \div d = n/d$). Consequently, according to Program **1.6**, the sign of the fraction itself will be positive when both n and d are of the same sign, and negative when n and d are of different signs, that is:

$$\frac{+n}{+d} = (+n) \div (+d) = +\frac{n}{d}$$

$$\frac{+n}{-d} = (+n) \div (-d) = -\frac{n}{d}$$

$$\frac{-n}{+d} = (-n) \div (+d) = -\frac{n}{d}$$

$$\frac{-n}{-d} = (-n) \div (-d) = +\frac{n}{d}$$

Thus it can be seen that

$$\frac{+n}{+d}, \quad \frac{-n}{-d}, \quad \text{and} \quad +\frac{n}{d}$$

are equivalent numerals for the same number, and that

$$\frac{+n}{-d}, \quad \frac{-n}{+d}, \quad \text{and} \quad -\frac{n}{d}$$

are also equivalent numerals for the same number. Moreover, the number symbolized by the first three fractions is the negative of that symbolized by the last three fractions, and conversely.

From this we may formulate a generalization that will be particularly helpful later on: If we think of any fraction as possessing three signs (the sign of the numerator, the sign of the denominator, and the sign of the fraction considered as a single numeral), then *any pair of these signs may be reversed without changing the number symbolized by the fraction.* Reversing one or all three signs, however, changes the number symbolized by the fraction to its "opposite" or negative.

EXERCISE 1-7

Compute the following quotients:

1. $(-4) \div (-2)$ **2.** $(-16) \div (+8)$

3. $(-42) \div (-6)$ **4.** $(-81) \div (+9)$

5. $(-32{,}256) \div (-48)$ **6.** $(+20{,}794) \div (-37)$

7. $(-4968) \div (-36)$

8. $(+6783) \div (+42)$

9. $(-\frac{3}{4}) \div (+\frac{2}{5})$

10. $(+\frac{3}{5}) \div (+\frac{4}{5})$

11. $(-\frac{7}{22}) \div (+\frac{2}{11})$

12. $(+8) \div (+\frac{3}{4})$

13. $(-1\frac{2}{3}) \div (-6\frac{2}{3})$

14. $(-4\frac{8}{19}) \div (+5\frac{1}{4})$

15. $(-51) \div (+3\frac{2}{5})$

16. $(-8.40) \div (+0.35)$

17. $(+2.627) \div (3.7)$

18. $(+29,664) \div (-0.36)$

19. $(-0.005832) \div (3.6)$

20. $(-0.52224) \div (64)$

Simplify:

21. $\dfrac{(+6) \times (-3)}{(-9)}$

22. $\dfrac{(-4) + (-3)}{(+14)}$

23. $\dfrac{(+5) - (+25)}{(-1) \times (-15)}$

24. $\dfrac{(-24) \div (-3)}{(-2)}$

25. $\dfrac{(+4) \div (+6)}{(-18)}$

26. $\dfrac{(-\frac{2}{3}) \div (+\frac{7}{6})}{(+\frac{3}{4}) \div (-\frac{3}{5})}$

Insert the proper sign in the empty parentheses to make the pairs of fractions equivalent.

27. $(-)\dfrac{-5}{+6} = (\ \ )\dfrac{+5}{+6}$

28. $(+)\dfrac{+2}{+3} = (\ \ )\dfrac{-2}{-3}$

29. $(+)\dfrac{-3}{-7} = (\ \ )\dfrac{-3}{+7}$

30. $(+)\dfrac{+15}{+10} = (\ \ )\dfrac{-3}{-2}$

31. $(-)\dfrac{+1}{-4} = (\ \ )\dfrac{+4}{-16}$

32. $(-)\dfrac{+9}{-16} = (\ \ )\dfrac{+36}{+64}$

12. SUMMARY

In this chapter we have discussed briefly the expansion of our collection of useful numbers, from the basic natural numbers to the negatives and to those which may be considered to express the ratio of two integers—referred to collectively as the rationals. Further expansions are in store. In Chapter VII, we shall introduce the *irrationals*, numbers which cannot be expressed as the ratio of two integers. Together the rationals and the irrationals make up a set of numbers known as the *real numbers*. In turn, the real numbers will be augmented by the *pure imaginary* numbers to form the larger set of *complex numbers*. In each instance, a new type of number is introduced to overcome a particular problem of closure. However, once complex numbers have been introduced, the last of the closure difficulties will have been resolved, and a number system for our type of algebra, the algebra of polynomials, will be complete.

REVIEW—CHAPTER I

Each of the following pairs of number expressions reflects one of the basic properties described in Section 3. Verify by performing the computations; carry out those within parentheses first.

Commutative—addition:

1. $47,235 + 98,674$; $98,674 + 47,235$

2. $74.683 + 193.24$; $193.24 + 74.683$

3. $0.00863 + 0.0324$; $0.0324 + 0.00863$

4. $\frac{3}{4} + \frac{5}{8} + \frac{7}{12}$; $\frac{7}{12} + \frac{5}{8} + \frac{3}{4}$

5. $\frac{2}{7} + \frac{5}{9} + \frac{2}{15} + \frac{1}{3}$; $\frac{1}{3} + \frac{5}{9} + \frac{2}{15} + \frac{2}{7}$

6. $4\frac{1}{4} + 3\frac{1}{2} + \frac{5}{6} + \frac{2}{9}$; $3\frac{1}{2} + \frac{5}{6} + \frac{2}{9} + 4\frac{1}{4}$

Associative—addition:

7. $(32,678 + 14,902 + 39,527) + 117,628$;
$(32,678 + 14,902) + (39,527 + 117,628)$

8. $(63.274 + 42.08 + 13) + 42.687$; $(63.274 + 42.08) + (13 + 42.687)$

9. $0.00362 + (0.4276 + 0.005)$; $(0.00362 + 0.4276) + 0.005$

10. $(\frac{7}{15} + \frac{2}{35} + \frac{9}{28}) + \frac{3}{10}$; $\frac{7}{15} + (\frac{2}{35} + \frac{9}{28} + \frac{3}{10})$

11. $(\frac{17}{30} + \frac{5}{24} + \frac{2}{9}) + (\frac{1}{6} + \frac{7}{15} + \frac{5}{12})$; $(\frac{17}{30} + \frac{5}{24}) + (\frac{2}{9} + \frac{1}{6}) + (\frac{7}{15} + \frac{5}{12})$

12. $(2\frac{1}{7} + 3\frac{1}{2}) + (\frac{5}{8} + \frac{3}{14})$; $2\frac{1}{7} + (3\frac{1}{2} + \frac{5}{8} + \frac{3}{14})$

Commutative—multiplication:

13. $52,326 \times 834$; $834 \times 52,326$

14. 62.84×1.036; 1.036×62.84

15. 0.00462×0.037; 0.037×0.00462

16. $\frac{3}{5} \times \frac{10}{13} \times \frac{26}{5}$; $\frac{26}{5} \times \frac{10}{13} \times \frac{3}{5}$

17. $\frac{5}{14} \times \frac{21}{25} \times \frac{6}{7}$; $\frac{6}{7} \times \frac{5}{14} \times \frac{21}{25}$

18. $3\frac{7}{8} \times 12\frac{3}{5} \times \frac{10}{11}$; $\frac{10}{11} \times 12\frac{3}{5} \times 3\frac{7}{8}$

Associative—multiplication:

19. $526 \times (324 \times 196)$; $(526 \times 324) \times 196$

20. $(42.6 \times 3.24) \times 1.61$; $42.6 \times (3.24 \times 1.61)$

21. $0.0463 \times (0.02 \times 0.056)$; $(0.0463 \times 0.02) \times 0.056$

22. $(\frac{3}{5} \times \frac{5}{22} \times \frac{11}{12}) \times \frac{2}{9}$; $\frac{3}{5} \times (\frac{5}{22} \times \frac{11}{12} \times \frac{2}{9})$

23. $(\frac{3}{4} \times \frac{9}{14} \times \frac{7}{11}) \times (\frac{3}{14} \times \frac{11}{18})$; $(\frac{3}{4} \times \frac{9}{14}) \times (\frac{7}{11} \times \frac{3}{14} \times \frac{11}{18})$

24. $(6\frac{2}{3} \times 5\frac{1}{4} \times \frac{5}{7}) \times 2\frac{3}{8}$; $(6\frac{2}{3} \times 5\frac{1}{4}) \times (\frac{5}{7} \times 2\frac{3}{8})$

Distributive:

25. $426 \times (783 + 268)$; $(426 \times 783) + (426 \times 268)$

26. $32.7 \times (4.68 + 7.9)$; $(32.7 \times 4.68) + (32.7 \times 7.9)$

27. $0.036 \times (0.0047 + 0.0003)$; $(0.036 \times 0.0047) + (0.036 \times 0.0003)$

28. $\frac{7}{9} \times (\frac{3}{4} + \frac{5}{7})$; $(\frac{7}{9} \times \frac{3}{4}) + (\frac{7}{9} \times \frac{5}{7})$

29. $\frac{6}{35} \times (\frac{7}{9} + \frac{5}{4} + \frac{3}{7})$; $(\frac{6}{35} \times \frac{7}{9}) + (\frac{6}{35} \times \frac{5}{4}) + (\frac{6}{35} \times \frac{3}{7})$

30. $3\frac{1}{8} \times (1\frac{2}{5} + 3\frac{1}{2} + 4\frac{2}{3})$; $(3\frac{1}{8} \times 1\frac{2}{5}) + (3\frac{1}{8} \times 3\frac{1}{2}) + (3\frac{1}{8} \times 4\frac{2}{3})$

Identify which of the basic properties is represented in each of the following:

31. $* \times p = p \times *$

32. $\# \times (\& + \%) = (\# \times \&) + (\# \times \%)$

33. $\& + ? = ? + \&$

34. $(f \times ?) \times \$ = f \times (? \times \$)$

35. $\& + (m + \#) = (\& + m) + \#$

Write the negative of:

36. 12   **37.** -7 **38.** $\frac{1}{4}$

39. -7.83 **40.** 0.035 **41.** $-3\frac{1}{2}$

42. Draw a number line and locate on it: $-\frac{3}{4}$, $+0.6$, 0, $-2\frac{1}{2}$, $+3.1$, $+\frac{3}{5}$, -0.25.

Add:

43. $\begin{array}{r} -468 \\ +397 \\ \hline \end{array}$ **44.** $\begin{array}{r} +36.4 \\ -72.7 \\ \hline \end{array}$ **45.** $\begin{array}{r} +4\frac{9}{16} \\ -3\frac{9}{12} \\ \hline \end{array}$

46. $\begin{array}{r} -18\frac{9}{10} \\ -13\frac{7}{8} \\ \hline \end{array}$ **47.** $\begin{array}{r} +273.41 \\ -\ \ 43.2 \\ \hline \end{array}$ **48.** $\begin{array}{r} +0.00294 \\ -1.473 \\ \hline \end{array}$

49. $\begin{array}{r} -38 \\ +47 \\ +19 \\ \hline \end{array}$ **50.** $\begin{array}{r} -192 \\ -237 \\ +444 \\ \hline \end{array}$ **51.** $\begin{array}{r} +492 \\ -381 \\ -117 \\ +162 \\ \hline \end{array}$

52. $\begin{array}{r} -4238 \\ +\ 742 \\ +3249 \\ -5327 \\ +5576 \\ \hline \end{array}$ **53.** $\begin{array}{r} -17.7 \\ -\ 8.3 \\ +20.3 \\ +\ 6.2 \\ \hline \end{array}$ **54.** $\begin{array}{r} -12.8 \\ +15.7 \\ -1.97 \\ \hline \end{array}$

55. $\begin{array}{r} -34.61 \\ +\ 3461 \\ -346.1 \\ \hline \end{array}$ **56.** $\begin{array}{r} +\ \frac{4}{5} \\ -1\frac{1}{16} \\ +\frac{7}{15} \\ \hline \end{array}$ **57.** $\begin{array}{r} -1\frac{5}{16} \\ -\frac{8}{11} \\ -\frac{5}{14} \\ \hline \end{array}$

58. $\begin{array}{r} -8\frac{3}{4} \\ +4\frac{9}{7} \\ +5\frac{5}{8} \\ -1\frac{3}{2} \\ \hline \end{array}$

Subtract the lower term from the upper:

59. $\begin{array}{r} -\ 6 \\ -18 \\ \hline \end{array}$ **60.** $\begin{array}{r} +1.7 \\ -1.5 \\ \hline \end{array}$ **61.** $\begin{array}{r} +83.7 \\ -64.2 \\ \hline \end{array}$

62. $\begin{array}{r} -37.672 \\ -54.791 \\ \hline \end{array}$

63. $\begin{array}{r} -53,627 \\ +42,874 \\ \hline \end{array}$

64. $\begin{array}{r} +\frac{5}{8} \\ -\frac{2}{5} \\ \hline \end{array}$

65. $\begin{array}{r} -\frac{8}{9} \\ -\frac{3}{4} \\ \hline \end{array}$

66. $\begin{array}{r} -\frac{34}{45} \\ +\frac{11}{12} \\ \hline \end{array}$

67. $\begin{array}{r} -47\frac{3}{8} \\ +24\frac{4}{5} \\ \hline \end{array}$

68. $\begin{array}{r} -16.37 \\ +2.482 \\ \hline \end{array}$

69. $\begin{array}{r} -14\frac{7}{5} \\ +18\frac{3}{2} \\ \hline \end{array}$

70. $\begin{array}{r} +500.006 \\ +\ \ 38.249 \\ \hline \end{array}$

Compute the products:

71. $(+10) \times (-13)$

72. $(-47) \times (-23)$

73. $(-\frac{2}{9}) \times (+81)$

74. $(-14) \times (-\frac{10}{7})$

75. $(-0.4) \times (+0.421)$

76. $(-4283) \times (-0.004)$

77. $(+98.76) \times (+0.02)$

78. $(-0.003) \times (+0.06) \times (-4.6)$

79. $(-\frac{1}{2}) \times (-7) \times (-8)$

80. $(+\frac{11}{9}) \times (-\frac{3}{2}) \times (-\frac{7}{4}) \times (+\frac{12}{7})$

81. $(-\frac{10}{7}) \times (-0.003) \times (-14) \times (-\frac{8}{15})$

82. $(-3.6) \times (-7.5 + 2.8 - 6.7 + 5.2)$

Compute the quotients:

83. $(-75) \div (-5)$

84. $(-4.5) \div (+12)$

85. $(+24.568) \div (-296)$

86. $(-389,906) \div (-439)$

87. $(-\frac{7}{6}) \div (+\frac{2}{5})$

88. $(-\frac{13}{9}) \div (+\frac{7}{3})$

89. $(-6\frac{1}{3}) \div (+10\frac{4}{9})$

90. $(-18.62) \div (+4.9)$

Simplify:

91. $\dfrac{(-56) - (+20)}{(-9) \times (-4)}$

92. $\dfrac{(+39) - (+44)}{(-5) \times (-5)}$

93. $\dfrac{(+40) \div (+\frac{2}{3})}{(-5)}$

94. $\dfrac{(-\frac{5}{8}) \div (+\frac{1}{6})}{(+\frac{14}{15}) \div (-\frac{9}{10})}$

Insert the proper sign in the empty parentheses to make the following pairs of fractions equivalent:

95. $(-)\dfrac{-28}{+40} = (\ \ )\dfrac{+28}{+40}$

96. $(+)\dfrac{+39}{+40} = (\ \ )\dfrac{-39}{-40}$

97. $(+)\dfrac{-11}{+77} = (\ \ )\dfrac{+1}{-7}$

98. $(\ \ )\dfrac{+49}{-21} = (\ \ )\dfrac{-7}{-3}$

II

ELEMENTARY OPERATIONS

1. TERMINOLOGY   A symbol that represents any of the numbers in some specified set of numbers is called a **variable**. Usually variables are expressed by letters of the alphabet. The specified set is the **domain of the variable**, often called the **replacement set**.

Among other things, variables make generalizations much easier to express. For example, if x (variable) represents any number in the set of rational numbers (domain of the variable), then it is true that:

$$\text{(a)} \quad 2 \times x = x + x \quad \text{and} \quad \text{(b)} \quad x - x = 0$$

In (a) we have expressed the generalization that twice any rational number is equal to the sum of that number and itself; (b) is the generalization that any rational number subtracted from itself is 0. Each generalization can be verified for specific numbers if we replace the variable, wherever it occurs, with a number from the domain of the variable. For instance, 7, $-\frac{1}{4}$, $2\frac{1}{2}$, and -3 are in the domain of the variable, and

$$2 \times 7 = 7 + 7 \qquad\qquad 2\frac{1}{2} - 2\frac{1}{2} = 0$$
$$2 \times (-\tfrac{1}{4}) = (-\tfrac{1}{4}) + (-\tfrac{1}{4}) \qquad (-3) - (-3) = 0$$

When a numeral, say 3, precedes a variable, say n, multiplication is implied: $3n = 3 \times n = 3 \cdot n$. Moreover, a succession of variables in algebra implies multiplication, e.g., $ab = a \times b$, and $2psm = 2 \times p \times s \times m$.

26

Algebraic expressions such as $3n$ or $5x$ are called **monomials**. Usually monomials consist of a numeral called the **numerical coefficient** and one or more variables:

$$\text{numerical coefficient} \overbrace{}^{23\,n} \overbrace{}_{6\,xy} \text{variable(s)}$$

When no numerical coefficient appears in a monomial, the numerical coefficient is understood to be 1:

$$s = 1s = 1 \times s$$
$$mn = 1mn = 1 \times m \times n$$

In the case of negative expressions, such as $-y$, the numerical coefficient is understood to be -1:

$$-y = -1y = -1 \times y$$

Technically, numerals alone, such as 6, -3, $+14$, are also classified as monomials. Monomials which differ by, at most, their numerical coefficients are called **like monomials**. For example, $4x$, $5x$, and x are like monomials because they agree in variable. On the other hand, $2x$ and $2y$ are not like monomials even though they agree in numerical coefficients.

Sums and differences of two monomials are called **binomials** (e.g., $m + n$, $3b + 2$, and $c - 5d$). Sums and differences of three monomials are called **trinomials** (e.g., $5x + 2y - 5$). Collectively, monomials, binomials, trinomials, and the sums and differences of four or more monomials are called **polynomials**.

2. SUMS AND DIFFERENCES OF MONOMIALS

The basic binary operations of algebra are the same as those of arithmetic: addition, subtraction, multiplication, and division. As in arithmetic, algebraic procedures for computing sums, differences, products, and quotients are normally stated and performed with symbols. The most elementary apply to monomials. The usual procedure for computing sums and differences of like monomials is based upon the distributive property. For example, if the like monomials are $8n$ and $5n$, their sum may be expressed as a binomial, $8n + 5n$, or more simply as a monomial, $13n$:

$$8n + 5n = \underbrace{(8 \times n) + (5 \times n) = (8 + 5) \times n}_{\text{distributive property}} = (13) \times (n) = 13n$$

Similarly for their difference:

$$8n - 5n = 8n + (-5n) = [(8) + (-5)] \times n = (3) \times (n) = 3n$$

We may program the procedure for computing sums and differences of like monomials as follows.

2.1

To add (subtract) like monomials:

Step 1. Add (subtract) the numerical coefficients of the monomials.

Step 2. Combine the sum (difference) of Step 1 and the common variable of the terms added (subtracted).

EXAMPLES

1. Add: $4x + x + 7x$.

Step 1.   Add the numerical coefficients: $4 + 1 + 7 = 12$.

Step 2.   Combine the common variable (x) and the sum of Step 1: $12x$ is the sum of $4x + x + 7x$.

2. Add: $3y + (-2y) + 4y + \frac{1}{2}y$.

Step 1.   Add: $3 - 2 + 4 + \frac{1}{2} = 5\frac{1}{2}$.

Step 2.   The common variable is y: $5\frac{1}{2}y$ is the desired sum.

3. Subtract: $(6y) - (2y)$.

Step 1.   Subtract the numerical coefficients: $6 - 2 = 4$.

Step 2.   The common variable is y: $4y$ is the desired difference.

4. Subtract: $(-3p) - (2p)$.

Step 1. $(-3) - (2) = (-3) + (-2) = -5$.

Step 2.   The common variable is p: $-5p$ is the desired difference.

5. Subtract: $(-6ab) - (-8ab)$.

Step 1. $(-6) - (-8) = (-6) + (+8) = 2$.

Step 2.   The common variable is ab: $2ab$ is the desired difference.

6. Add: $(4p) + (3q)$.

The monomials are unlike; hence the simplest expression of their sum is the binomial: $(4p + 3q)$.

7. Subtract: $(6a) - (5b)$.

The monomials are unlike; hence the simplest expression of their difference is the binomial: $(6a - 5b)$.

EXERCISE 2-1

Add:

1. $\begin{array}{r} +14x \\ +\ 7x \\ \hline \end{array}$ **2.** $\begin{array}{r} -18y \\ -\ 6y \\ \hline \end{array}$ **3.** $\begin{array}{r} +13p \\ -\ 6p \\ \hline \end{array}$

4. $+\frac{7}{8}m$
$-\frac{2}{3}m$

5. $+31s$
$-65s$

6. $-42k$
$-57k$

7. $-5.6x$
$+1.8x$

8. $-\frac{1}{2}z$
$+\frac{1}{4}z$

9. $-16m$
$+94n$

10. $+1.8r$
$-2.2r$

11. $+14f$
$+\ 6f$
$-\ 8f$

12. $-27x$
$-18x$
$+\ 6x$

13. $+1.8m$
$-1.6x$
$-1.8m$

14. $-37y$
$+14y$
$+19y$

15. $-42a$
$-36a$
$-51a$

16. $+14x$
$+13y$
$+17x$

17. $-\frac{2}{3}x$
$+\frac{3}{4}x$
$-\frac{1}{6}x$

18. $+0.15s$
$+0.12s$
$-0.20s$

19. $+1.6m$
$-1.7m$
$-1.9m$

20. $-\ 1.2b$
$-\ 1.4b$
$+10.7a$

Subtract; the lower term is the subtrahend:

21. $+14r$
$+\ 7r$

22. $-18m$
$-\ 6m$

23. $+1.2k$
$+1.4k$

24. $+16a$
$-37b$

25. $+17c$
$-12c$

26. $-\frac{3}{5}t$
$+\frac{1}{3}t$

27. $-27d$
$-34e$

28. $+2.6d$
$-1.8d$

29. $+\frac{1}{2}x$
$-\frac{3}{4}x$

30. $-44p$
$-44p$

Simplify:

31. $4xy + 3xy$

32. $3ab - 2ac$

33. $5m + 3m - 8m$

34. $3xy - 5xy - 7xy$

35. $3a - 5a + 4k$

36. $2m - m - \frac{1}{2}m$

37. $3.2s - 5.6s - 0.7s$

38. $\frac{1}{2}d - \frac{3}{4}d + \frac{5}{8}d$

39. $5xy + 2xy - 7xy$

40. $5x + 0.5x + 0.05x$

41. $3x + 0.3x - 0.03x$

42. $\frac{1}{2}x - 7x + \frac{1}{4}x$

43. $\frac{1}{5}ap - \frac{1}{10}ap - \frac{1}{100}ap$

44. $-3q - 6p + 9q$

3. POWERS

Since we agree, in algebra, to write $3 \times n$ as $3n$, and $m \times n$ as mn, then to be consistent, $n \times n$ is nn. In this last case, we have a special type of product, one in which both factors are the same. A product in which all factors are identical is called a **power** of the repeated factor.

Instead of listing all of the factors (as is necessary when they are all different), a highly useful shorthand notation may be employed: a numeral written above and to the right of a factor indicates the number of times which that factor is used in the product. Thus:

$$5 \times 5 \times 5 \times 5 = 5^4$$

$$a \times a \times a = a^3$$

$$b \times b \times b \times b = b^4$$

$$ab \times ab \times ab = a \times b \times a \times b \times a \times b$$
$$= a \times a \times a \times b \times b \times b$$
$$= a^3 b^3$$

In such expressions, the repeated factor is called the **base**; the upper-level numeral that indicates the number of times the factor is used is called the **exponent**.

As with numerical coefficients, when no exponent is attached to an expression, whether numerical or literal, the exponent is understood to be 1:

$$a = a^1; \qquad 3 = 3^1; \qquad ab = a^1 b^1; \qquad 4d = 4^1 d^1$$

4. PRODUCTS OF POWERS HAVING THE SAME BASE

If, as we have seen in the previous section,

$$a^4 = a \times a \times a \times a \quad \text{and} \quad a^3 = a \times a \times a$$

then

$$a^4 \times a^3 = (a \times a \times a \times a) \times (a \times a \times a)$$
$$= a \times a \times a \times a \times a \times a \times a = a^7$$

Thus we could have obtained the same resulting exponent (7) by simply adding the exponents of the two like-based factors:

$$a^4 \times a^3 = a^{4+3} = a^7$$

2.2

To multiply powers having the same base:

Step 1. Add the exponents of the like-based factors.

Step 2. Write as the desired product the common base with an exponent equal to the sum of the exponents found in Step 1.

EXAMPLES

1. Find the product of $a^4 \times a^3 \times a^7$.

Step 1.   Add the exponents: $4 + 3 + 7 = 14$.

Step 2.   The common base is a; the product is a^{14}.

2. Find the product of $a^3b^2 \cdot a^2b^2 \cdot a^4b$. [*Note:* When all factors are not of the same base, powers of the same base are multiplied together independently.]

$$a^3b^2 \cdot a^2b^2 \cdot a^4b^1 = a^3 \cdot a^2 \cdot a^4 \cdot b^2 \cdot b^2 \cdot b^1$$
$$= a^{3+2+4}b^{2+2+1}$$
$$= a^9b^5$$

5. PRODUCTS OF MONOMIALS

We noted previously that a monomial may be considered to be a product in itself, e.g., $8pq = 8 \times p \times q$. Computing the product of several monomials is essentially a matter of reorganizing and collecting factors according to the associative and commutative properties. For instance, to multiply $6ab \times 2ab^2c$:

$$6ab \times 2ab^2c = (6 \times a \times b) \times (2 \times a \times b \times b \times c)$$
$$= 6 \times a \times b \times 2 \times a \times b \times b \times c$$
$$= 6 \times 2 \times a \times a \times b \times b \times b \times c$$
$$= 12 \times a^2 \times b^3 \times c$$
$$= 12a^2b^3c$$

2.3

To multiply several monomials:

Step 1. Multiply the numerical coefficients of the several monomials for the numerical coefficient of the product.

Step 2. Multiply the variables of the several monomials for the variable of the product.

EXAMPLES

1. Multiply: $(4a^2b) \times (-2ac^2) \times (bc^3)$.

Step 1.   Numerical coefficients: $(4) \times (-2) \times (1) = -8$.

Step 2.   Variables: $(a^2b) \times (ac^2) \times (bc^3) = a^2 \times b^1 \times a^1 \times c^2 \times b^1 \times c^3$
$$= a^{2+1}b^{1+1}c^{2+3} = a^3b^2c^5.$$

The product is $-8a^3b^2c^5$.

2. Multiply: $(-3a^2b) \times (-2a^3b) \times (4)$.

Step 1. $(-3) \times (-2) \times (+4) = +24$

Step 2. $(a^2b) \times (a^3b) = a^2 \cdot b^1 \cdot a^3 \cdot b^1 = a^5b^2$

The product is $24a^5b^2$.

3. Multiply: $(6xy^3) \times (-3xy) \times (-x^3y)$.

Step 1. $(+6) \times (-3) \times (-1) = +18$

[*Note:* $-x^3y$ may be thought of as $-1x^3y$.]

Step 2. $(xy^3) \times (xy) \times (x^3y) = x^1 \cdot y^3 \cdot x^1 \cdot y^1 \cdot x^3 \cdot y^1 = x^5y^5$

The product is $18x^5y^5$.

EXERCISE 2-2

Express the product for each of the following as a monomial.

1. $a^3 \cdot a^2$ **2.** $b^4 \cdot b^3 \cdot b$ **3.** $d^2 \cdot d \cdot d^{10}$

4. $x^3 \cdot x^5 \cdot x^7$ **5.** $m^2 \cdot m^3 \cdot m^5$ **6.** $c^2 \cdot c^2 \cdot d$

7. $a^2b^2 \cdot a^4b^2 \cdot ab^3$ **8.** $a^3b^2 \cdot a^5 \cdot b^7 \cdot a$

9. $x^3y \cdot x^2y^3 \cdot x^4$ **10.** $m^2n \cdot mn^2 \cdot m^3 \cdot n^3$

11. $5x$ **12.** $-3x$ **13.** $5n$
$\underline{-2}$ $\underline{4}$ $2n$

14. $-7x$ **15.** $4y$ **16.** $5xy$
$\underline{-3x}$ $\underline{-2x}$ $6x$

17. $-3xy^2$ **18.** $-8a^2b$ **19.** $(4x)(-3x)$
$\underline{5x^2y}$ $\underline{-3bx}$

20. $(7x)(-3y)$ **21.** $(-12x)(3a)$ **22.** $(\frac{3}{4}x)(\frac{2}{3}y)$

23. $(\frac{4}{5}m)(-\frac{1}{2}n)$ **24.** $(-4x)(-3y)(2x)$

25. $(-3x)(-2b)(3x)$ **26.** $(4m)(-3m)(-7m^2)$

27. $(3st)(3st)(3st)$ **28.** $(4ab)(3bc)(2ac)$

29. $(-2mn)(-3mp)(-4np)$ **30.** $(\frac{3}{5}xy)(\frac{2}{9}xy^2)(\frac{5}{8}x^2y^2)$

31. $(\frac{7}{8}ab)(-\frac{2}{3})(-\frac{4}{7}b)$ **32.** $(-x^3)(-y^5)(x^2)(m)$

33. $(0.6x)(-0.3y)(0.2x)$ **34.** $(0.07xy)(-0.03xy^2)(-0.001x^2)$

6. QUOTIENTS OF POWERS HAVING THE SAME BASE

Since

$$a^5 = a \times a \times a \times a \times a \quad \text{and} \quad a^3 = a \times a \times a$$

then

$$a^5 \div a^3 = \frac{a^5}{a^3} = \frac{a \times a \times a \times a \times a}{a \times a \times a} = a^2 \quad (a \neq 0)$$

and

$$a^3 \div a^5 = \frac{a^3}{a^5} = \frac{a \times a \times a}{a \times a \times a \times a \times a} = \frac{1}{a^2} \quad (a \neq 0)$$

In both of these examples, we could have obtained the exponent of the quotient by subtracting the smaller exponent from the larger exponent of the two original terms. Although later, in Chapter VII, we shall find it advantageous to define negative exponents, it will be useful for now to handle problems such as those above with a two-part program for dividing powers having the same base.

2.4(a)

To divide powers having the same base when the exponent of the dividend is greater than the exponent of the divisor:

Step 1. Subtract the exponent of the divisor from the exponent of the dividend.

Step 2. Write as the desired quotient the common base (zero excepted) with an exponent equal to the difference of the exponents found in Step 1.

EXAMPLES

1. Divide b^4 by b^3 $(b \neq 0)$.

Step 1.   Subtract exponents: $4 - 3 = 1$.

Step 2.   The quotient is b^1 (or b).

2. Divide: $a^9 \div a^3$ $(a \neq 0)$.

Step 1. $9 - 3 = 6$

Step 2.   The quotient is a^6.

2.4(b)

To divide powers having the same base when the exponent of the divisor is greater than the exponent of the dividend:

Step 1. Subtract the exponent of the dividend from the exponent of the divisor.

Step 2. Write as the desired quotient a fraction whose numerator is 1 and whose denominator is the common base (zero excepted) with an exponent equal to the difference of the exponents found in Step 1.

EXAMPLES

1. Divide k^5 by k^8 $(k \neq 0)$.

Step 1. $8 - 5 = 3$

Step 2.   The quotient is $1/k^3$.

2. Divide: $x^3 \div x^7$ $(x \neq 0)$.

$$x^3 \div x^7 = \frac{1}{x^{7-3}} = \frac{1}{x^4}$$

7. QUOTIENTS OF MONOMIALS

Computing the quotient of two monomials essentially involves (a) expressing the dividend monomial as the numerator of a fraction and the divisor as the denominator of that fraction, and (b) reducing that fraction to lowest terms by dividing out (cancelling) common factors. In arithmetic, this is a familiar process:

$$60 \div 84 = \frac{60}{84} = \frac{\cancel{2} \times \cancel{2} \times \cancel{3} \times 5}{\cancel{2} \times \cancel{2} \times \cancel{3} \times 7} = \frac{5}{7}$$

2.5

To divide one monomial by another:

Step 1. Express dividend and divisor as the numerator and denominator of a fraction, respectively.

Step 2. Simplify the fraction by dividing out common factors.

EXAMPLES

1. Divide $36a^3b^2c$ by $9ab^2$.

Step 1. $\dfrac{36a^3b^2c}{9ab^2}$

Step 2. $\dfrac{2 \cdot 2 \cdot \cancel{3} \cdot \cancel{3} \cdot \cancel{a} \cdot a \cdot a \cdot \cancel{b} \cdot \cancel{b} \cdot c}{\cancel{3} \cdot \cancel{3} \cdot \cancel{a} \cdot \cancel{b} \cdot \cancel{b}} = 4a^2c$

2. Divide $-6a^3bc^2$ by $2ab^4$.

$$\frac{\overset{-3a^2}{\cancel{-6a^3}bc^2}}{\underset{b^3}{\cancel{2ab^4}}} = \frac{-3a^2c^2}{b^3}$$

3. Divide: $(-25x^3) \div (-5x^2yz)$.

$$\frac{\overset{+5\ x^1}{\cancel{-25x^3}}}{\cancel{-5x^2yz}} = \frac{5x}{yz}$$

EXERCISE 2-3

Find the quotient for each of the following:

1. $2^5 \div 2^3$ **2.** $2^2 \div 2^3$ **3.** $2^4 \div 2^7$

4. $x^8 \div x^4$ **5.** $x^8 \div x^2$ **6.** $x^7 \div x^9$

7. $(xy)^3 \div (xy)^5$ **8.** $(ab)^2 \div (ab)$ **9.** $(-a)^7 \div (-a)^5$

10. $(-2)^3 \div (-2)^6$ **11.** $14a \div (-2)$ **12.** $-15x \div 5$

13. $-24a^2 \div 6a$ **14.** $18x^4 \div 2x$ **15.** $25p^2q \div (-5pq)$

16. $(-84a^3b^2) \div (7a^2b^2)$ **17.** $(26a^3b) \div (-13a^3b)$

18. $40x^2y^3z \div 16x^2y$ **19.** $32a^2y \div \frac{1}{4}ay^3$

20. $\frac{3}{4}a^2b \div 4ab^2$ **21.** $\frac{2}{5}ab^3 \div \frac{1}{3}ac$

22. $(-87a^2b^3c) \div (3bc^2)$ **23.** $16a^2bx \div \frac{2}{3}ax^2$

24. $0.04ab^3 \div 0.01ab^4$ **25.** $(-0.006xy^2) \div (0.002xy^3)$

26. $(-9t^2v^3) \div (21t^2vx^2)$ **27.** $(-13a^2b^5) \div (-26a^6b)$

28. $\frac{2}{3}x^2y^7 \div \frac{5}{6}xy^6$ **29.** $(-\frac{3}{7}x^2yz) \div (\frac{5}{14}xz^2)$

30. $(-0.032m^2n) \div (0.04pq)$ **31.** $4\frac{1}{2}x^2y^7 \div 2\frac{1}{2}xy^5$

32. $3\frac{2}{3}m^2y^7 \div 0.3my^8$ **33.** $(-80ab^4) \div (-0.16ab^5)$

34. $(7\frac{3}{8}x^4y^2z^3) \div (-9\frac{1}{4}xy^5z)$

8. SUMS AND DIFFERENCES OF POLYNOMIALS

For technical reasons of classification, monomials are considered to be included among the polynomials, i.e., the monomials are a subset of the polynomials. From a computational viewpoint, however, it is better to discuss the operations on monomials first and separately, as we have done, and then to discuss the operations on the rest of the polynomials. This is so because performing operations on the polynomials depends for the most part upon operations on the monomials. For example, computing the sum of two polynomials is essentially a matter of adding the like monomial parts of the polynomials:

$$
\begin{aligned}
(3a + 2b - 4c) + (2a - b + c) &= 3a + 2b - 4c + 2a - b + c \\
&= 3a + 2a + 2b - b - 4c + c \\
&= (3a + 2a) + (2b - b) + (-4c + c) \\
&= (5a) + (b) + (-3c) \\
&= 5a + b - 3c
\end{aligned}
$$

2.6

To add polynomials:

Step 1. Arrange each polynomial addend so that like monomials are in the same column.

Step 2. Add each column separately.

EXAMPLES

1. Add: $(3a + b - c) + (4a - 3b + 6c) + (2a + b + c)$.

Step 1.   Arrange addends:

$$
\begin{array}{l}
3a + b - c \\
4a - 3b + 6c \\
2a + b + c
\end{array}
$$

Step 2.   Add by columns: $(+)$ _____

$$9a - b + 6c = 9a - b + 6c \quad \text{(sum)}$$

2. Add: $(4p + 6q + 11r) + (3p - 2r) + (3q - 4r)$.

Step 1.   Arrange addends:

$$
\begin{array}{l}
4p + 6q + 11r \\
3p - 2r \\
 + 3q - 4r
\end{array}
$$

Step 2.   Add by columns: $(+)$ _____

$$7p + 9q + 5r = 7p + 9q + 5r \quad \text{(sum)}$$

As with monomials, the difference of two polynomials is readily found by adding the negative of the subtrahend to the minuend. The **negative of a polynomial** is a polynomial in which the signs of the terms of the original polynomial have been reversed. For example, the negative of $(2a - b + c)$ is $(-2a + b - c)$. That each is a negative of the other can be verified if we add the two polynomials together by Program **2.6** and note that their sum is zero. Thus, to subtract $(2a - b + c)$ from $(3a + 2b - 4c)$, we have:

$$
\begin{aligned}
(3a + 2b - 4c) - (2a - b + c) &= 3a + 2b - 4c + (-2a + b - c) \\
&= 3a + 2b - 4c - 2a + b - c \\
&= 3a - 2a + 2b + b - 4c - c \\
&= (a) + (3b) + (-5c) \\
&= a + 3b - 5c
\end{aligned}
$$

2.7

To subtract polynomials:

Step 1. Arrange the minuend and subtrahend so that like monomials are in the same column.

Step 2. Reverse the signs of the monomials of the subtrahend.

Step 3. Add the polynomial of Step 2 to the minuend.

EXAMPLES

1. Subtract: $(6x - 4y) - (3x + 8y)$.

Step 1.   Arrange minuend and subtrahend:

$$\begin{array}{r} +6x - 4y \\ +3x + 8y \\ (-) \hline \end{array}$$

Step 2.   Reverse the signs of subtrahend: $+3x + 8y \rightarrow -3x - 8y$

Step 3.   Add by columns:

$$\begin{array}{r} +6x - 4y \\ -3x - 8y \\ (+) \hline +3x - 12y \quad \text{(difference)} \end{array}$$

2. Subtract $(4a - 2b + c)$ from $(3a - 4b + 8c)$.

$$\begin{array}{r} +3a - 4b + 8c \\ +4a - 2b + c \\ (-) \hline \end{array} \rightarrow \begin{array}{r} +3a - 4b + 8c \\ -4a + 2b - c \\ (+) \hline -a - 2b + 7c \quad \text{(difference)} \end{array}$$

3. Subtract from $(4x - 3y)$ the binomial $(3x - 2z)$.

$$\begin{array}{r} +4x - 3y \\ +3x - 2z \\ (-) \hline \end{array} \rightarrow \begin{array}{r} +4x - 3y \\ -3x + 2z \\ (+) \hline +x - 3y + 2z \quad \text{(difference)} \end{array}$$

EXERCISE 2-4

Add:

1. $\begin{aligned} 6xy - 2xy^2 \\ -3xy - 17xy^2 \\ 4xy + 12xy^2 \\ -xy - xy^2 \end{aligned}$

2. $\begin{aligned} 3x - 4p \\ -2x \\ 3x - 3p \\ - 4p \end{aligned}$

3. $\begin{aligned} 7xy - 3y + 2z \\ - 4y + 7z \\ 3xy + 5y + 2z \\ -2xy - 8z \end{aligned}$

4. $\begin{aligned} x + 3y - 12p^2 \\ -x + 5y \\ - 5y + 17p^2 \\ 3x - 4p^2 \end{aligned}$

5. $(4x - 3y); (3d + 4x); (3y + 2z); (5z - 8x)$

6. $(3x - 2y); (4x - 3y + 2z); (4z - 6x); (2y + 5z)$

7. $(0.3x - 2y); (0.6x + 0.7y); (-0.8x - 1.4y)$

8. $(\frac{2}{3}x - \frac{1}{2}y); (\frac{3}{4}x - \frac{1}{3}y); (-\frac{5}{6}x - \frac{3}{2}y)$

Subtract the second polynomial from the first:

9. $(3x - 2y); (6x - 7y)$

10. $(3x - 2y); (2y - 3x)$

11. $(3x - 4y); (6y + 12)$

12. $(x + 3p); (3p + 2)$

13. $(x - \frac{1}{2}y); (-\frac{3}{4}y)$

14. $(0.6a - 0.3b); (-0.4b + 0.7a)$

15. $(3 + 0.6a)$; $(-5 + 0.2b)$ **16.** $(17a - 4b)$; $(-12c + 3d)$

17. $(3x + 4y - 7)$; $(2x - 2y + 5)$ **18.** $(3x - 7y)$; $(2p + 3x - 15y)$

19. $(0.1a - 3.4b - 2.7c)$; $(6d - 3.2a - 0.5c)$

Simplify:

20. $3x - 2y + 3x - 5x + 3y - x - 3y$

21. $5a - 3b + 2a - 6c - 7a - 15b - 4c$

22. $6ab - 0.6b + 3.2ab - 0.7b + 4ab - 0.1b$

23. $\frac{1}{2}x - \frac{3}{4}y + \frac{2}{3}x + \frac{1}{4}y - \frac{1}{5}x$

24. $2g - 5h + 0.3g - \frac{1}{2}h + \frac{1}{5}g - 0.4h$

25. Add $(3x - 2y)$ to itself and then subtract the sum from $(6y - 7x)$.

26. Subtract $(3a - \frac{1}{2}b)$ from the sum of $(3x - a)$ and $(2b - 3x)$.

27. Increase $(4x - 7)$ by 3 and subtract from $(5x - 12)$.

28. Subtract $6x - 3y$ from 1.

29. Subtract $4x - 3y$ from 0.

30. What must be added to $(3x - 2y)$ to make the sum $(4y + z)$?

9. PRODUCTS OF POLYNOMIALS

The usual program for computing the product of a monomial and a polynomial is based upon the distributive property:

$$a \times (b + c) = (a \times b) + (a \times c)$$

as can be noted in the following examples.

2.8

To multiply a polynomial by a monomial:

Step 1. Multiply each term of the polynomial by the monomial.

Step 2. Add the products of Step 1 for the desired product.

EXAMPLES

1. Multiply: $6a \times (3b - 2a)$, or $6a(3b - 2a)$.

Step 1. $(6a) \times (3b) = 18ab$

$(6a) \times (-2a) = -12a^2$

Step 2.   Add the products of Step 1: $18ab - 12a^2$.

2. Multiply: $5a^3b(3a^2 - 2b + 4)$.

Step 1. $(5a^3b)(3a^2) = 15a^5b$

$$(5a^3b)(-2b) = -10a^3b^2$$

$$(5a^3b)(+4) = 20a^3b$$

Step 2. Add products: $15a^5b - 10a^3b^2 + 20a^3b$.

3. Multiply: $-3a^3(4c - 2a^4 - 6ab^2)$.

Step 1. $(-3a^3)(4c) = -12a^3c$

$$(-3a^3)(-2a^4) = 6a^7$$

$$(-3a^3)(-6ab^2) = 18a^4b^2$$

Step 2. Add products: $-12a^3c + 6a^7 + 18a^4b^2$.

4. Multiply: $-c(a^3 - 2b^2 + d)$.

$$\begin{aligned}
-c(a^3 - 2b^2 + d) &= (-c)(a^3) + (-c)(-2b^2) + (-c)(d) \\
&= (-ca^3) + (2cb^2) + (-cd) \\
&= -ca^3 + 2cb^2 - cd
\end{aligned}$$

Computing the product of two polynomials usually involves a repeated application of the distributive property. For instance, in $(a + b) \times (c + d + e)$, we consider the second factor—for the moment—as a single term:

$$(a + b) \times (c + d + e) = [a \times (c + d + e)] + [b \times (c + d + e)]$$

Now applying the distributive property a second time, within the bracketed expressions:

$$\begin{aligned}
[a \times (c + d + e)] &+ [b \times (c + d + e)] \\
&= [(a \times c) + (a \times d) + (a \times e)] + [(b \times c) + (b \times d) + (b \times e)] \\
&= [ac + ad + ae] + [bc + bd + be] \\
&= ac + ad + ae + bc + bd + be
\end{aligned}$$

Essentially, this is what we do when we multiply two multidigit numbers in arithmetic, say 23×312:

$$\begin{aligned}
23 \times 312 &= (20 + 3) \times (300 + 10 + 2) \\
&= [20 \times (300 + 10 + 2)] + [3 \times (300 + 10 + 2)] \\
&= [(20 \times 300) + (20 \times 10) + (20 \times 2)] \\
&\quad + [(3 \times 300) + (3 \times 10) + (3 \times 2)] \\
&= [6000 + 200 + 40] + [900 + 30 + 6] \\
&= [6240] + [936] \\
&= 7176
\end{aligned}$$

But in arithmetic, a shorter way to obtain this same result has been developed, the familiar:

$$\begin{array}{r}
312 \\
23 \\
(\times) \ \overline{} \\
936 \\
624 \quad \text{(in effect: 6240)} \\
\hline
7176
\end{array}$$

In algebra there is a corresponding short way by which to find the product of two polynomials, as given in the following program.

2.9

To multiply two polynomials:

Step 1. Write one of the polynomial factors directly above the other.

Step 2. Multiply each monomial of the bottom factor with each monomial of the top factor and arrange the products in columns according to likeness.

Step 3. Add the products of monomials of Step 2 for the desired product.

EXAMPLES

1. Find the product: $(2x + 1) \times (3x + 2)$.

Step 1. $\qquad \begin{cases} 3x + 2 \\ 2x + 1 \end{cases}$

$(\times) \overline{}$

Step 2. $\qquad \begin{cases} 6x^2 + 4x \\ \quad 3x + 2 \end{cases}$

Step 3. $\qquad 6x^2 + 7x + 2$

[*Note:* Since there is no "carrying" in algebraic multiplication, the computation may be performed from left to right. This usually helps in organizing the partial product terms by likeness.]

2. Multiply $(3a - 4c)$ by $(2a + 3y)$.

Step 1. $\qquad \begin{cases} 3a - 4c \\ 2a + 3y \end{cases}$

$(\times) \overline{}$

Step 2. $\qquad \begin{cases} 6a^2 - 8ac \\ + 9ay - 12cy \end{cases}$

Step 3. $\qquad 6a^2 - 8ac + 9ay - 12cy$

3. Multiply: $(4x^2 - 3x)(2x + 7)$.

Step 1. $\qquad \begin{cases} 4x^2 - 3x \\ 2x + 7 \end{cases}$

$(\times) \overline{}$

Step 2. $\qquad \begin{cases} 8x^3 - 6x^2 \\ + 28x^2 - 21x \end{cases}$

Step 3. $\qquad 8x^3 + 22x^2 - 21x$

4. Multiply: $(4a - 2b + 3c)(2a - b - 4c)$.

Step 1. $\qquad \begin{cases} 4a - 2b + 3c \\ 2a - b - 4c \end{cases}$

$(\times) \overline{}$

Step 2. $\qquad \begin{cases} 8a^2 - 4ab + 6ac \\ - 4ab + 2b^2 - 3bc \\ - 16ac + 8bc - 12c^2 \end{cases}$

Step 3. $\qquad 8a^2 - 8ab - 10ac + 2b^2 + 5bc - 12c^2$

EXERCISE 2-5

Multiply:

1. $(3x)(4x + 2y - z)$

2. $(-x)(4a - 2x + 3y)$

3. $(3ab)(-2a + 3b - ab)$

4. $(2x - 3b + 4c)(8a)$

5. $(-3ax)(-2a + 5x - 6)$

6. $(2a)(a - 2 + a + 3)$

7. $(-2a^2c)(3a - 4c - 2ac)$

8. $(-8a^2)(3b^2 - 4a - 2b)$

9. $(-pqr)(3p + 2qr + 7)$

10. $(-6ab^2c^3)(3a^2bc - 3ab + 3ac)$

11. $2x - 7$
$5x + 2$

12. $5a - 2x$
$a + 3x$

13. $m - n$
$2p - m$

14. $3x - 2y + 6$
$x - y$

15. $3x^2 - 2x + 7$
$3x - 5$

16. $6x^2 - 3xy + 5y^2$
$x - 2y$

17. $5a - 2a^2 - 3a^3$
$a^2 - 2a$

18. $-8d^2 + 4cd - 6c^2$
$-d + 2c$

19. $a^5b - 6a^4b^2 + b^3$
$3ab - 4ab^2$

20. $3x^2 - 2x + 1$
$5x^2 - 3x + 7$

21. $6m^2 - 5mn + n^2$
$2m^2 + 3mn - 6n^2$

22. $4a^2 - 2b + c$
$3a - 3b + c$

23. What is the area of a rectangle $(3x + 2)$ ft by $(4x - 7)$ ft?

24. What would be the sum if $(s - 3t)$ were added repeatedly $(2s - 3t)$ times?

25. What polynomial, when divided by $(x - y)$, will yield a quotient $x^2 + xy + y^2$?

26. What is the product if one factor is a increased by b and the other is a decreased by b?

10. SYMBOLS OF GROUPING

Parentheses ( ), brackets [ ], and braces { } are the usual symbols of grouping used in algebra. They indicate to the reader that certain algebraic expressions are to be treated as a single expression. For instance, if we wish to multiply the sum of $a + b$ by c, we may express the product as $c \cdot (a + b)$ or $c(a + b)$. In effect, these symbols of grouping can be looked upon as the punctuation marks of our algebraic language. In the English language, we know it is possible for the same sequence of words to have two totally different meanings depending upon how the words are grouped. Consider the two statements, which differ only by a comma:

(a) "I am ready to eat, Sam."
(b) "I am ready to eat Sam."

For an arithmetic parallel, note:

(a′) Three times five, plus six.
(b′) Three times five plus six.

Without punctuating parentheses, $3 \times 5 + 6$ must serve both (a′) and (b′). With punctuating parentheses, it is possible to make clear the distinction between the two:

(a′) $(3 \times 5) + 6 = 15 + 6 = 21$
(b′) $3 \times (5 + 6) = 3 \times 11 = 33$

In the removal of symbols of grouping, cases in which a minus sign precedes a grouped expression, such as $-(a - b)$, sometimes cause difficulty. It can be easily resolved if we remember the role that the unwritten numeral one (1) plays in multiplicative situations (e.g., $6 = 1 \times 6 = \frac{6}{1}$, $1 \times a = a^1 = a$, etc.). Here, too, the "1" part of the numeral -1 goes unwritten, but its minus sign is written; thus, by the distributive property:

$$-(a - b) = -1 \times (a - b)$$
$$= (-1)(a) + (-1)(-b)$$
$$= -a + b$$

In other words, the negative of $(a - b)$ is $-(a - b)$, and the latter is equivalent to $(-a + b)$.

When there are several symbols of grouping involved, the convention is to begin with the innermost, and work outward.

EXAMPLES

1. $\{-3a[4b - a(c - d)] + 14ab\}$
   $\{-3a[4b - ac + ad] + 14ab\}$
   $\{-12ab + 3a^2c - 3a^2d + 14ab\}$
   $3a^2c - 3a^2d + 2ab$

2. $-3\{a + [6(a - 2b)] + b - [7(a + 3b)]\}$
   $-3\{a + [6a - 12b] + b - [7a + 21b]\}$
   $-3\{a + 6a - 12b + b - 7a - 21b\}$
   $-3\{-32b\} = 96b$

The insertion of parentheses or other symbols of grouping is the opposite of the foregoing. It must be remembered that prefixing a grouped expression with a minus sign negates it. Therefore, the sign of each term of the enclosed expression must be reversed if it is to remain equivalent to the original expression.

EXAMPLES

1. $-6 + 2x - 4y \overset{?}{=} -(6 - 2x + 4y)$.

   To check, *remove* the parentheses from the right-hand expression:

$$-(6 - 2x + 4y) = (-1)(+6) + (-1)(-2x) + (-1)(+4y)$$
$$= -6 + 2x - 4y$$

2. $-3x^2 + 4x - 7y = -(3x^2 - 4x + 7y)$
$$= -[3x^2 - (4x - 7y)]$$

EXERCISE 2-6

Remove all signs of grouping and simplify:

1. $6 + (3 + 4) - 5$ **2.** $8 - (3 - 2) + (4 - 6)$

3. $5(4 - 6) - 2(3 + 1)$ **4.** $5 - [6(3 - 5) + 2]$

5. $-[3(2 - 5) - (-6 + 2)]$ **6.** $4\{[6 - 7(3 - 2 - 4) + 8] + 3\}$

7. $20 - \{3[4(6 - 7 + 15) - 18]\}$

8. $\{2 - 3\}\{6 - [5 + (2 - 1)] + 4\}\{-7[3 - (6 - 6)]\}$

9. $(4a + 3x) + (5a - 7x)$ **10.** $(2p - 3t) - (p + 4t)$

11. $4p - (k - 2x) + 8k$ **12.** $7s - [4s + (4s + 7)] + 7$

13. $-(3a - 5b + 3c) - (a + 7b - 4c)$ **14.** $(2x - y) + (4x + 2y) - (6x + 3y)$

15. $(5x + 3y) - [(2y - 6x) - (2x + 8y)]$

16. $-[(2a - b) + (5b + 6a)] - (8a + 3b)$

17. $[(4m - s) - (2m + 5s)] + (m - 4s)$

18. $[5 - (2a - 4)] - [a - (6a - 3)]$

19. $6x - \{2x - [5x - (3x - 8) - 8] + 4\}$

20. $4(2y + 6) + 3(4y - 5)$

21. $4(3k + t) - 2(5k + 4t)$

22. $7(-5c + 4) - (-2c + 7)$

23. $7f - \{3f - [5 - (4f - 2) + 6f] - 5\}$

24. $-3[2(5a - 4b) - 6a] - 4b$

Rewrite the expressions below, enclosing the last three terms in parentheses, preceded by a minus sign:

25. $2x + 6y - 3t - 8$ **26.** $4x - 6y + 3z + 5$

27. $ab - c + gk - h$ **28.** $-st + q - a - c - d$

11. QUOTIENTS OF POLYNOMIALS

Division is said to be *distributive from the right:*

$$(6 + 4) \div 2 \overset{?}{=} (6 \div 2) + (4 \div 2)$$
$$10 \quad \div 2 \overset{?}{=} \quad 3 \quad + \quad 2$$
$$5 = 5$$

But division is *not distributive from the left*:

$$2 \div (6 + 4) \overset{?}{=} (2 \div 6) + (2 \div 4)$$
$$2 \div 10 \overset{?}{=} \tfrac{1}{3} + \tfrac{1}{2}$$
$$\tfrac{1}{5} \neq \tfrac{5}{6}$$

We take advantage of the property of right distributivity to develop the following program.

2.10

To divide a polynomial by a monomial:

Step 1. Divide each term of the polynomial by the monomial.

Step 2. Add the quotients of Step 1 for the desired quotient.

EXAMPLES

1. Divide $(6a^4b^3 - 3a^2b^3)$ by $3ab$.

Step 1.
$$\frac{\overset{2}{\cancel{6}} \, \overset{3}{\cancel{a^4}} \, \overset{2}{\cancel{b^3}}}{\cancel{3} \, \cancel{a} \, \cancel{b}} = 2a^3b^2$$

$$\frac{\overset{-1}{-\cancel{3}} \, \overset{1}{\cancel{a^2}} \, \overset{2}{\cancel{b^3}}}{\cancel{3} \, \cancel{a} \, \cancel{b}} = (-1)(ab^2) = -ab^2$$

Step 2.   The desired quotient is $2a^3b^2 - ab^2$.

2. Divide: $(18x^4y - 15x^2y^3) \div 3x^3y^2$.

Step 1.
$$\frac{\overset{6}{\cancel{18}} \, \overset{1}{\cancel{x^4}} \, \cancel{y}}{\cancel{3} \, \underset{1}{\cancel{x^3}} \, \cancel{y^2}} = \frac{6x}{y}$$

$$\frac{\overset{-5}{-\cancel{15}} \, \cancel{x^2} \, \overset{1}{\cancel{y^3}}}{\underset{1}{\cancel{3}} \, \cancel{x^3} \, \cancel{y^2}} = \frac{-5y}{x} = -\frac{5y}{x}$$

Step 2.   The desired quotient is $\dfrac{6x}{y} - \dfrac{5y}{x}$.

3. Compute: $(8x^3 + 4x^2y + 16xy^2) \div 4x$.

$$(8x^3 + 4x^2y + 16xy^2) \div 4x = (8x^3 \div 4x) + (4x^2y \div 4x) + (16xy^2 \div 4x)$$
$$= \frac{\overset{2x^2}{\cancel{8x^3}}}{\cancel{4x}} + \frac{\overset{x}{\cancel{4x^2y}}}{\cancel{4x}} + \frac{\overset{4}{\cancel{16xy^2}}}{\cancel{4x}}$$
$$= 2x^2 + xy + 4y^2$$

The usual long-division algorithm of arithmetic parallels algebraic division involving polynomials. For instance:

$$\begin{array}{r} 14 \\ 12\overline{)168} \\ 12 \\ \hline 48 \\ 48 \\ \hline \end{array} \rightarrow \begin{array}{r} 10 + 4 \\ 10 + 2\overline{)100 + 60 + 8} \\ 100 + 20 \\ \hline 40 + 8 \\ 40 + 8 \\ \hline \end{array} \rightarrow \begin{array}{r} t + 4 \\ t + 2\overline{)t^2 + 6t + 8} \\ t^2 + 2t \\ \hline 4t + 8 \\ 4t + 8 \\ \hline \end{array}$$

When t is replaced by 10, the three computations above are essentially the same.

2.11

To divide one polynomial by another:

Step 1. Arrange both polynomials in descending powers of the same variable.

Step 2. Find the first term of the quotient by dividing the first term of the dividend by the first term of the divisor.

Step 3. Multiply the divisor by the quotient term of Step 2.

Step 4. Subtract the product of Step 3 from the dividend; bring down the next term of the original dividend to form the new dividend.

Step 5. Repeat the loop or sequence of Steps 2, 3, and 4 on the new dividend; keep repeating this loop of steps until the exponent of the first term of any remainder is less than the exponent of the first term of the divisor.

EXAMPLES

1. Divide $(8x + 6x^2 + 2)$ by $(2x + 2)$.

Step 1. $(6x^2 + 8x + 2) \div (2x + 2)$

Step 2. $\dfrac{6x^2}{2x} = 3x$

Step 3.

$$2x + 2\overline{)6x^2 + 8x + 2} \qquad \begin{array}{r} 3x + 1 \end{array}$$

$(3x)(2x + 2) \longrightarrow 6x^2 + 6x$

Step 4.   Subtract and bring down $+2$ $\longrightarrow$ $2x + 2$
$\qquad\qquad\qquad\qquad\qquad 2x + 2$

Step 5. $\begin{cases} \dfrac{2x}{2x} = 1 \\ (1) \times (2x + 2) = 2x + 2 \\ (2x + 2) - (2x + 2) = 0 \end{cases}$

2. Divide: $(2x + 6x^2 - 20) \div (7 + 3x)$.

Step 1. $(6x^2 + 2x - 20) \div (3x + 7)$

Step 2. $\dfrac{6x^2}{3x} = 2x$

Step 3.
$$3x + 7\overline{)6x^2 + 2x - 20}$$
$$\underline{6x^2 + 14x}$$

Step 4. Subtract and $\longrightarrow$ bring down -20

$$-12x - 20$$
$$\underline{-12x - 28}$$
$$+ 8$$

Step 5.
$$\begin{cases} \dfrac{-12x}{3x} = -4 \\ (-4)(3x + 7) = -12x - 28 \\ (-12x - 20) - (-12x - 28) = +8 \end{cases}$$

(remainder)

As in arithmetic, when a remainder occurs, it may be treated simply as a remainder or added to the quotient as a fraction whose denominator is the divisor. Thus:

$$(6x^2 + 2x - 20) \div (3x + 7) = \begin{cases} 2x - 4, \text{ R } (+8) \\ \quad\quad \text{or} \\ 2x - 4 + \dfrac{8}{3x + 7} \end{cases}$$

3. Divide $(x^4 - y^4)$ by $(x - y)$.

$$x - y\overline{)\begin{array}{l} x^3 + x^2y + xy^2 + y^3 \\ \overline{x^4 - y^4} \end{array}}$$
$$\underline{x^4 - x^3y}$$
$$x^3y - y^4$$
$$\underline{x^3y - x^2y^2}$$
$$x^2y^2 - y^4$$
$$\underline{x^2y^2 - xy^3}$$
$$xy^3 - y^4$$
$$\underline{xy^3 - y^4}$$

EXERCISE 2-7

Divide; express quotients in simplest terms:

1. $(12a^2 - 18a^4) \div (6a^2)$

2. $(t^3 - t^2) \div (-t^2)$

3. $(x - 6x^2y^2) \div (x^2)$

4. $(5xyz - 10x^2z) \div (-5xz)$

5. $(12k^2n^3 - 28k^3n^2) \div (4k^2n^2)$

6. $(-12a^2b^3 + 48a^4b^4) \div (-6a^2b^3)$

7. $(-21a^3b - 12a^2b^2 + 3ab) \div (-3ab)$

8. $(2x^4 - 6x^3 + 10ax) \div (-4x^3)$

9. $\dfrac{-18x^3y + 33x^2y - 27xy^3}{3x^2}$

10. $\dfrac{28x^2yz^3 - 21x^3y + 14x^3y^2z}{-7x^2y}$

11. $\dfrac{25a^3b^2c^2 - 20a^2b^3c^2 - 15a^2b^2c^4}{-5a^2b^2c^2}$

12. $\dfrac{30x^4y - 5x^2y^2 + 12xy^2}{-5x^2y^2}$

13. $\dfrac{a^3 - 3b^2 - 6a^5}{-1/(5a^2)}$

14. $\dfrac{1.2ab^2 - 0.9a^3b - 1.5b^3}{0.3b}$

15. $\dfrac{m^2 - \frac{1}{2}mn - \frac{3}{4}m^2n^2}{\frac{1}{4}m}$

16. $\dfrac{3.6x^2y - 5.2xy^3 + 3.0xy}{-0.4x^2y^2}$

17. $(6a^2 + 11a + 3) \div (3a + 1)$

18. $(12x^2 + 7xy - 12y^2) \div (4x - 3y)$

19. $(8x - 8x^2 + 6) \div (2x - 3)$

20. $(2y^2 - 17yz + 35z^2) \div (y - 5z)$

21. $(53a^2 + 15a^3 - 8 - 30a) \div (3a - 2)$

22. $(4x^3 + 1 - 3x) \div (2x - 1)$

23. $(10x^2 + 1 + 11x) \div (2x + 3)$

24. $(a^3 - a^2 + a - 1) \div (a + 1)$

25. How may $(x - 2)$'s are there in $(x^3 - 8)$?

26. If $(2x - 4)$ similar machines cost $(4x^2 - 9x + 2)$ dollars, what does one machine cost?

REVIEW—CHAPTER II

Add:

1. $-3.9x$
$\underline{+4.6x}$

2. $-\frac{5}{8}m$
$\underline{+\frac{3}{5}m}$

3. $+42x$
$-18x$
$\underline{+\ 7x}$

4. $-6.3a$
$-7.2a$
$\underline{+4.1a}$

5. $-103y$
$+\ 27y$
$\underline{-\ 63y}$

6. $-20.6m$
$-1.72m$
$\underline{-3.04m}$

7. $-\frac{3}{8}a$
$+\frac{2}{3}a$
$\underline{-\frac{3}{5}a}$

8. $+.624m$
$-1.83m$
$\underline{+24.3m}$

9. $-3.6a$
$+4.2a$
$\underline{-3.1b}$

10. $-\frac{3}{10}r$
$+\ \frac{2}{5}r$
$+\ \frac{1}{2}r$
$\underline{-\ \frac{3}{8}r}$

Subtract; the lower term is the subtrahend:

11. $+623a$
$\underline{-418a}$

12. $-6.37x$
$\underline{+2.42x}$

13. $+\ 9.62m$
$\underline{+10.37m}$

14. $+62p$
$\underline{+37q}$

15. $-\ \frac{5}{8}x$
$\underline{-3\frac{1}{4}x}$

Simplify:

16. $7.9x - 3.4x + 2 - 8.7x + 1$

17. $96x - 362x - 421x + 37x$

18. $\frac{3}{4}x - \frac{2}{3}x + \frac{5}{8}x + \frac{1}{2}x - \frac{5}{12}x$

19. $7x - 0.7x + 0.07x$

20. $4\frac{1}{4}x - 3\frac{1}{5}x - 2\frac{1}{2}x + 16$

Compute the products:

21. $a^7 \cdot a^3 \cdot a^5$

22. $a^3b^2 \cdot ab^3 \cdot ab \cdot a^2b \cdot a$

23. $(-6) \times (14xy)$

24. $(-0.3xy)(15x^2y)$

25. $(\frac{1}{3}ab)(63a^2b^2)$

26. $(-\frac{3}{5}x)(-\frac{2}{3}x)$

27. $(0.03a^2b)(-0.02ab)$

28. $(-6x)(-2y)(3x)(-y)$

29. $(0.3m)(-6m)(0.01m)(-0.02)$

30. $(-2ab)(-3ab^2)(2a^2b)(ab)$

31. $(\frac{3}{4}xy)(\frac{2}{3}xyz)(-\frac{1}{5}x)(\frac{5}{7})$

32. $(-0.02ab)(-3a^2b)(-4ab^2)$

33. $(x^4)(-y^3)(-x^2)(y)$

34. $(3\frac{1}{4}x)(-2\frac{1}{2}y)(2\frac{3}{8}x)$

35. $(0.02x)(-0.01y)(-0.0003x^2)$

36. $(-0.02x)(3\frac{1}{4}x)(-16x^3)$

Compute the quotients:

37. $x^9 \div x^3$

38. $x^{12} \div x^{15}$

39. $(xy)^3 \div (xy)$

40. $(-3)^4 \div (-3)^6$

41. $-18x \div 6$

42. $24x^2 \div 2x^2$

43. $(-76a^2b^3) \div (19ab^2)$

44. $320x^2y^4z^5 \div 16xyz^3$

45. $\frac{5}{8}a^2b \div \frac{1}{4}ab$

46. $(-\frac{3}{4}x^2y) \div (4x)$

47. $0.08ab^4 \div 0.02ab^2$

48. $(-6m^2n^2) \div (21m^2n^2p)$

49. $\frac{3}{5}x^2y^3 \div \frac{7}{10}xy^3$

50. $(-0.025a^2b^3) \div (-0.05a^3b^4)$

51. $(-3\frac{3}{8}m^2n) \div 3\frac{1}{3}mn^3$

52. $(-36a^2b^3c) \div (-0.12ab^4)$

Add:

53. $\quad 32xy - 6xy^2$
$\quad\quad 14xy + 9xy^2$
$\quad\quad -3xy - 7xy^2$
$\quad\quad \underline{\quad xy + \ xy^2}$

54. $\quad x + 2y - \ z$
$\quad\quad -x \quad\ + \ z$
$\quad\quad\quad\quad 3y - 4z$
$\quad\quad \underline{2x - 2y}$

55. $(4x - 2y)$; $(3x - 7y + 2z)$; $(5z - 3y)$; $(2x + 2y)$

56. $(\frac{3}{4}x - \frac{1}{3}y)$; $(\frac{2}{5}x + \frac{3}{7}y)$; $(-\frac{1}{2}x + \frac{3}{4}y)$; $(\frac{2}{3}y - \frac{1}{2}x)$

Subtract the second polynomial from the first:

57. $(4x - 7y)$; $(9y - 4x)$

58. $(3a + b^2)$; $(b^2 - 5a)$

59. $(0.3a + 0.7b)$; $(0.5a - 0.2b)$

60. $(30a - 2b)$; $(30c + 2d)$

61. $(0.3a - 4.1b + 2.6c)$; $(7.1c - 3.2d)$

Simplify:

62. $14ab - 0.3b + 0.62ab - 0.4b + 2.1ab$

63. $\frac{3}{4}x - \frac{2}{3}y + \frac{3}{8}xy + \frac{1}{2}x - \frac{1}{5}y$

64. $4m - 0.6h - \frac{2}{3}m + \frac{1}{5}h - 0.2h$

65. Subtract $5x - 8y$ from -1.

66. Subtract $9a - 2b$ from 0.

67. What must be added to $(7m - 3k)$ to make the sum $(5a + 2k)$?

Multiply:

68. $(-3xy)(4a - 2x + 7)$

69. $(-6x - 3x + 2)(-3x)$

70. $(3m)(a - 3m + 2 + 4a)$

71. $(-9a^3)(3a^2 - 4a + 2m)$

72. $(-1.6a^2b)(-0.1a + 4b - 0.3c)$ **73.** $(3a - 4x)(2a - 7x)$

74. $(3x - 2y)(3x - 3y + 7)$ **75.** $(x - 5y)(6x - 3y + 2z)$

76. $(-a + 4d)(-6a^3 + 3a^2 + 2a)$ **77.** $(a^2 + b - c)(a + b + c)$

78. $(3x^2 + 4x - 2)(12x^2 - 3x + 1)$

79. What would be the sum if $(3a - \frac{1}{2}c)$ were added repeatedly $(2a - d)$ times?

80. What is the product if one factor is x increased by y, and the other is y increased by x?

Remove all signs of grouping and simplify:

81. $9 - [3(5 + 2) + 6]$

82. $3\{[8 - 2(6 - 3 + 7) + 4] - 7\}$

83. $\{9 - 11\}\{7 - [6 + (2 - 4) + 5]\}\{-3[2 - (6 - 6)]\}$

84. $(3m - 6t) - (m + 2t)$

85. $8a - [5a + (3a - 2)] + 7$

86. $(5a - b) + (3a - 7b) - (2a + 3b)$

87. $-[(5x - 3y) + (4y - 2x)] - (2y - 3x)$

88. $4m(3a + 2b) - a(4m - 6b)$

89. $0.5(4a - 6b) - 0.8(5a + 1.5b)$

90. $12m - \{3m - [7 - (4m + 2) + 3m] - 8\}$

91. Write the equivalent of $2z + 6m - 4n + 2$ with the last three terms in parentheses, preceded by a minus sign.

Divide; express quotients in simplest terms:

92. $(15a^2b - 18a^4) \div (3a^2)$

93. $(m - 12m^2n^3) \div (mn)$

94. $(24a^2b^3 - 16ab^4) \div (4ab^2)$

95. $(-14x^2y + 21xy^2 - 63x^2y^3) \div (-7xy)$

96. $(36x^2y^3 + 12x^2y - 15x^2y^2) \div (-3xy)$

97. $(102x^4y^4z^2 - 51x^3y^2z^3 + 85x^2y^2z^2) \div (-17x^2y^2z^2)$

98. $(x^5 - 2x^2 - 3) \div \left(-\frac{1}{3a}\right)$

99. $(a^4 - a^3 - 3a^2) \div (-\frac{1}{2}a)$

100. $(10a^2 - 19a + 6) \div (2a - 3)$

101. $(21x^2 - xy - 10y^2) \div (3x + 2y)$

102. $(4a^3 - 13a^2 + 11a - 2) \div (4a - 1)$

103. $(8x^2 + 10x - 65) \div (2x + 7)$

104. How many $(x - 3)$'s are there in $(x^3 - 27)$?

III

FIRST-DEGREE EQUATIONS
AND INEQUALITIES

1. EQUATIONS   An **equation** is a mathematical sentence which states that two number expressions, called **members**, are equal. When one or both members of an equation contain a variable, and when the equation is true for all numbers in the domain of the variable—i.e., the replacement set—the equation is called an **identity**. For example,

$$x^2 + 2x + 1 = (x + 1)^2$$

is an identity. No matter which number in the domain replaces the variable, x, a true sentence results. For replacements of, say, 3 and -2:

$$
\begin{aligned}
& & x^2 & \quad + 2x & + 1 &= (x + 1)^2 \\
x = 3: & & 3^2 & \quad + 2(3) & + 1 &\overset{?}{=} (3 + 1)^2 \\
& & 9 & \quad + 6 & + 1 &\overset{?}{=} (4)^2 \\
& & & & 16 &= 16 \\
x = -2: & \quad (-2)^2 & + 2(-2) & + 1 &\overset{?}{=} (-2 + 1)^2 \\
& \quad 4 & + (-4) & + 1 &\overset{?}{=} (-1)^2 \\
& & & 1 &= 1
\end{aligned}
$$

[*Note:* Unless otherwise stated, in this chapter the

50

domain of the variable is assumed to be the set of rational numbers.]

When an equation is true for some but not all numbers in the domain or replacement set, the equation is called a **conditional equation**. For example:

$$x + 2 = 7$$

$$
\begin{aligned}
x = -4: &\quad (-4) + 2 = 7 \quad \text{(false)} \\
x = 0: &\quad (0) + 2 = 7 \quad \text{(false)} \\
x = 3: &\quad (3) + 2 = 7 \quad \text{(false)} \\
x = 5: &\quad (5) + 2 = 7 \quad \text{(true)} \\
x = 6: &\quad (6) + 2 = 7 \quad \text{(false)}
\end{aligned}
$$

Those numbers in the replacement set that make the equation a true sentence are called **solutions** of the equation. Solutions are said to *satisfy* an equation. For an identity, all numbers in the replacement set satisfy the equation. For a conditional equation, only some numbers in the replacement set satisfy the equation.

EXERCISE 3-1

For each equation, the domain of the variable—the set of replacements for the variable— is specified. Test each possible replacement to decide which makes the equation a true sentence.

1. $4 + x + 7 = 2 + 15$ $\quad \{2, 4, 6, 8, 10\}$

2. $3 + x - 2 + 5 = 7 - 9$ $\quad \{-2, -4, -6, -8, -10\}$

3. $3 + 2(x) = x + 5$ $\quad \{1, 2, 3, 4, 5, 6\}$

4. $2x + 3 = 6 - x$ $\quad \{-2, -1, 0, 1, 2\}$

5. $3 + 2x - 1 = x + 2$ $\quad \{-2, -1, 0, 1, 2\}$

6. $x + 3 = 2x + 4$ $\quad \{-2, -1, 0, 1, 2\}$

7. $\frac{1}{2} + x = \frac{5}{6}$ $\quad \{\frac{1}{2}, \frac{1}{3}, \frac{1}{4}, \frac{1}{5}\}$

8. $3x + \dfrac{1}{x} = 8 - \dfrac{3}{x}$ $\quad \{0, 1, 2, 3, 4\}$

9. $\dfrac{4}{x} + 3 = x + \dfrac{1}{x}$ $\quad \{-2, -1, 0, 1, 2\}$

10. $2x + \dfrac{3}{x} = 5x$ $\quad \{-2, -1, 0, 1, 2\}$

11. $\dfrac{x + 7}{x} = x + \dfrac{1}{x}$ $\quad \{0, 1, 2, 3, 4\}$

12. $\dfrac{(x \cdot x) - x}{x + 1} = 2x$ $\quad \{-4, -3, -2, -1, 0, 1\}$

2. SOLVING FIRST-DEGREE EQUATIONS IN A SINGLE VARIABLE

When an equation involves a single variable having an exponent of 1, the equation is known as a **first-degree equation** in a single variable. Because the graph of a first-degree equation is a straight line—as we shall see in Chapter IX—first-degree equations are also called **linear equations**.

Two equations are said to be **equivalent** if they have the very same solutions. The following operations on members of equations always result in a new equation equivalent to the original equation:

1. Addition or subtraction of the same number (which may be represented by the variable) to both members.
2. Multiplication and division of both members by the same number (except zero and terms involving the variable).

EXAMPLES

1. Given the equation: $x - 4 = 6$
Add 4 to each member: $x - 4 + 4 = 6 + 4$
Simplify each member: $x = 10$
The equation $x = 10$ has the same solution as $x - 4 = 6$; this means that the two equations are equivalent. However, the solution is immediately evident in $x = 10$, while in $x - 4 = 6$ it is not.

2. Given the equation: $x + 2 = 7$
Subtract 2 from each member:
$$x + 2 - 2 = 7 - 2$$
Simplify each member: $x = 5$
The equation $x = 5$ is equivalent to $x + 2 = 7$. The solution is immediately evident in $x = 5$, while in $x + 2 = 7$ it is not.

3. Given the equation: $\frac{1}{2}x = 6$
Multiply each member by 2: $2(\frac{1}{2}x) = 2(6)$
Simplify each member $x = 12$
The equation $x = 12$ is equivalent to $\frac{1}{2}x = 6$. The solution is immediately evident in $x = 12$, while in $\frac{1}{2}x = 6$ it is not.

4. Given the equation: $4x = 12$
Divide each member by 4:

$$\frac{4x}{4} = \frac{12}{4}$$

Simplify each member: $x = 3$
The equation $x = 3$ is equivalent to $4x = 12$. The solution is immediately evident in $x = 3$, while in $4x = 12$ it is not.

EXERCISE 3-2

For each of the following equations, a single operation on both members will result in an equivalent equation, but one in which the solution is immediately evident. Find the solution for each.

1. $x - 3 = 7$

2. $x + 4 = 6$

3. $2x = 12$

4. $\frac{1}{3}x = 5$

5. $x - 8 = 2\frac{1}{3}$

6. $x + 3 = 6\frac{5}{8}$

7. $5x = 0.25$

8. $\frac{1}{2}x = \frac{3}{4}$

9. $\frac{3}{4}x = 24$

10. $x + 2 = -6$

11. $x - 3 = -2\frac{1}{4}$

12. $\frac{2}{3}x = -3$

13. $6 = \frac{2}{5}x$

14. $-2x = -9$

15. $-\frac{1}{3}x = 2$

16. $x - 3.62 = 5.91$

17. $x + 6.27 = 4.28$

18. $\frac{x}{25} = 0.03$

19. $\frac{x}{0.03} = -8$

20. $\frac{x}{0.27} = 0.63$

21. $0.25x = 12$

22. $1.37x = -13.7$

23. $-2.6x = 0.26$

24. $-3.62x = -0.00362$

3. PROGRAM FOR SOLVING FIRST-DEGREE EQUATIONS

There is no established pattern for the sequence of equivalent equations that one might follow in solving an equation. A generally useful tactic is to work toward an equation which has the terms containing the variable congregated on one side of the equality sign (either side), and the remaining terms on the other side.

EXAMPLES

1. Solve for x: $7x - 2 = 4x + 7$.

Subtract $4x$ from both members:

$$7x - 2 - \mathbf{4x} = 4x + 7 - \mathbf{4x}$$
$$3x - 2 = 7$$

Add 2 to both members:

$$3x - 2 + \mathbf{2} = 7 + \mathbf{2}$$

$$3x = 9$$

Divide both members by 3:

$$\frac{3x}{\mathbf{3}} = \frac{9}{\mathbf{3}}$$

$$x = 3$$

2. Solve for x: $2x - 12 = 4x - 8$.

Subtract $4x$ from both members:

$$2x - 12 \mathbf{- 4x} = 4x - 8 \mathbf{- 4x}$$

$$-2x - 12 = -8$$

Add 12 to both members:

$$-2x - 12 \mathbf{+ 12} = -8 \mathbf{+ 12}$$

$$-2x = 4$$

Divide both members by -2:

$$\frac{-2x}{\mathbf{-2}} = \frac{4}{\mathbf{-2}}$$

$$x = -2$$

[*Note:* It is not necessary to keep the variables on the left side of the equality sign. In this example, it might have been simpler to maintain the variable on the right side of the equality sign:

$$2x - 12 = 4x - 8$$

Subtract $2x$: $-12 = 2x - 8$

Add 8: $-4 = 2x$

Divide by 2: $-2 = x$]

A program for solving first-degree equations is given below. Note that what we find after applying Step 2 is called a "possible solution." Not all equations have solutions, no matter what set of numbers is made available as possible replacements (e.g., $x + 2 = x$ and $3/x = 0$ are equations which have no solutions). What is determined at the end of Step 2 of the program is a number about which this can be said: *If the original equation has a solution, it must be that number* (or for higher-degree equations, the solutions must be among these numbers). However, this does not assure that the number in question is a solution to the equation any more than the fact that all Presidents of the United States have been males implies that one's great-grandfather was a President because he was a male. Any doubt about a number being indeed a solution can be cleared if we substitute it for the variable in the original equation; if it is a solution, a true sentence results. Thus, Step 3 has more significance than being simply a check against computational errors, as many regard it.

3.1

To solve a first-degree equation:

Step 1. If the equation involves fractional coefficients, multiply both members of the equation by the least common denominator (**LCD**) of the fractions; remove all parentheses and simplify.

Step 2. As necessary, add to, subtract from, divide both members equally to produce an equivalent equation in which one member contains only the variable with a coefficient of $+1$. The other member is a possible solution of the equation.

Step 3. Substitute the possible solution of Step 2 for the variable in the given equation; it is a solution if it satisfies the given equation.

EXAMPLES

1. Solve for x: $5x - \frac{5}{12} = \frac{2}{3} + 3x$.

Step 1.   LCD $= 15$: $15(5x - \frac{2}{15}) = 15(\frac{2}{3} + 3x)$

$$75x - 2 = 10 + 45x$$

Step 2.   Add 2 to each member; subtract $45x$ from each member; simplify:

$$75x - 2 + 2 - 45x = 10 + 45x + 2 - 45x$$

$$30x = 12$$

Divide each member by 30:

$$\frac{30x}{30} = \frac{12}{30}$$

$$x = \tfrac{12}{30} = \tfrac{2}{5} \quad \text{(possible solution)}$$

Step 3.   Verify by substituting $\frac{2}{5}$ for x wherever it appears in the given equation:

$$5x - \tfrac{2}{15} = \tfrac{2}{3} + 3x$$

$$x = \tfrac{2}{5}: \quad 5(\tfrac{2}{5}) - \tfrac{2}{15} \overset{?}{=} \tfrac{2}{3} + 3(\tfrac{2}{5})$$

$$2 - \tfrac{2}{15} \overset{?}{=} \tfrac{2}{3} + \tfrac{6}{5}$$

$$\tfrac{30}{15} - \tfrac{2}{15} \overset{?}{=} \tfrac{10}{15} + \tfrac{18}{15}$$

$$\tfrac{28}{15} = \tfrac{28}{15}$$

Thus $\frac{2}{5}$ is the solution of $5x - \frac{2}{15} = \frac{2}{3} + 3x$.

2. Solve for y: $\frac{5}{6}(y + 1) = \frac{1}{2}y - \frac{1}{4}(2y + 5)$.

Step 1.   LCD $= 12$:

$$12[\tfrac{5}{6}(y + 1)] = 12[\tfrac{1}{2}y - \tfrac{1}{4}(2y + 5)]$$

$$10(y + 1) = 6y - 3(2y + 5)$$

$$10y + 10 = 6y - 6y - 15$$

$$10y + 10 = -15$$

Step 2.   Subtract 10 from both members:

$$10y + 10 - 10 = -15 - 10$$

$$10y = -25$$

Divide each member by 10:

$$\frac{10y}{10} = \frac{-25}{10}$$

$$y = \frac{-25}{10} = -\frac{5}{2}$$

Step 3.   Verify by replacing y in the given equation with $-\frac{5}{2}$:

$$\tfrac{5}{6}(y + 1) = \tfrac{1}{2}y - \tfrac{1}{4}(2y + 5)$$

$$y = -\tfrac{5}{2}: \qquad \tfrac{5}{6}(-\tfrac{5}{2} + 1) \overset{?}{=} \tfrac{1}{2}(-\tfrac{5}{2}) - \tfrac{1}{4}[2(-\tfrac{5}{2}) + 5]$$

$$\tfrac{5}{6}(-\tfrac{3}{2}) \overset{?}{=} \tfrac{1}{2}(-\tfrac{5}{2}) - \tfrac{1}{4}(-5 + 5)$$

$$-\tfrac{5}{4} \overset{?}{=} -\tfrac{5}{4} - \tfrac{1}{4}(0)$$

$$-\tfrac{5}{4} = -\tfrac{5}{4}$$

Thus $-\frac{5}{2}$ is the solution.

EXERCISE 3-3

Solve the following equations.

1. $3x + 8 = 12 + x$

2. $3x - 12 = 2x - 10$

3. $3x + 7 = 6x$

4. $4x - 11 = 2x - 7$

5. $3x - 8 = 12x - 7$

6. $4x - 3 = 3(6 - x)$

7. $7x = -2(x + 9)$

8. $(2a - 9) - (a - 3) = 0$

9. $4 - (2 - b) = 2 + 2b$

10. $(1 - t) - 6 = 3(7 - 2t)$

11. $3x + 4(3x - 5) = 12 - x$

12. $3(6a - 5) = 7(3a + 10)$

13. $4.6 + 2x + 1 = 3.8 + 5$

14. $6(x - 5) = 15 + 5(7 - 2x)$

15. $\frac{1}{2}x - x = 12$

16. $a - 5 = \frac{1}{4}a$

17. $\frac{p}{3} - \frac{7}{6} + \frac{2p}{5} = \frac{3p}{4}$

18. $\frac{1}{3}(x + 2) - \frac{1}{2}(x + 8) = -2$

19. $\frac{5}{7}y - y - \frac{5}{3} = \frac{1}{21}(3y + 1)$

20. $1 - \frac{1}{15}(a + 7) = -\frac{1}{12}(3a - 2)$

21. $(a - 1) - (a + 4) = (2a - 5)$

22. $\frac{x}{2} + \frac{3x + 1}{5} = \frac{x + 3}{10}$

23. $\frac{2x + 3}{4} - \frac{3x + 2}{12} = 1$

24. $\frac{x + 1}{3} - \frac{x + 3}{5} = x - \frac{3x - 2}{3}$

25. Is 1 the only solution of

$$\frac{x - 3}{5} + 2x = \frac{11x - 3}{5}?$$

What are equality sentences of this type called?

26. Without solving the equation, decide whether $\frac{1}{2}$ is a solution of

$$(x - 2) + \frac{x - 3}{3} = \frac{x + 7}{15}.$$

4. LITERAL EQUATIONS

Equations containing several variables are sometimes called *literal equations*; for example:

$$2x - p = 4$$

$$mx + 3 = 7b$$

$$a = lw$$

Formulas of science, engineering, business, etc., are usually equations of this type. For instance:

$$I = Prt \qquad \text{(simple interest)}$$

$$A = \tfrac{1}{2}h(b_1 + b_2) \quad \text{(area of a trapezoid)}$$

$$C = \tfrac{5}{9}(F - 32) \qquad \text{(Fahrenheit-centigrade conversion)}$$

To solve such equations for a certain variable means to treat the equation as though it were one in a single variable (the variable to be solved for) and the other variables were constants. The equation is then solved in the manner of any other equation in a single variable.

EXAMPLES

1. Solve $2x - p = 4$ for x.

Here x is treated as the variable, p and 4 as constants.

$$2x - p = 4$$

$$2x = 4 + p$$

$$x = \frac{4 + p}{2} \quad \text{(possible solution)}$$

Verification:

$$2x - p = 4$$

$$x = \frac{4 + p}{2}: \qquad 2\left(\frac{4 + p}{2}\right) - p \overset{?}{=} 4$$

$$4 + p - p \overset{?}{=} 4$$

$$4 = 4$$

$\dfrac{4 + p}{2}$ is the solution.

2. Solve $2x - p = 4$ for p.

Here p is treated as the variable, x and 4 as constants.

$$2x - p = 4$$
$$-p = 4 - 2x$$
$$p = 2x - 4 \quad \text{(possible solution)}$$

Verification:

$$2x - p = 4$$
$$p = 2x - 4: \quad 2x - (2x - 4) \overset{?}{=} 4$$
$$2x - 2x + 4 \overset{?}{=} 4$$
$$4 = 4$$

$2x - 4$ is the solution.

3. Solve $A = \frac{1}{2}h(b_1 + b_2)$ for b_1.

$$A = \frac{1}{2}h(b_1 + b_2)$$
$$2A = hb_1 + hb_2$$
$$2A - hb_2 = hb_1$$
$$\frac{2A - hb_2}{h} = b_1 \quad \text{(possible solution)}$$

Verification:

$$A = \frac{1}{2}h(b_1 + b_2)$$
$$b_1 = \frac{2A - hb_2}{h}: \quad A \overset{?}{=} \frac{1}{2}h\left[\left(\frac{2A - hb_2}{h}\right) + b_2\right]$$
$$A \overset{?}{=} \frac{1}{2}h\left[\frac{2A - hb_2 + hb_2}{h}\right]$$
$$A \overset{?}{=} \frac{1}{2}h\left[\frac{2A}{h}\right]$$
$$A = A$$

$\dfrac{2A - hb_2}{h}$ is the solution.

EXERCISE 3-4

Solve for x.

1. $5x = b$

2. $4a^2x = 16$

3. $x + 6 = d$

4. $6x + 3m = 9m$

5. $2x - 6b = 8a$

6. $3x + c = 2x$

7. $3a - x = x - 4b$

8. $\dfrac{2x + 3b}{7} = \dfrac{x + b}{3}$

9. $\frac{1}{2}(x - 6a) = \frac{1}{3}(6a - x)$

10. $\frac{3x + 2y}{3} = \frac{x + 4y}{6}$

11. $3(cx - 2ac) = c(x - 4a)$

12. $2(px + pd) - p(2d - x) = 0$

13. $\frac{6x - 5a}{2} = \frac{5x + 2a}{8} + x$

14. $5a(bx - 3bc) = 7b(ax - 5ac)$

15. $0.3x + 0.5a = 1.8x + 4.5b$

16. $\frac{3}{4}\left(\frac{x}{m} + 4\right) = \frac{2}{3}\left(\frac{x}{m} - 1\right)$

17. Solve $y = mx + b$ (equation of a straight line) for b; for x; for m. When $m = \frac{1}{2}$, $b = 5$, and $y = 7$, what is x?

18. Solve $P_2 = P_1 T_2 / T_1$ (Gay-Lussac's law of gases) for P_1; T_1; T_2. When $P_1 = 15$ lb, $T_1 = 300°$A, and $T_2 = 280°$A, what is P_2?

19. Solve $S = \frac{n}{2}(a + l)$ (sum of arithmetic progression) for n; a; l. When $n = 8$, $a = 2$, and $S = 48$, what is l?

20. Solve $\frac{x^2}{a^2} + \frac{y^2}{b^2} = 1$ (equation of an ellipse) for x^2 and for y^2. When $x = 2$, $a = 4$, and $b = 2$, what is y^2?

5. INEQUALITIES

An **inequality** is also a mathematical sentence, one which states that two number expressions (members) are unequal; for example:

$$3 + x > 5$$
$$6 < x + 3$$

As with equations, two inequalities are equivalent if they have the same solutions, though inequalities generally have a great many more solutions than do equations. Let us assume that the replacement set for the first inequality above is the set of integers. Because every integer greater than 2 will satisfy the inequality, all integers greater than 2 are said to belong to the **solution set**. Thus, for

$$3 + x > 5$$

the solution set is

$$\{3, 4, 5, 6, 7, \ldots\}$$

since

$$3 + 3 > 5$$
$$3 + 4 > 5$$
$$3 + 5 > 5$$

etc.

For the inequality

$$x + 2 < 4$$

when the replacement set is the set of integers, the solution set is

$$\{\ldots, -3, -2, -1, 0, 1\}.$$

[*Note:* The brace symbol is standard notation in mathematics for listing the members of a set. More on sets may be found in Supplementary Unit C, pages 289–301.]

EXAMPLES

1. What is the solution set for $8 + x > 12$ if the replacement set is the set of integers?

The integer 4 is not in the solution set because $8 + 4$ *equals* 12; it is not *greater than* 12. Integers less than 4 are not in the solution set either, because when each is added to 8, the resulting sum is less than 12. Integers greater than 4, however, are in the solution set because, when each is added to 8, the resulting sum is greater than 12. So, the set of integers greater than 4, or $\{5, 6, 7, 8, 9, \ldots\}$, satisfies $8 + x > 12$.

2. What is the solution set for $x + 3 < 6$ if the replacement set is the set of integers?

All integers less than 3, or $\{\ldots, -4, -3, -2, -1, 0, 1, 2\}$, satisfy the inequality $x + 3 < 6$.

3. What is the solution set for $5 \leq x - 1$ if the replacement set is the set of integers?

The symbol $\leq$ means "is less than or equal to." The sentence, $5 \leq x - 1$, is really a double or compound sentence which states "$5 < x - 1$ or $5 = x - 1$." The solution set for such compound sentences contains all those numbers which make *either* sentence true. Thus, substituting 6 for x makes $5 = x - 1$ true, and the integers, $7, 8, 9, 10, \ldots$, make $5 < x - 1$ true when each replaces the variable. So

$$\{6, 7, 8, 9, 10, \ldots\}$$

is the solution set for $5 \leq x - 1$.

4. What is the solution set for $2x \geq -6$ if the replacement set is the set of integers?

The symbol $\geq$ means "is greater than or equal to." The compound sentence $2x \geq -6$ is equivalent to the two sentences, $2x = -6$ and $2x > -6$. The solution set for $2x \geq -6$ is

$$\{-3, -2, -1, 0, 1, \ldots\}$$

Had the replacement set in the preceding four examples been the set of rational numbers instead of the set of integers, the various solution sets would have contained many more numbers. Because such solution sets cannot be readily tabulated, a graph is frequently used, as illustrated in the following examples.

EXAMPLES

1. Graph the solution set for $8 + x > 12$ if the domain of the variable is the set of rational numbers.

   $8 + x > 12$:

   The graph is based on the number line for rational numbers. The circled dot at 4 on the number line means that 4 is *not* in the solution set. The thickened half-line to the right of 4 means that all rational numbers greater than 4 are in the solution set.

2. Graph the solution set for $x + 3 < 6$ if the replacement set is the set of rational numbers.

   $x + 3 < 6$: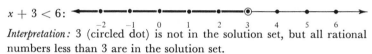

   Interpretation: 3 (circled dot) is not in the solution set, but all rational numbers less than 3 are in the solution set.

3. Graph the solution set for $5 \leq x - 1$ if the replacement set is the set of rational numbers.

   $5 \leq x - 1$: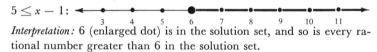

   Interpretation: 6 (enlarged dot) is in the solution set, and so is every rational number greater than 6 in the solution set.

4. Graph the solution set for $2x \geq -6$ if the domain of the variable is the set of rational numbers.

   $2x \geq -6$:

   Interpretation: -3 (enlarged dot) is in the solution set, and so is every rational number greater than -3 in the solution set.

EXERCISE 3-5

Test each member in the replacement set accompanying the inequality. Tabulate the solution set using brace notation, e.g., {1, 2, 3}.

1. $x < 5$ {3, 4, 5, 6, 7}

2. $y > 2$ $\{0, 1, 2, 3, 4, 5\}$

3. $m > 2$ $\{0, 1, 2, 3, 4, 5\}$

4. $n < 5$ $\{1, 2, 3, 4, 5, 6\}$

5. $t > -4$ $\{-6, -5, -4, -3, -2\}$

6. $x < -1$ $\{-3, -2, -1, 0, 1\}$

7. $a < -6$ $\{-8, -7, -6, -5, -4\}$

8. $p > 0$ $\{-3, -2, -1, 0, 1, 2\}$

9. $y < 20$ $\{0, 5, 10, 15\}$

10. $n > -2$ $\{-3, -2, -1, 0, 1, 2, 3\}$

11. $g \geq -1$ $\{-2, -1, 0, 1, 2, 3\}$

12. $k \leq 5$ $\{6, 2, 4, 5, -3, 0\}$

13. $x \leq -3$ $\{-10, -5, 0, 5, 10\}$

14. $d \geq 0$ $\{-5, 0, 5, 10, \ldots\}$

Graph the solution set for each inequality. The replacement set is the set of rational numbers.

15. $x > 3$ **16.** $x < 5$

17. $x < -3$ **18.** $x > -4$

19. $x \geq 6$ **20.** $x \leq 5$

21. $x \leq 3$ **22.** $x \geq 1$

23. $x \geq -2$ **24.** $x \leq -7$

25. $x \leq 0$ **26.** $x \geq -1$

6. SOLVING FIRST-DEGREE INEQUALITIES IN A SINGLE VARIABLE

Simple inequalities, such as those of the Examples in Section 5, can often be solved by inspection. More complicated inequalities are not so easily solved. As with equations, however, it is possible to proceed through a series of equivalent but simpler inequalities until we arrive at an inequality for which the solution set is obvious.

The following operations invariably result in an inequality that is equivalent to the original inequality, i.e., they both have identical solution sets.

1. Addition or subtraction of the same number (which may be in the form of the variable) to both members.
2. Multiplication or division of both members by the same positive number (variable excepted).
3. Multiplication or division of both members by the same negative

number (variable excepted) when the sense of the inequality is reversed.

[*Note*: Two inequalities have the same *sense* when their symbols of inequality both "point" in the same direction. For instance, $x < 4$ and $y < 8$ have the same sense; $a > 6$ and $b < 2$ are opposite in sense.]

EXAMPLES

1. Solve: $8 + x > 12$.

Subtract 8 from both members:

$$8 + x - \mathbf{8} > 12 - \mathbf{8}$$

Simplify:

$$x > 4$$

The inequality $x > 4$ is equivalent to $8 + x > 12$. The solution set for $x > 4$ is immediately evident, but not for $8 + x > 12$. (Compare with Examples 1 in Section 5.)

2. Solve: $5 \le x - 1$.

Add 1 to both members:

$$5 + \mathbf{1} \le x - 1 + \mathbf{1}$$

Simplify:

$$6 \le x$$

(Compare with Examples 3 in Section 5.)

[*Note:* $6 \le x$ and $x \ge 6$ are equivalent.]

3. Solve: $2x \ge -6$.

Divide both members by 2:

$$\frac{2x}{\mathbf{2}} \ge \frac{-6}{\mathbf{2}}$$

$$x \ge -3$$

(Compare with Examples 4 in Section 5.)

4. Solve: $3 - 2x < 5$.

Subtract 3 from both members:

$$3 - 2x - \mathbf{3} < 5 - \mathbf{3}$$

$$-2x < 2$$

Divide by -2 (reverse the sense or direction of the inequality):

$$\frac{-2x}{\mathbf{-2}} > \frac{2}{\mathbf{-2}}$$

$$x > -1$$

An alternative procedure for solving such inequalities may be programmed as follows.

3.2

To solve a first-degree inequality in a single variable:

Step 1. Transform the given inequality into an equation by replacing the $<$ or $>$ sign with an $=$ sign.

Step 2. Solve the equation of Step 1.

Step 3. Choose two trial numbers, one greater and one less than the solution found in Step 2, and substitute each for the variable in the given inequality.

    (a) If the greater trial number satisfies the inequality, then its solution set consists of those numbers in the replacement set that are greater than the solution found in Step 2.

    (b) If the lesser trial number satisfies the inequality, then its solution set consists of those numbers in the replacement set that are less than the solution found in Step 2.

EXAMPLES

1. Solve: $3 - 2x < 5$.

Step 1. Replace $<$ with $=$:

$$3 - 2x = 5$$

Step 2. Solve for x:

$$-2x = 5 - 3$$
$$-2x = 2$$
$$x = -1$$

Step 3. Select two trial numbers, one greater than -1 (say, 0) and one less than -1 (say, -3), and substitute each for the variable in the given inequality.

$$3 - 2x < 5$$

$$x = 0: \qquad 3 - 2(0) < 5$$
$$3 - 0 < 5$$
$$3 < 5 \quad \text{(true)}$$

$$x = -3: \qquad 3 - 2(-3) < 5$$
$$3 + 6 < 5$$
$$9 < 5 \quad \text{(false)}$$

The greater trial number satisfies the given inequality; therefore, the solution set for the given inequality is $x > -1$, or

(Compare with Example 4, p. 63.)

2. Solve: $2x - 2 > 4x + 6$.

Step 1. $2x - 2 = 4x + 6$

Step 2. $2x - 4x = 6 + 2$

$$-2x = 8$$

$$x = -4$$

Step 3. Trial numbers: $-5(< -4)$ and 1 (> -4) .

$$2x - 2 > 4x + 6$$

$$x = -5: \quad 2(-5) - 2 > 4(-5) + 6$$

$$-12 > -14 \quad \text{(true)}$$

$$x = 1: \quad 2(1) - 2 > 4(1) + 6$$

$$0 > 10 \quad \text{(false)}$$

Solution: $x < -4$, or

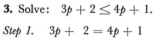

3. Solve: $3p + 2 \le 4p + 1$.

Step 1. $3p + 2 = 4p + 1$

Step 2. $3p - 4p = 1 - 2$

$$-p = -1$$

$$p = 1$$

Step 3. Trial numbers: 0 and 2.

$$3p + 2 \le 4p + 1$$

$$p = 0: \quad 0 + 2 \le 0 + 1$$

$$2 \le 1 \quad \text{(false)}$$

$$p = 2: \quad 3(2) + 2 \le 4(2) + 1$$

$$8 \le 8 + 1$$

$$8 \le 9 \quad \text{(true)}$$

Solution: $p \ge 1$, or

[*Note:* $p = 1$ is included in the solution set because 1 satisfies the equality part of the given inequality.]

Solution sets may also be expressed by what is called **set-builder notation**. For example,

$\{x \mid x < 3, x \in I\}$ means "the set of all integers less than 3."

$\{m \mid m > 2, m \in R\}$ means "the set of all rational numbers greater than 2."

The symbolism, taken apart and expressed literally, may be read:

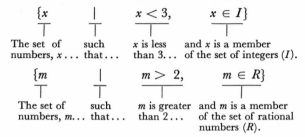

The solution sets for the three examples on pages 64 and 65 may therefore be expressed in set-builder notation as follows:

Example 1: $\{x \mid x > -1, x \in R\}$

Example 2: $\{x \mid x < -4, x \in R\}$

Example 3: $\{p \mid p \geq 1, p \in R\}$

EXERCISE 3-6

Solve the following inequalities. When the specified replacement set is the set of intergers (I), give the solution set in regular set notation, e.g., $\{1, 2, 3, \ldots\}$. When the replacement set is the set of rational numbers (R), graph the solution set.

1. $x + 1 < 6$ $\{x \in I\}$

2. $6 < x - 1$ $\{x \in I\}$

3. $p + 3 \geq 4$ $\{p \in I\}$

4. $5 > t - 6$ $\{t \in I\}$

5. $x + 3 \leq 0$ $\{x \in I\}$

6. $3x < 12$ $\{x \in I\}$

7. $144 \leq 12m$ $\{m \in I\}$

8. $\frac{1}{2}k \leq -5$ $\{k \in I\}$

9. $p \leq -2$ $\{p \in R\}$

10. $x \geq 0$ $\{x \in R\}$

11. $3x > -15$ $\{x \in R\}$

12. $14 \leq \frac{1}{2}x$ $\{x \in R\}$

13. $2a - 3 \leq 5a$ $\{a \in R\}$

14. $3z - 4 > 2z - 9$ $\{z \in R\}$

15. $k + 3 < 2k + 1$ $\{k \in I\}$

16. $6 - x > 2x + 5$ $\{x \in I\}$

17. $2p - 3p + 7 \geq 0$ $\{p \in R\}$

18. $0 < 2x - 3 + 4x$ $\{x \in R\}$

19. $3m - 2 + m \leq 5m + 6$ $\{m \in R\}$

20. $5p - 2 + 7p > 4p - 6 + 2p$ $\{p \in R\}$

21. Use set-builder notation to express the solution sets for Exercises 1, 5, 11, 13, and 15.

REVIEW—CHAPTER III

From the specified replacement set, select the member that will make the equality true.

1. $\dfrac{2}{x} + x = 2x - 3\frac{1}{2}$ $\{0, 1, 2, 3, 4\}$

2. $\dfrac{3}{x} + 2x = 3x + 2$ $\{-5, -4, -3, -2\}$

3. $\dfrac{2}{x} + x = \dfrac{3}{x} - 1.5$ $\{\frac{1}{5}, \frac{1}{4}, \frac{1}{3}, \frac{1}{2}, 1\}$

4. $\dfrac{2}{x^2} + 3x = x^2 + 2\frac{1}{2}$ $\{0, 1, 2, 3, 4\}$

5. $2x + 4x^2 = 204x^2$ $\{1, 0.1, 0.01, 0.001, 0.0001\}$

6. $x^2 + 2x - 3 = 0$ $\{-1, -2, -3, -4\}$

Solve for x.

7. $2x - 4 + 3x - 2x = 6 + x + 7x$

8. $4x - 2 + 3x + 7 = 4 - 3x + 2 - 5x$

9. $3x + 2 = 4(x - 7) + 6$

10. $3(x - 5) = 2(3x - 2) - 2$

11. $4(2x - 3) = 7(x - 4 + x)$

12. $2(7 - 3x) + 10 = 5(2x + 3 - 6 + 2x)$

13. $4(2x + 5) + 3(x - 7) = x - 1$

14. $2(x - 3 + x) - x + 3 = 5(x - 2 + x - 7)$

15. $3(x + 2) - 4(x - 7) + 3(x - 5) = 2(x - 1)$

16. $\dfrac{x - 2}{5} + \dfrac{3x - 2}{6} = \dfrac{x + 8}{15}$

17. $\frac{1}{3}(4 - x) - \frac{3}{5}(5 - 2x) = \frac{2}{3}(x - 7)$

18. $\frac{2}{3}(2 - 3x) + x - \frac{3}{4}(x - 5) = 2 + \frac{1}{6}(2x + 1)$

19. $m - 3(2x - 3m) = 2(3x - m)$

20. $a(3p + x) - am = 2(ap - ax)$

21. $6a = 4(a - 3x) + 5(a + x)$

22. $3a - 2x = \dfrac{x - 8a}{6}$

23. $\frac{5}{6}(c + 3x) = \frac{3}{4}(x - 2c)$

24. $\dfrac{c - 4x}{3} = \dfrac{2c - x}{5} + x$

25. $2(m + 3x) - 2(4 - x) = 3(2x - 1) + 2x$

26. $\frac{1}{3}(x - 5a + 2) = \frac{3}{4}(2x - 5a + 2) + \frac{5}{12}(5a - 2)$

27. $\dfrac{2}{5}\left(\dfrac{2x}{a} + 3\right) = \dfrac{3}{4}\left(\dfrac{2x}{a} - 4\right)$

28. Solve $A = \dfrac{h}{2}(b_1 + b_2)$ (area of a trapezoid) for h; for b_2. When $A = 36$, $b_1 = 10$, and $h = 4$, what must be the value of b_2?

Solve for the variable; the replacement set is the set of integers.

29. $x - 1 < 6$ **30.** $y + 5 \geq 6$

31. $-4 > t + 5$ **32.** $m - 4 \geq 0$

33. $3x \leq -15$ **34.** $100 \geq 10a$

35. $\frac{1}{3}p + 2 > 8$ **36.** $-6 + \frac{1}{2}x < 0$

Solve for the variable; the replacement set is the set of rational numbers.

37. $k \geq -4$ **38.** $x < 0$

39. $2x > -4$ **40.** $3m + 2 \leq -4$

41. $-x + 3 > 0$ **42.** $2m - 3 \leq 4m - 5$

43. $0 \leq 4x - 2 - 7x$ **44.** $3y - \frac{1}{2} \geq y - 4 + 3y$

IV

PROBLEM SOLVING
BY EQUATION

Algebra is truly an international language, one that is especially well-suited for expressing quantitative and spatial relationships. Like any language, it possesses its own symbolism, punctuation, and grammar (rules of operation), and it conveys its ideas in words, phrases, and sentences. To solve problems by algebra, it is necessary to translate them into the language of algebra, then apply solution techniques peculiar to that language. In this chapter, we direct our attention to a selection of problems which may be translated into algebraic sentences of equality and then solved.

An algebraic sentence, like any similarly complex sentence in English, is made up of phrases; these phrases in turn are made up of terms which correspond to the data of the problem. Thus, before considering algebraic sentences, we need to consider the matter of translating the information of the problem, both what is known and what is sought (unknown), into accurate algebraic phrases.

Ordinarily, most problems solvable by first-degree equations involve either comparisons of more and less, or of ratios. Phrases of more and less can be expressed by sums and differences, using a letter (variable) to hold the place of an unspecified or unknown number. For instance, if x represents some unknown number, then

1. TRANSLATING FROM ENGLISH TO ALGEBRA

69

$x + 7$ or $7 + x$ can be used to symbolize the number that is "7 greater than the unknown," or perhaps the number "x more than 7." The "difference between 12 and the unknown," or perhaps "12 less than the unknown," may be expressed algebraically as $x - 12$.

Phrases of ratio and measurement involving multiple and submultiple units can be stated in terms of products and quotients. For instance, if n represents an unknown number, then $7n$ represents the number "seven times the size of n"; similarly, $\frac{1}{7}n$ or $n/7$ represents the number "$\frac{1}{7}$ the size of n", or perhaps "one of seven equal parts of n."

Algebraic phrases, like those in English, can also be compounded. For instance, $5(x + 7)$ represents a number "five times as great as the unknown number (x) increased by 7"; $\frac{1}{3}(x - 2)$ represents a number that is "$\frac{1}{3}$ the size of an unknown number which has been diminished by 2." And $(5x + 10x)/100$ might represent the worth in dollars of any group of coins containing the same number (x) of nickels as dimes.

Translating the statement of a problem into accurate algebraic language is absolutely essential to successful problem solving in algebra.

EXERCISE 4-1

Translate the following statements into algebraic language; use x, unless otherwise indicated, to represent the unknown.

1. A number 5 more than the unknown.

2. A number 17 less than the unknown.

3. A number twice the size of the unknown.

4. A number one-fourth the size of the unknown.

5. A number 3 more than twice the unknown.

6. A number half the size of 3 more than the unknown.

7. A number 40% of the size of the unknown.

8. A number 40% larger than the unknown.

9. The remainder after the unknown has been reduced by 6.

10. The remainder after the unknown has been subtracted from 6.

11. The remainder after the unknown has been reduced by one-sixth of itself.

12. 17 divided by double the unknown.

13. The sum of two numbers, a and b, decreased by half their product.

14. The other number when one number is x and their sum is p.

15. John's age (x) 17 years from now.

16. Cost of x items at p dollars per item.

17. One-fifth of twice the unknown reduced by 20.

18. One-fifth of twice the unknown, reduced by 20.

19. The number of minutes in x hours.

20. The number of weeks in x days.

Translate the following statements into algebraic language. Use x to represent the unknown and the $=$ sign to represent such predicates as "is", "is the same number as," and "equals." Solve the resulting equation for x.

21. The unknown decreased by 7 is 15.

22. The unknown increased by 8 is 19.

23. Five less than twice the unknown is the same number as 16.

24. Three times the unknown is 8 greater than the unknown.

25. Ten divided by the unknown equals 2.

26. Sixty is the same number as 4 greater than twice the unknown.

27. Ten increased by 8 and the unknown is equal to twice the unknown.

28. One-third of the unknown equals the unknown diminished by 6.

29. Two-thirds of the unknown, increased by 4, is 8.

30. Two-thirds of the unknown increased by 4 is 8.

2. SOLVING PROBLEMS BY EQUATIONS

The first step in solving a problem by equation is to decide upon the unknown to be sought from the situation described in the statement of the problem. Then *represent that unknown by a letter symbol*, which subsequently will become a variable in an equation.

The second step is to *identify within the problem some equivalence relationship which involves known data and the unknown* of the problem. Every problem capable of solution by an equation must contain—either stated or implied—such an equivalence. The following two problems are illustrative.

PROBLEM WITH THE EQUIVALENCE RELATIONSHIP STATED: What number when doubled and then increased by 3 is 27?

If we substitute the word "equals" for the "is" and translate the phrase before it, using x to represent the unknown but sought-for number, we get

x represents the unknown number.

$2x$ represents the unknown number doubled.

$2x + 3$ represents the doubled unknown number increased by 3.

$2x + 3 = 27$ is the equation.

PROBLEM WITH THE EQUIVALENCE RELATIONSHIP IMPLIED: What is the altitude of a triangle that has an area of 36 and a base of 6?

In this problem it is necessary to know, either as a matter of fact or by formula, that the area of a triangle is numerically equal to one-half the product of its base and altitude measures. Thus, if we let x represent the number associated with the unknown altitude, then

$$(\text{area}) \quad 36 = \tfrac{1}{2}(6)(x) \qquad \left.\vphantom{\begin{array}{c}a\\b\\c\end{array}}\right\} \text{ the equation}$$

with (altitude) pointing to x and (base) pointing to 6.

The third step in problem solving by equation has already been demonstrated in the two foregoing illustrations: *incorporate the unknown and the knowns into an algebraic statement of equivalence.* This is the heart of the solution of the problem, for it provides an equation by which the unknown can become known.

The fourth step is purely mechanical: *solve the equation for the variable which represents the unknown.*

The final step is to verify by *substituting the solution of the equation back into the problem* rather than into the equation. Substitution of the solution back into the equation simply verifies the computational work and will not detect any error in the setup of the equation, the major source of error in the solution of problems by equation. Morevoer, not every solution of an equation is certain to be a solution of the problem.

4.1

To solve a problem by equation:

Step 1. Determine from the statement of the problem that which is sought —the unknown—and represent it by a letter symbol.

Step 2. Identify some equivalence relationship which involves both known data and the unknown of the problem.

Step 3. State algebraically (by equation) the equivalence relationship of Step 2.

Step 4. Solve the equation of Step 3 for the variable.

Step 5. Verify by substituting the solution found in Step 4 back into the problem.

EXAMPLES

1. A board 12 ft long is cut in two pieces so that one piece is three times the length of the shorter piece. How long is the shorter piece?

Step 1.   What is sought? The length of the shorter piece. Let x represent the length of the shorter piece.†

Step 2.   Identify an equivalence relationship:

length of short piece + length of long piece = total length

Step 3.   State algebraically (by equation) the equivalence relationship of Step 2: If x represents the length of the shorter piece, then the length of the longer piece, which is three times the length of the shorter piece, can be represented as $3x$. Therefore:

length of short piece + length of long piece = total length

$$x \qquad + \qquad 3x \qquad = 12 \text{ (correspond-ing equation)}$$

Step 4.   Solve the equation of Step 3:

$$x + 3x = 12$$
$$4x = 12$$
$$x = 3$$

Length of short piece: 3 ft.

Step 5.   Verify by going back to the problem and reasoning: If the length of the short piece is 3 ft, then the length of the long piece, which is three times that of the short piece, must be 3×3 ft, or 9 ft; the sum of the lengths of the short and the long piece is 3 ft + 9 ft, or 12 ft, which should equal the length of the board (12 ft). Since it does, the solution of the equation is verified in the problem.

2. Given three consecutive integers. If the sum of the two smaller integers is decreased by three times the largest, the result is -37. What are the integers?

Step 1.   Three consecutive integers are sought. Let x represent the smallest of these.

Step 2.   Equivalence relationship:

$$\text{integer}_1 + \text{integer}_2 - 3(\text{integer}_3) = -37$$

Step 3.   If the smallest integer is represented by x, the next consecutive integer will be represented by one greater than x, or $(x + 1)$; the integer after the second will be two greater than the smallest, or one greater than the second integer; so $(x) + 2$ or $(x + 1) + 1$ or $(x + 2)$ represents the third integer:

---

†More precisely, "Let x represent the *number of feet* in the length of the shorter piece." The variable represents only numbers: not lengths, but the *number* of units of measure in the length; not rates, but the *number* of miles per hour or the *number* of feet per second, etc. For simplicity of statement, however, we assume the reader understands that, when we say "Let x represent *something*," we mean that x represents the number associated with that *something*.

$$\text{integer}_1 + \text{integer}_2 - 3(\text{integer}_3) = -37$$
$$x \quad + (x+1) - 3(x+2) \quad = -37 \text{ (corresponding equation)}$$

Step 4.   Solve the equation of Step 3:

$$x + (x+1) - 3(x+2) = -37$$
$$x + x + 1 - 3x - 6 = -37$$
$$x + x - 3x = -37 + 6 - 1$$
$$x = 32$$

If $x = 32$, then $x + 1 = 33$ and $x + 2 = 34$. Hence, integer$_1$ is 32, integer$_2$ is 33, and integer$_3$ is 34.

Step 5.   The sum of the two smaller integers is $32 + 33$ or 65; if 65 is diminished by 3 times the larger $(3 \times 34 = 102)$, the result should be -37: $65 - 102 = -37$. The solution checks in the problem.

3. A merchant sells a mixture of olive oil and corn oil for salad dressing. If olive oil is priced at \$0.32 a pint and corn oil at \$0.28 a pint, how much of each should he use to set the price for ten pints of the mixture at \$2.94?

Let x represent the number of pints of corn oil. Then $10 - x$ will represent the number pints of olive oil. The equivalence relationship is:

$$\left[\begin{matrix}\text{pints of}\\ \text{olive oil}\end{matrix} \times \begin{matrix}\text{price per}\\ \text{pint}\end{matrix}\right] + \left[\begin{matrix}\text{pints of}\\ \text{corn oil}\end{matrix} \times \begin{matrix}\text{price per}\\ \text{pint}\end{matrix}\right] = \$2.94$$

$$[(10 - x) \times (\$0.32)] + [(x) \times (\$0.28)] = \$2.94$$

The corresponding equation is:

$$(10 - x)(0.32) + (x)(0.28) = 2.94$$
$$3.2 - 0.32x + 0.28x = 2.94$$
$$-0.04x = -0.26$$
$$x = \frac{-0.26}{-0.04} = \frac{26}{4}$$
$$x = 6\tfrac{1}{2}$$

If $x = 6\tfrac{1}{2}$, then

$$10 - x = 10 - 6\tfrac{1}{2} = 3\tfrac{1}{2}$$

Hence

$$\text{pints of corn oil: } 6\tfrac{1}{2}$$
$$\text{pints of olive oil: } 3\tfrac{1}{2}$$

Verification:

$$6\tfrac{1}{2} \text{ pts corn oil @ } \$0.28 = \$1.82$$
$$3\tfrac{1}{2} \text{ pts olive oil @ } \$0.32 = \$1.12$$
$$10 \text{ pts} \qquad\qquad\qquad \$2.94$$

4. A coin bank holds nickels, dimes, and quarters. If it contains three times as many dimes as nickels and two more quarters than dimes, and if the total value of its contents is $4.90, how many of each coin are there in the bank?

Let x represent the number of nickels. Then $3x$ represents the number of dimes, and $3x + 2$ represents the number of quarters. The equivalence relationship is:

$$\left(\begin{array}{c}\text{No. of}\\\text{nickels}\end{array} \times 5¢\right) + \left(\begin{array}{c}\text{No. of}\\\text{dimes}\end{array} \times 10¢\right) + \left(\begin{array}{c}\text{No. of}\\\text{quarters}\end{array} \times 25¢\right) = 490¢$$

The corresponding equation is:

$$[(x)(5)] + [(3x)(10)] + [(3x + 2)(25)] = 490$$
$$5x + 30x + 75x + 50 = 490$$
$$110x = 440$$
$$x = 4$$

If $x = 4$, then $3x = 12$ and $3x + 2 = 14$; hence

$$\text{nickels: } 4$$
$$\text{dimes: } 12$$
$$\text{quarters: } 14$$

Verification:

$$\begin{array}{lcl}4 \text{ nickels} & = & \$0.20 \\ 12 \text{ dimes} & = & 1.20 \\ 14 \text{ quarters} & = & 3.50 \\ \hline & & \$4.90\end{array}$$

5. A stream flows at the rate of 4 miles per hour. A launch takes 3 hours to make a trip to a certain town downstream, but it takes 7 hours to return because of the adverse current. What is the speed of the launch in still water?

Let x represent the speed of the launch in still water. Then $x + 4$ represents the speed of the launch going downstream, and $x - 4$ represents the speed of the launch going upstream. The basic equivalence for this problem will be in terms of distance:

$$\text{distance down} = \text{distance back}$$
$$(\text{rate down}) \times (\text{time down}) = (\text{rate back}) \times (\text{time back})$$
$$(x + 4) \quad \times \quad (3 \text{ hr}) \quad = \quad (x - 4) \quad \times \quad (7 \text{ hr})$$

The corresponding equation is:

$$3(x + 4) = 7(x - 4)$$
$$3x + 12 = 7x - 28$$
$$40 = 4x$$

$$10 = x$$

The speed of the launch in still water: 10 mph.

Verification: If the launch moves at 10 miles per hour in still water, with the current it "makes good" $10 + 4$, or 14 miles per hour. A 3-hour trip at this rate must be 3×14, or 42 miles long. Coming back, against the current, the launch makes good only $10 - 4$ or 6 miles per hour. At this rate, a 7-hour trip must be 7×6, or 42 miles long. Hence,

distance down (42 miles) = distance up (42 miles)

and the solution of the equation is verified in the problem.

3. NUMERICAL APPROACH TO PROBLEM SOLVING

Often difficulty in setting up an equation for a problem can be resolved by the following "numerical" technique. It is based upon the fact that the variable in the equation is simply a placeholder for the yet-to-be-determined value of the unknown. So instead of representing that unknown with a letter symbol, we assume some plausible number for the unknown, and then reason through the problem as we would in the verification (Step 5 of Program **4.1**). Except by accident, of course, the assumed number will be incorrect. However, if the *reasoning* is correct, substituting a letter symbol for the assumed number leads to a valid basic equation.

EXAMPLES

1. Bill is now twice the age of John. Seven years ago the sum of their ages was 28. What are their ages now?

Numerical Approach: Assume that John is now 12 years old; that would make Bill, who is twice John's age, 24 years old. Seven years ago John would have been $12 - 7$, or 5, and Bill would have been $24 - 7$, or 17. According to the data of the problem, the sum of their ages seven years ago would have been 28. But

$$(12 - 7) + (24 - 7) = 28 \quad \text{(false)}$$

However, if we now introduce x instead of the assumed 12 as John's present age (and $2x$ instead of Bill's 24) in the false sentence, we obtain a valid basic equation:

$$(x - 7) + (2x - 7) = 28$$
$$3x - 14 = 28$$
$$3x = 42$$
$$x = 14$$

If $x = 14$, then $2x = 28$. John is 14 at present, and Bill is 28.

2. If half the sum of two consecutive even integers is subtracted from twice the smaller integer, the result will be 23. What are the integers?

Numerical Approach: Assume the smaller even integer is 18; the next consecutive even integer would then be 20. Half their sum would be $\frac{1}{2}(18 + 20) = 19$, which according to the data of the problem, when subtracted from twice of the smaller ($2 \times 18 = 36$) should equal 23. But

$$(2 \times 18) - \tfrac{1}{2}(18 + 20) = 23 \quad \text{(false)}$$

However, by substituting x for the smaller integer instead of the assumed 18, and $x + 2$ for the next consecutive even integer instead of the assumed 20, we are led to the appropriate equation:

$$2(x) - \tfrac{1}{2}[(x) + (x + 2)] = 23$$
$$2x - \tfrac{1}{2}[2x + 2] = 23$$
$$2x - x - 1 = 23$$
$$x = 24$$

If $x = 24$, then $x + 2 = 26$. The smaller desired integer is 24 and the next consecutive even integer is 26.

4. DIMENSION PROBLEMS

A most useful aid in solving dimension problems is a drawing or sketch of the situation posed by the problem. Drawing it to exact scale is not necessary, but the given as well as the desired or unknown dimensions—the latter represented by a letter—should be inserted as on a blueprint or plan. From this we try to find some equivalence relationship (often it is a geometric relationship) which involves both the unknown and known data. We then state this equivalence relationship in algebraic language, and solve.

EXAMPLES

1. A board is 12 ft long and is to be cut into three pieces so that the second piece is twice the size of the first piece, and the third piece is three times the size of the second piece. Find the length of the three pieces of board.

Let x represent the length of the first piece. Then $2x$ represents the length of the second piece; and $3(2x)$ represents the length of the third piece.

The equivalence relationship is:

length of first piece + length of second piece

+ length of third piece = total length

The equation is:

$$x + 2x + 3(2x) = 12$$
$$x + 2x + 6x = 12$$
$$9x = 12$$
$$x = \tfrac{12}{9} = 1\tfrac{1}{3}$$

If $x = 1\tfrac{1}{3}$, then $2x = 2\tfrac{2}{3}$ and $3(2x) = 8$. Hence the length of the first piece is 1 ft 4 in., the second piece 2 ft 8 in., and the third piece 8 ft 0 in.

Verification: The length of the three pieces add to 12 ft: 1 ft 4 in. + 2 ft 8 in + 8 ft 0 in. = 12 ft 0 in. So the conditions of the problem are met by this solution.

2. The perimeter of a rectangle, whose length is $2\tfrac{1}{2}$ times its width, is 21 ft. Find the dimensions of the rectangle.

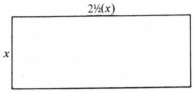

Let x represent the width of the rectangle. Then $2\tfrac{1}{2}(x)$, or $\tfrac{5}{2}x$, represents the length of the rectangle. The equivalence relationship is:

perimeter of a rectangle = 2 × (length + width)

The equation is:

$$21 = 2(x + \tfrac{5}{2}x)$$
$$21 = 2x + 5x$$
$$3 = x$$

If $x = 3$, then $\tfrac{5}{2}x = 7\tfrac{1}{2}$. The width of the rectangle is 3 ft and its length is $7\tfrac{1}{2}$ ft.

Verification: The distance around the rectangle, which is its perimeter, is the sum:

$$3 \text{ ft} + 7\tfrac{1}{2} \text{ ft} + 3 \text{ ft} + 7\tfrac{1}{2} \text{ ft} = 21 \text{ ft.}$$

So the conditions of the problem are met by this solution.

(See Example 1, p. 72, for another illustration of a dimension problem.)

EXERCISE 4-2

1. A piece of wire 36 ft long is cut so that one piece is 12 ft longer than the other. Find the dimensions of the two pieces.

2. The perimeter of a rectangle, whose width is $\frac{3}{8}$ its length, is 176 ft. Find its length.

3. A circular plate is to be painted so that the first sector has three times the area of the second, and the third sector has twice the area of the other two sectors combined. Find the central angle for each of the sectors.

4. The perimeter of a triangle is 78 in.; the longest side is twice the length of the shortest side and 12 in. longer than the other side. Find the length of the longest side.

5. A triangle sits on top of a square and forms a five-sided figure in which two sides are equal and the three other sides are equal. If the perimeter of the triangle is 22 in. and the perimeter of the complete five-sided figure is 38 in., find the length of the edge of the square.

6. A board is 41 in. long and is to be cut into three pieces. One piece is to be twice the length of the smallest piece and $\frac{1}{3}$ the length of the longest piece. Allowing $\frac{1}{4}$ in. loss for the sawdust and the sanding of each saw cut, determine the final length of each piece.

7. A man paid $1044 to have his rectangular lot, whose perimeter is 600 ft, fenced in. If fencing the 80-ft front part cost him 30¢ more per foot than that of the back and sides, how much did he pay per foot for the front fencing?

8. Find the sizes of the three angles of a triangle if the smallest is half the middle-sized angle, and the largest is 27° larger than three times the size of the middle-sized angle.

9. One dimension of a rectangular lot is $\frac{3}{4}$ the other. If the perimeter becomes 680 ft when each dimension is extended 30 ft, what are the original dimensions?

10. A bus starting from Glendale each morning makes a trip to Avalon, and from Avalon it goes 26 mi to Barret. From Barret it returns to Glendale over a route that is 9 mi longer than the first leg of the trip. If the total trip is 205 mi, what is the distance of this last leg?

11. A triangle sits atop a square. The altitude of the triangle is 4 in. shorter than twice its base, and a side of the square has the same dimension as the base of the triangle. If the area of the square is 20 sq in. greater than the area of the triangle, what is the altitude of the triangle?

12. A rectangular garden plot is $1\frac{3}{5}$ times as long as it is wide. If its length is decreased by 2 ft and its width is increased by 2 ft, the perimeter stays the same, but the area increases by 20 sq ft. Find the original dimensions of the plot.

5. INTEGER PROBLEMS

Number sequences and component parts of the standard Hindu-Arabic numerals for integers underlie many integer problems. The most frequent number sequences are the consecutive integers, consecutive even integers, and consecutive odd integers. By consecutive integers is meant a set of integers arranged in the same order as would occur in counting. Examples of sets of consecutive integers are 5, 6, 7, 8; 18, 19; and 126, 127, 128, 129, 130. Since each integer in sequence differs from the previous integer by 1, such a sequence may be represented algebraically as:

$$n, n + 1, n + 2, n + 3, \ldots$$

where n is the smallest of the integers under consideration.

Each integer may be further classified as being odd or even. Even integers are those which are exactly divisible by 2, while odd integers invariably have a remainder of 1 when divided by 2. The nonnegative even integers are $0, 2, 4, 6, \ldots$, and the nonnegative odd integers are $1, 3, 5, 7, \ldots$. Since, in both cases, one integer differs from the next by 2, both sequences may be expressed algebraically as $n, n + 2, n + 4, n + 6$, etc., where n represents the smallest odd or even integer, as the case may be.

The standard numeral aspect of integer problems usually relates to the way we express the integers in the decimal (tens) system of number notation. For instance, the numeral for eighty-six is 86, which implies 8 tens $+$ 6 ones, or $(8 \times 10) + (6 \times 1)$. If we reverse the order of the two digits to 68, we have a numeral that represents a number eighteen less than the other, for 68 implies 6 tens $+$ 8 ones, or $(6 \times 10) + (8 \times 1)$.

EXAMPLES

1. There are three consecutive odd integers whose sum is 5 more than twice the largest. What are the integers?

   Let x represent the smallest odd integer. Then $x + 2$ represents the middle odd integer, $x + 4$ represents the largest odd integer, and $2(x + 4) + 5$ represents 5 more than twice the largest odd integer. The equivalence relationship is:

   integer$_1$ + integer$_2$ + integer$_3$ = twice the largest, increased by 5

   The equation is:

$$(x) + (x + 2) + (x + 4) = 2(x + 4) + 5$$
$$x + x + 2 + x + 4 = 2x + 8 + 5$$
$$3x + 6 = 2x + 13$$
$$3x - 2x = 13 - 6$$
$$x = 7$$

   If $x = 7$, then $x + 2 = 9$ and $x + 4 = 11$. The integers are 7, 9, 11.

   Verification: 7, 9, 11 are three consecutive odd integers. Their sum is 27 which *is* 5 more than twice the largest, 2×11, or 22. So the conditions of the problem are met by this solution.

2. The digits of a two-digit numeral are consecutive integers. The number expressed when these two digits are reversed differs from the original by an amount equal to the sum of the digits. What are the digits?

Let x represent the smaller of the two consecutive digits. Then:

$x + 1$ represents the larger of the two consecutive digits;

$(x + 1)$ tens $+ (x)$ ones represents the larger of the two numbers expressed by these digits;

(x) tens $+ (x + 1)$ ones represents the smaller of the two numbers expressed by these digits; and

$(x + 1) + x$ represents the sum of the two digits.

The equivalence relationship is:

larger number $-$ smaller number $=$ sum of digits of either

$[(x + 1)$ tens $+ (x)$ ones$] - [(x)$ tens $+ (x + 1)$ ones$] = (x + 1) + x$

The equation is:

$$[10(x + 1) + 1(x)] - [10(x) + 1(x + 1)] = (x + 1) + x$$
$$10x + 10 + x - 10x - x - 1 = x + 1 + x$$
$$10 - 1 = 2x + 1$$
$$8 = 2x$$
$$4 = x$$

If $x = 4$, then $x + 1 = 5$. The smaller digit is 4 and the larger is 5.

Verification: The two integers that have in their standard numeral the digits 4 and 5 are 45 and 54. Their difference is $54 - 45$, or 9, which is also the sum of the digits. So the conditions of the problem are met by this solution.

(See Example 2, p. 73. for another illustration of an integer problem.)

EXERCISE 4-3

1. The sum of three consecutive even integers is 54. What are the integers?

2. The sum of eight consecutive odd integers is 112. What are the integers?

3. What four consecutive multiples of 3 have a sum of 102?

4. The tens digit of a two-digit numeral is 2 more than twice the other digit. If the sum of these digits is 8, what is the two-digit numeral?

5. What is the three-digit Hindu-Arabic number expression whose hundreds digit is 3 greater than the tens digit and 3 less than the ones digit, and the sum of whose digits is 12?

6. Find four consecutive odd integers such that the sum of the first three, when subtracted from three times the fourth, leaves a remainder of 12.

7. There are four consecutive integers whose sum is 2. Which are they?

8. Four integers in an ordered sequence differ one from the next by 5. Twice the sum of the middle pair exceeds the sum of the largest three of them by 65. What integers are in this sequence?

9. The digits of a standard two-digit numeral representing an odd number differ by 5. If the tens digit is tripled and added to four times the ones digit, the sum is

5 greater than the next consecutive odd number. What is the original odd number?

10. The tens digit of a standard two-digit numeral is 2 less than the ones digit. The value of the numeral when the digits are reversed is 5 more than twice its original value. What is the original numeral?

11. An integer has a pair of digits in its Hindu-Arabic numeral which differ by 2. Twice the difference between the two numbers that can be represented by this pair of digits exceeds the smaller of the two numbers by 1. What are the digits?

12. An integer has three consecutive digits in its numeral. Eight times the hundreds digit added to seven times the ones digit yields a sum that is 190 less than the original integer. What is the integer?

6. MIXTURE PROBLEMS

Problems usually classified as "mixture problems" often have the key data given in terms of their component parts. For instance, 20 lbs of 60¢ coffee is worth less than 15 lbs of $1 coffee because the former, though greater in bulk, has a total worth of only $12, while the latter, smaller in bulk, is worth more at $15. This judgment could not be made on the basis of either weight or pound-worth alone, but only by considering the products of these factors. Similarly, 100 cc of 60% alcohol solution contains more pure alcohol than 150 cc of 30% alcohol solution. As usual, in setting up the equivalence relationship and equation, it is necessary that both members of the equation refer to the same thing.

EXAMPLES

1. A grocer has some coffee worth 70¢ a pound and some worth 80¢ a pound. How much of each must he mix together to produce a blended mixture of 40 lbs worth 76¢ a pound?

Let x represent the number of pounds of 70¢ coffee to be used in the 40-lb mixture. Then $40 - x$ represents the number of pounds of 80¢ coffee to be used in the 40-lb mixture, and 40×76¢ is the total worth of the 40-lb mixture. The equivalence relationship is:

$$\begin{bmatrix} \text{worth of the 70¢ coffee} \\ \text{in the mixture} \end{bmatrix} + \begin{bmatrix} \text{worth of the 80¢ coffee} \\ \text{in the mixture} \end{bmatrix} = \begin{bmatrix} \text{worth of the} \\ \text{total mixture} \end{bmatrix}$$

The equation is:

$$70x + (40 - x)(80) = (40)(76)$$
$$70x + 3200 - 80x = 3040$$
$$160 = 10x$$
$$16 = x$$

If $x = 16$, then $40 - x = 24$. In the mixture there should be 16 lbs of 70¢ coffee and 24 lbs of 80¢ coffee.

Verification: The worth of the total mixture is 40×76¢, or $30.40. The worth of the 16 lbs of 70¢ coffee in the mixture is 16×70¢, or $11.20; the worth of the 24 lbs of 80¢ coffee in the mixture is 24×80¢, or $19.20. Since the worth of the parts of the mixture ($11.20 + $19.20) equals the worth of the total mixture ($30.40), the conditions of the problem are met by this solution.

2. A chemist has 15 oz of 4% acid solution. How much 20% acid solution must he add to bring the total solution up to 10% strength?

Let x represent the number of ounces of 20% solution to be added. Then:
   $15 + x$ represents the number of ounces in the resulting solution;
   4% of 15 is the number of ounces of pure acid in the original solution;
   20% of x is the number of ounces of pure acid that will be added when x ounces of 20% solution is added; and
   10% of $(15 + x)$ is the number of ounces of pure acid in the resulting 10% solution.
The equivalence relationship is:

   pure acid in original solution + pure acid in added solution

   $\qquad\qquad\qquad\qquad$ = pure acid in resulting solution

The equation is:

$$(0.04)(15) + (0.2)(x) = 0.1(15 + x)$$
$$0.6 + 0.2x = 1.5 + 0.1x$$
$$0.1x = 0.9$$
$$x = 9$$

Thus, 9 oz of the 20% acid solution must be added.

Verification: In the 15 oz of 4% solution there is 0.04×15 or 0.6 oz of pure acid; in the 9 oz of 20% acid solution that is added, there are 0.20×9 or 1.8 oz of pure acid. This amounts to $0.6 + 1.8$ or 2.4 oz of pure acid in $15 + 9$ or 24 oz of solution; $2.4 \div 24 = 0.1 = 10\%$, the strength of the resulting solution. So the conditions of the problem are met by this solution of the equation.

(See Example 3, p. 74, for another illustration of a mixture problem.)

EXERCISE 4-4

1. A grocer has some tea that sells for 50¢ a pound and some that sells for 80¢ a pound. To make a 45-lb blend of the two that will sell for 60¢ a pound, what proportions should he use?

2. A nurse must make a 20% alcohol solution by diluting 10 oz of 25% solution. How much water must she add?

3. How much water may be added to 1 gallon of lemonade without reducing its strength by more than 20%?

4. The manager of a candy store wishes to put a mixture of hard candy on sale for 49¢ a pound. If he plans on a 70-lb mixture, using candy that usually sells for 29¢ per pound and others that usually sell for 59¢ per pound, how many pounds of the cheaper candy must he use?

5. A nut mixture of peanuts and cashews costs a merchant 80¢ a pound. If he pays 65¢ a pound for the peanuts and 90¢ a pound for the cashews, what percent peanuts must be in the mixture?

6. A candy dealer mixes chocolates that normally sell for 90¢ a pound with others that sell for 60¢ a pound. If he has 40 lbs of the cheaper candy he wants to use up in an assortment that will sell for 80¢ a pound, how many pounds of the expensive candy should he use?

7. A druggist must make 20 oz of 12% argyrol from his supply of 5% and 15% solutions. How much of the 5% solution should he use?

8. Two alloys, one 1 part silver and 5 parts copper, and the other 3 parts silver and 1 part copper, are mixed to form 350 lbs of an alloy that is equal in silver and copper content. How many pounds of the first alloy should be used?

9. How much 24% butterfat cream must be mixed with 4% butterfat milk to make 10 gallons of a lighter cream that is 20% butterfat?

10. A can of evaporated milk contains 0.2% fat, and a certain brand of whole milk has 3.9% fat. How much whole milk must be mixed with the contents of a 13-oz can of evaporated milk to make a 2.0% fat mixture?

11. A 24-qt automobile radiator contains a mixture that is 25% alcohol and 75% water. How much water must be drained off and replaced with an 85% alcohol solution to bring the alcoholic content in the radiator up to 30%?

12. A chemistry student has 50 cc of 30% sulphuric acid. How much of this solution must he remove and replace with pure sulphuric acid to bring the acid strength up to 58%?

7. UNIFORM MOTION PROBLEMS

Basic to all uniform motion problems is the relationship:

$$\text{distance} = \text{rate} \times \text{time} \quad \text{or} \quad d = rt$$

From this we may derive its associated relationships:

$$\text{rate} = \text{distance} \div \text{time} \quad \left(r = \frac{d}{t} \right)$$

and

$$\text{time} = \text{distance} \div \text{rate} \quad \left(t = \frac{d}{r} \right)$$

When analyzing a uniform motion problem for some equivalence relationship that will lead to an equation for solving the problem, we must keep in mind that the members of the equation must represent equal measures of the same kind, such as equal distances, equal times, or equal rates.

EXAMPLES

1. An airplane flies 980 mi in $3\frac{1}{2}$ hrs against a steady headwind blowing at 20 mph. How fast would the plane be flying in still air?

Let x represent the rate of the plane in still air. Then $x - 20$ represents the rate of the plane against the wind, and $980/3\frac{1}{2}$ represents the rate of the plane during flight. The equivalence relationship is: rate = rate. The equation is:

$$x - 20 = \frac{980}{3\frac{1}{2}}$$

$$x - 20 = 280$$

$$x = 300$$

The plane's rate of speed in still air would be 300 mph.

Verification: The rate in still air (300 mph) reduced by the rate of the headwind (20 mph) implies the plane can make only 280 mph. Flying at that speed for $3\frac{1}{2}$ hrs, the plane would be able to make good $3\frac{1}{2} \times 280$ or 980 mi. So the conditions of the problem are met by this solution.

2. A truck leaves the highway terminal and heads for New York, averaging 40 mph. A mistake in shipping orders is noted, and 24 min after the truck leaves, a car is dispatched to overtake the truck. If the car averages 50 mph, how long will it take it to catch up to the truck?

Let x represent the time in hours that the car travels. Then:
$x + \frac{24}{60}$ represents the time in hours that the truck travels;
$40(x + \frac{24}{60})$ represents the distance $(r \times t)$ the truck travels; and
$50(x)$ represents the distance $(r \times t)$ the car travels.
The equivalence relationship is:

$$\text{distance for car} = \text{distance for truck}$$

The equation is:

$$50x = 40(x + \tfrac{24}{60})$$

$$50x = 40x + 16$$

$$10x = 16$$

$$x = 1.6$$

It will take the car 1.6 hr, or 1 hr 36 min, to overtake the truck.

Verification: When the car overtakes the truck, the truck will have been traveling for 1 hr 36 min + 24 min, or 2 hrs, exactly; at 40 mph, the truck would have made good 80 mi. To travel those 80 mi at 50 mph, the car would have to travel for 80 ÷ 50 or 1.6 hrs. So the conditions of the problem are met by this solution.

(See Example 5, p. 75 for another illustration of a uniform motion problem.)

EXERCISE 4-5

1. A boat travels 63 mi upstream in $4\frac{1}{2}$ hrs against a steady current of 4 mph. How fast would the boat be moving if there were no current?

2. A speeding automobile is going 55 mph, and a highway trooper, 2 mi behind, is chasing it at a speed of 70 mph. How long will it take the trooper to overtake the speeding car if they both maintain these speeds?

3. A boy rides his bike 6 mi from his home to the bus stop at the rate of 9 mph. He arrives just in time to catch the city bus which averages 30 mph. If the boy spent 1 hr and 28 min traveling from home to city, how far did he travel by bus?

4. Two airplanes pass each other going in opposite directions at 0602 on the clock. One is traveling at a speed of 270 mph and the other at 210 mph. At what time will they be 100 mi apart?

5. Two automobiles leave towns 470 mi apart at the same time; they travel toward each other, one at a rate of speed that is 20 mph faster than the other. If they meet in 5 hrs, what is the speed of the faster automobile?

6. A pair of hikers, 18 mi apart, begin at the same time to hike toward each other. If one walks at a rate that is 1 mph faster than the other, and if they meet 2 hr later, how fast is the slower hiker walking?

7. Able ran a 440-yd race in 59.2 sec, and Baker came in second with a time of 60.4. Assuming that each runner ran his race at a uniform speed throughout the race, how far back was Baker when Able crossed the finish line?

8. A boy rode with his father from home to town at the average rate of 40 mph and walked back at the average rate of 4 mph. If the total riding and walking time was 1 hr and 39 min, how far was it from home to town?

9. A driver averaged 50 mph on the turnpike and 30 mph on secondary roads. If a trip of 185 mi over turnpike and secondary roads took 4 hrs and 30 min, how many miles were over the turnpike?

10. An automobile 19 ft long overtakes a truck that is 25 ft long, traveling at 35 mph. At what rate must the automobile move in order to pass the truck completely in 3 sec?

11. A man drives the first mile of a 2-mi stretch at the average speed of 30 mph. What speed must he average in the second mile in order to average 40 mph overall?

12. At how many minutes after 4 p. m. will the minute hand of the clock be even with the hour hand?

8. INVESTMENT PROBLEMS

When a person borrows money, he pays a rental called *interest* for the use of the money he has borrowed, somewhat in the way one pays rent for the use of a house or apartment. When someone lends money, he receives this interest as compensation. Countless business transactions each day revolve around the borrowing and lending of money, each with an agreed-upon rate of interest. Just as the comparison of distance to time produces a *rate* of speed—a measure of motion—the comparison of the rental or interest for the use of a certain sum of money to that sum of money produces a *rate* of interest—a measure of the cost of borrowing money. From the lender's point of view, the rate of interest is a measure of the return his loaned money is producing.

EXAMPLES

1. A man invested part of $4000 at 3% and the rest at 5%. If he expects a return of $180 a year on his investment, how much has he invested at each rate?

   Let x represent the amount of money invested at 3%. Then:
   $4000 - x$ represents the amount of money invested at 5%;
   3% of x is the return each year on 3% investments; and
   5% of $(4000 - x)$ is the return each year on 5% investments.
   The equivalence relationship is:

   3% investment return + 5% investment return = total return

   The equation is:

   $$(0.03)(x) + (0.05)(4000 - x) = 180$$
   $$0.03x + 200 - 0.05x = 180$$
   $$-0.02x = -20$$
   $$x = 1000$$

   If $x = 1000$, then $4000 - x = 3000$. The man has $1000 invested at 3% and $3000 invested at 5%.

   Verification: $1000 invested at 3% will produce an annual return of 0.03×1000, or $30; $3000 invested at 5% will produce an annual return of 0.05×3000, or $150. Together the two investments will produce an annual return of $30 + $150, or $180. So the conditions of the problem are met by this solution.

2. A man pays $1200 interest annually for a sum of money. Part he borrows at 3%, and the balance, which exceeds the other part by $4000, costs

him 8% by the time he pays insurance and special fees. How much money has he borrowed at the 3% rate?

Let x represent the number of dollars borrowed at 3%. Then:

$x + 4000$ represents the number of dollars borrowed at 8%;

3% of x is the cost of sum borrowed at 3%; and

8% of $(x + 4000)$ is the cost of sum borrowed at 8%.

The equivalence relationship is:

cost of part at 3% + cost of part at 8% = total cost of borrowed money

The equation is:

$$(0.03)(x) + (0.08)(x + 4000) = 1200$$

$$0.03x + 0.08x + 320 = 1200$$

$$0.11x = 880$$

$$x = 8000$$

The man has borrowed $8000 at the 3% rate.

Verification: $8000 borrowed at 3% costs 0.03×8000, or $240, per year; $4000 more, or $12,000, borrowed at 8% costs $0.08 \times 12,000$, or $960, per year. Together these two loans cost a total of $240 + $960, or $1200. So the conditions of the problem are met by this solution.

EXERCISE 4-6

1. Mr. Wallace invests part of his $7000 at 4% and the rest at 7%. If his annual return is $400, how much has he invested at each rate?

2. An investor can borrow money on his credit rating at 4% and can invest it at $5\frac{3}{4}$%. How much should he borrow in order to realize a clear profit of $875 a year?

3. Mr. Walton has three times as much money invested in 4% bonds as he does in a stock that pays 7%. If his annual return from both investments is $456, how much has he invested in the bonds?

4. Last year a man had invested $3 in a safe 4% bond for every $1 he had invested in a speculative stock that paid him 7%. If his annual return on these investments was $209, how much must he have had invested in the stock?

5. Part of a $10,000 trust fund is invested at 3%, and the balance at 5%. The return from these two investments is the equivalent of $4\frac{1}{4}$% on the total investment. What amount is invested at 3%?

6. A man has $2800 invested at one rate and $4600 at another rate that is 1% more than the former. If his total return for both investments is $342, what are the two rates?

7. Mr. Miller has invested $6000 at one rate of interest and another $4000 at half that rate. If his annual return is $360, what are the two rates?

8. Mrs. Carey has $9000 invested in various bonds. She has twice as much invested in a $4\frac{1}{2}\%$ issue as she does in a 3% issue; the balance is invested in a 5% issue. How much money has she invested in each of these bond issues if her annual return is $420?

9. A man pays $955 annually for borrowed money for which his house is used as security. He borrows part through a first mortgage at $5\frac{1}{2}\%$ and the rest, which is $6000 less than the first part, by means of a second mortgage at 7%. How much does he borrow on the first mortgage?

10. For every two dollars invested in a company at 4%, the investor is allowed to invest another dollar in a 6% investment. If a person desires a yield of $700 per year from such an investment arrangement, how much should be invested altogether?

11. Mr. Jones had invested part of his $6000 at 3% and the remainder at 5%. A reversal of these investments would have resulted in a return of $48 less. How much did he invest at 3%?

12. A man has $8000 invested, part at 4% and the rest at 6%. Had he switched these investments he would have earned $60 more. What amounts did he have invested at these two rates?

9. MISCELLANEOUS PROBLEMS

There is an almost limitless variety of possible problems solvable by first-degree equations in a single variable. The types covered in Sections 4 through 8 in this chapter are among the more frequently encountered. Among others having a long tradition in algebra courses are age problems, coin problems and lever problems. There is a sampling of these in the exercises that follow.

EXERCISE 4-7

1. John is 8 years older than Sue, but 2 years ago his age was twice her age then. How old is John now?

2. A mother is twice the age of her son now, but in 4 years she will be 4 times as old as her son was 9 years ago. How old is the mother now?

3. A boy 8 years old has a father who is 28 years old. In how many years will the father be twice the son's age?

4. In 6 years Jane will be $1\frac{2}{3}$ her age 4 years ago. How old is Jane now?

5. A clerk gave a customer 25 coins, all quarters and dimes. If the amount of these coins totaled $3.85, how many of each coin did he give?

6. Admission to a college play is set at $1, $2.50, and $3. There are 7 times as

many of the cheapest seats as there are of the most expensive, and 38 more middle-priced seats than expensive seats. If a sellout will realize $220, how many of each type of seat are there?

7. A man at the Mexican border had 48 pieces of paper currency, all $1 (U.S.) and 1 peso (Mexican) notes. If the rate of exchange is 12.5 pesos for each $1 (U.S.), and the value of the money was the equivalent of $9.36 (U.S.), how many pesos did he have?

8. A man paid his restaurant bill of $1.85 with change: nickels, dimes, and quarters. If there were six times as many quarters as dimes and four more nickels than dimes, how many of each must he have used?

9. Two boys sit at opposite ends of a seesaw. The boy who weighs 120 lbs sits 8 ft from the fulcrum (point of balance) and the other boy, who weighs 96 lbs, sits at a point that just balances the seesaw. What is the distance between the two boys?

10. Jack (162 lbs) and Joe (126 lbs) sit at opposite ends of a seesaw that is 16 ft long and balance it. How far must the fulcrum be from Jack?

11. How heavy a rock can a man of 180 lbs move with a crowbar that is 8 ft long if the fulcrum, or point of leverage, is 9 in. from the end of the bar and the rock?

12. In a mechanics experiment, three weights are to be distributed so as to balance a yardstick on a knife-edge or fulcrum. If a weight of 24 oz is placed 8 in. to the left of the fulcrum, and a 15-oz weight is placed 6 in. to the right of the fulcrum, where should a 6-oz weight be placed to balance the yardstick?

REVIEW—CHAPTER IV

Translate the following statements into algebraic language; use x for the unknown, unless otherwise indicated:

1. A number p less than the unknown.

2. A number one-ninth the size of twice the unknown.

3. A number half the size of the unknown diminished by k.

4. A number 78% larger than the unknown.

5. A number 125% greater than the unknown.

6. The remainder after the unknown has been taken from 22.

7. The other number when the larger is x and their positive difference is p.

8. The average rate of speed when the distance is $3x$ and the time is $x - 7$.

9. One-sixth of twice the unknown, reduced by m.

10. The number of dozens in x.

11. The unknown decreased by 22 is the same number as six times the unknown.

12. Three less than the unknown is a number that is three-quarters of the unknown.

13. Seven decreased by the unknown and 12 is 5 less than the negative of the unknown.

14. Two-thirds of the unknown is the equivalent of the unknown increased by p.

15. A number 3 more than 7, when divided by the unknown, is $1\frac{1}{2}$ more than the unknown.

Solve:

16. The perimeter of a rectangle is 148 in., and the length of one side is 1 in. short of being $1\frac{1}{2}$ times the length of the other side. What are the dimensions of the rectangle?

17. A circular "pie" graph is to be divided into four sectors. One sector is to have twice the central angle of the first, $1\frac{1}{2}$ times the central angle of the second, and 75° less than the central angle of the third. Find the measurements of the four central angles.

18. A line 7 ft long is to be cut into three pieces, and each finished piece is to have a knot, which takes up 2 in. of line, tied on each end. If one finished piece is to be twice the length of another and 2 in. shorter than the third, what should be the unknotted lengths of the three pieces of line?

19. Find the sizes of the angles of a triangle if one is 19° larger than the middle-sized angle, and the third is 19° less than that angle.

20. An isosceles (two legs equal) triangle has a base which is $\frac{3}{4}$ the length of one of its legs. If each of the legs is increased by 4 in., and the base by 4 in., the perimeter of the new triangle is 60 in. Find the dimensions of the original triangle.

21. When opposite sides of a square are increased by 7 ft, and the other pair of opposite sides are decreased by 2 ft, the result is a rectangle whose area is 31 sq ft more than that of the square. What is the length of side of the square?

22. Seven consecutive integers add to 728. Which are they?

23. A numeral has two digits. The tens digit is $2\frac{1}{2}$ times the size of the ones digit, and the sum of the digits is 7. What is the numeral?

24. Reversing the digits, which differ by 2, of a two-digit numeral and subtracting the number it represents from that of the first yields a difference of 18. What are the digits?

25. Five integers, in ordered sequence, differ one from the next by 7. The sum of the four smallest is 10 less than the largest. What are the integers?

26. Given a standard two-digit numeral. When 4 times the tens digit is added to 3 times the ones digit (which is 6 less than the tens digit) the result is 4 times the value of the number expressed by reversing the original digits. What are the digits?

27. Two different numbers can be represented by a pair of digits that differ by 3, and the difference between these two numbers is 3 less than twice the sum of the digits. Which is the larger of the two numbers in question?

28. A confectioner has some candy that normally sells for $1.90 a pound and some that sells for 70¢ a pound. How much of the expensive candy must he use to produce 20 lbs of a mixture to sell for $1 a pound?

29. What volumes of 10% and 4% solutions must be mixed together to yield 81 cc of 6% solution?

30. A basketball team scored 86 points in a game in which they scored 16 more field goals than foul shots. (Field goal = 2 points; foul shot = 1 point.) How many field goals were scored?

31. An 8-gallon tank contains a 40% salt solution. How much of the solution must be drawn off and replaced with pure water in order to bring the salt concentration down to 25%?

32. Thirty-six ounces of 1 part gold, 8 parts copper alloy is to be mixed with an alloy containing 5 parts gold, 3 parts copper to make an alloy containing 7 parts gold and 5 parts copper. How much of the first alloy must be used?

33. Two hikers, Al and Pete, are carrying loads of 60 lbs and 70 lbs, respectively. How many pounds must be removed from Al's load and added to Pete's so that Al will then be carrying a load that is just $\frac{3}{4}$ of Pete's?

34. Two automobiles leave towns 333 mi apart and travel toward each other, one averaging 10 mph faster than the other. If they meet in $4\frac{1}{2}$ hrs, what must have been the average speed of the slower automobile?

35. A driver averaged 52 mph on one leg of a 206-mi trip, and 40 mph on the second. If the whole trip took $4\frac{1}{4}$ hrs, how many miles did he cover at the faster speed?

36. A room can be painted by a fast workman in 12 hrs, and by a slow workman in 16 hrs. How long would it take a slow and a fast workman to paint the room working together?

37. A little girl runs to the store at the rate of 9 mph and walks back at the rate of 3 mph. If her total traveling time is 10 min, how far away is the store?

38. A passenger train 225 ft long overtakes a freight train 555 ft long and traveling at a speed of 32 mph on a parallel track. If it takes exactly 1 min for the passenger train to pass the freight train completely, how fast must the passenger train be moving?

39. When the hands of a clock register 5:15, how long will it be before the hands are together?

40. An investor can borrow money on his good credit at $4\frac{1}{4}$% and can invest these same funds at $5\frac{3}{4}$% elsewhere. If he earned $600 that way last year, how much did he borrow?

41. A man has $10,000 to invest. He invests $3000 at 6% and $2500 at $4\frac{1}{2}$%. What rate should he seek in investing the balance so that his total return is $585 annually?

42. Mr. Wilkins has $42,000 invested in two real estate holdings which pay him 8% and 11%. What must be the amount invested in the 11% property if his total yield is $3690?

43. A man was able to borrow money on his house at 5% and a quarter of that amount on his furniture at 6%. If his total interest charge is $234 a year, how much must he have borrowed against the furniture?

44. Mr. Allen had part of $40,000 invested at 4% and the rest at 6%. Had he switched these investments, his yield would have been $400 more. How much must he have had invested at 6%?

45. Two men each invested $8000. The first invested part at 4% and the balance at 5%. The second invested $1000 more at $2\frac{1}{2}$% than the first did at 4%, and the rest at 10%. If both realized the same return, how much must the first man have had invested at 4%?

46. A father, who is 36, has two sons, one twice the age of the other. In three years, the sum of all three ages will be 63. How old are the sons now?

47. An uncle is 3 times the age of his nephew, and in 3 years he will be 3 times as old as his niece is now. How old is the uncle if the total of the 3 ages is now 66?

48. A woman has 3 times as many quarters as dimes in her purse. If the quarters were dimes and the dimes were dollars, she would be richer by $1.80. How much money does she have in the purse?

49. A man spent $3.20 for 4¢, 7¢, and 30¢ stamps. If he bought $2\frac{1}{2}$ times as many 4¢ stamps as 7¢ stamps, and his 30¢ stamps numbered $\frac{1}{7}$ of the combined number of the other two types, how many 7¢ stamps did he buy?

50. Two boys together weigh 126 lbs. If they can balance a seesaw when one sits 6 ft from the fulcrum and the other 8 ft from it on the other side, what are their respective weights?

V

SPECIAL PRODUCTS
AND FACTORING

1. INTRODUCTION

Often in mathematics we find it advantageous to express a product of several algebraic factors as a sum. This is referred to as **expanding.** Equally often we find it advantageous to do the reverse—express a sum of terms as the product of several factors. This is referred to as **factoring.** The distributive property,

$$a(b + c) = ab + ac$$

provides the fundamental bridge for passing back and forth between product and sum form. A general way to expand the product of several factors to an equivalent sum is to apply the multiplication procedure, Program **2.9**. The division procedures, Programs **2.10** and **2.11**, make possible the factoring of a sum (the dividend) into a product of two factors (the divisor and the quotient).

It is useful to learn how to expand and factor certain types of algebraic expressions on sight. In this chapter, we make a study of "special products," which not only provides shortcuts for expanding frequently used product types, but also affords basic clues and insights into the inverse process of factoring.

[*Note:* Whether a polynomial is factorable or not depends upon the domain of the coefficients. *In this chapter,* a polynomial with integral coefficients is considered factorable if it can be expressed as the product of two or more factors whose numerical coefficients are integers.]

94

2. COMMON FACTOR

The procedure for multiplying a polynomial by a monomial (multiply each term of the polynomial by the monomial and add the products) was given by Program **2.8**. The result is invariably a polynomial in which each term contains a common factor, namely, the monomial. Inversely, this suggests a way of changing such polynomials to equivalent product form.

5.1

To factor a polynomial whose terms contain a common factor:

Step 1. Inspect the prime factors† of each term and identify all prime factors which are common to each term.

Step 2. Divide the polynomial by the product of the common prime factors of Step 1. The divisor and quotient are a pair of factors of the given polynomial.

EXAMPLES

1. Factor $8x^2 + 12xy + 4x$.

Step 1. $\quad 8x^2 = 2 \cdot 2 \cdot 2 \cdot x \cdot x$

$\qquad\quad 12xy = 2 \cdot 2 \cdot 3 \cdot x \cdot y$

$\qquad\quad 4x = 2 \cdot 2 \cdot x$

The common factors are 2, 2, and x; $2 \cdot 2 \cdot x = 4x$ is their product.

Step 2. $\quad \dfrac{2x + 3y + 1}{4x)\,8x^2 + 12xy + 4x}$

$\qquad\quad 8x^2 + 12xy + 4x = (4x)(2x + 3y + 1)$

2. Factor $4x^2 - 2x + 6xy^2$.

Step 1. $\quad 4x^2 = 2 \cdot 2 \cdot x \cdot x$

$\qquad\quad 2x = 2 \cdot x$

$\qquad\quad 6xy^2 = 2 \cdot 3 \cdot x \cdot y \cdot y$

Common factors are 2 and x; $2 \cdot x = 2x$.

Step 2. $\quad \dfrac{2x - 1 + 3y^2}{2x)\,4x^2 - 2x + 6xy^2}$

$\qquad\quad 4x^2 - 2x + 6xy^2 = (2x)(2x - 1 + 3y^2)$

---

†A prime factor of a number is one which has for exact integral divisors only itself and one. For our purposes we shall consider a prime factor of an algebraic expression to be one which is not further "factorable" (except for the trivial products of itself and one, and the negative of itself and -1) according to the criteria noted in Section 1.

3. Factor $-15x^3 + 45x^2y - 30x^2$.

Step 1. $15x^3 = 3 \cdot 5 \cdot x \cdot x \cdot x$

$45x^2y = 3 \cdot 3 \cdot 5 \cdot x \cdot x \cdot y$

$30x^2 = 2 \cdot 3 \cdot 5 \cdot x \cdot x$

Common factors are 3, 5, x, x; $15x^2$ is their product.

Step 2.   (When the numerical coefficient of the first term is negative, the coefficient of the common factor is usually made negative, which makes the coefficient of the first term in the other factor positive.)

$$-15x^2 \overline{) -15x^3 + 45x^2y - 30x^2} \quad \begin{matrix} x & - & 3y & + & 2 \end{matrix}$$

$$-15x^3 + 45x^2y - 30x^2 = (-15x^2)(x - 3y + 2)$$

EXERCISE 5-1

Write products for each of the following directly:

1. $(-3)(a - 6)$ **2.** $a(x - y)$

3. $4c(a^2 - c^2)$ **4.** $5x(a - b + c)$

5. $3xy(x - y + z)$ **6.** $-(4x - 3y + 2z)$

7. $-p^2(p^3 + p^2 - 3p)$ **8.** $-ab(ab - ab^2 + a^2b)$

9. $-0.4xy(0.2x - 0.3y - 0.5xy)$ **10.** $3m(n^2 + 2mn - m^2)$

11. $-5(4a^3 - 3a^2 + 2a + 1)$ **12.** $-\frac{1}{2}ab(\frac{3}{4}a^3b - \frac{2}{3}a^2b^2 - \frac{8}{9}ab^3)$

Factor:

13. $5x - 10y$ **14.** $6x - 2x^2$

15. $3a^2 - 12a$ **16.** $27a^2p^2 + 3ap$

17. $34pq - 51p$ **18.** $6a - 36a^2$

19. $14a^2xy - 2ax^2y$ **20.** $x^3 - 6x^2$

21. $2a^2 + a^2 - 3a$ **22.** $g^5 + 3g^4 - 2g^2$

23. $2mn - 4m + 6n$ **24.** $2x^2 - 4xy + 12y^2$

25. $2x^3y + 4x^2y^2 + 16xy^3$ **26.** $a^6 - 2a^4 + 3a^2$

27. $a^3b^2c^4 + a^2b^3c^3 + a^4b^2c^3$ **28.** $\frac{1}{4}a^2 - \frac{3}{8}a^2b + \frac{7}{16}ab^2$

29. $9x^2y - 0.6xy^2 + 0.3xy$ **30.** $2a^5 - 4a^4 + 8a^3 + 10a$

31. $h^2x^2 - 8px^2 + 4x^2y + 6x^2$ **32.** $24x^2y - 36xy^2 + 12xy$

3. SQUARE OF A BINOMIAL

A binomial has been defined as the sum or difference of two monomials, e.g., $(3a + b)$ or $(x - 2y)$. To **square** a number or expression means to use it as a factor twice or, simply, to multiply it by itself. Thus, we may express the square of $(3a + b)$ as

$$(3a + b)(3a + b) \quad \text{or} \quad (3a + b)^2$$

and the square of $(x - 2y)$ as

$$(x - 2y)(x - 2y) \quad \text{or} \quad (x - 2y)^2$$

By Program **2.9** (multiplication of two polynomials), we can compute the product of two such factors (their **expansion**) and detect a pattern that will allow us to write the product directly. Note in the following illustrative expansions, carried out according to Program **2.9**, that:

The square of the sum of two terms is equal to the sum of the squares of each term and twice the product of the two terms.

The square of the difference of two terms is equal to the sum of the squares of each term less twice the product of the two terms.

$$(3a + b)^2 = \begin{cases} \begin{array}{r} 3a + b \\ 3a + b \\ \hline (\times) \\ 9a^2 + 3ab \\ 3ab + b^2 \\ \hline 9a^2 + 6ab + b^2 \end{array} \end{cases} = (3a)^2 + (b)^2 + 2(3a \cdot b)$$

$$(x - 2y)^2 = \begin{cases} \begin{array}{r} x - 2y \\ x - 2y \\ \hline (\times) \\ x^2 - 2xy \\ - 2xy + 4y^2 \\ \hline x^2 - 4xy + 4y^2 \end{array} \end{cases} = (x)^2 + (2y)^2 - 2(x \cdot 2y)$$

The following program accomplishes the same result, though the sequence of terms in the expansion differs slightly from that stated in the foregoing italicized generalizations.

5.2

To square a binominal:

Step 1. Square the first term.

Step 2. Double the product of the first and second terms (consider the sign between the two terms as belonging to the second term).

Step 3. Square the second term.

Step 4. Express the desired product as the sum of the results of Steps 1, 2, and 3.

EXAMPLES

1: Square $4a + 3c$ [or "Expand $(4a + 3c)^2$"].

Step 1.   Square the first term:

$$(4a)^2 = 16a^2$$

Step 2.   Double the product of the two terms:

$$2[(4a) \cdot (3c)] = 2[12ac] = 24ac$$

Step 3.   Square the second term:

$$(3c)^2 = 9c^2$$

Step 4.   Express the results of Steps 1, 2, and 3 as a sum:

$$(4a + 3c)^2 = 16a^2 + 24ac + 9c^2$$

2. Expand $(2a - b)^2$.

Step 1. $(2a)^2 = 4a^2$

Step 2. $2[(2a)(-b)] = 2[-2ab] = -4ab$

Step 3. $(-b)^2 = b^2$

Step 4.   Add: $4a^2 - 4ab + b^2$

3. Expand $(ab - c)^2$.

Step 1. $(ab)^2 = a^2b^2$

Step 2. $2[(ab) \cdot (-c)] = 2[-abc] = -2abc$

Step 3. $(-c)^2 = c^2$

Step 4. $a^2b^2 - 2abc + c^2$

4. Square 53.

Express 53 as $(50 + 3)$:

$$53^2 = (50 + 3)^2 = (50 + 3)(50 + 3)$$
$$= (50)^2 + 2(50)(3) + (3)^2$$
$$= 2500 + 300 + 9 = 2809$$

These examples illustrate that the result of squaring a binomial is always a **perfect square trinomial**: a three-term polynomial in which two of the terms (usually the first and third) are perfect squares, and the other (middle) term is (disregarding the sign) twice the product of the square roots[†] of the other two terms. Furthermore, when the binomial is the sum of two terms, the sign of the middle term is invariably positive; when the original binomial is the difference of two terms, the sign of the middle term is invariably negative. From this we derive the following program.

5.3

To factor a perfect square trinomial:

Step 1. Find the square roots for each of the two terms that are perfect squares.

Step 2. Write two identical binomial factors whose terms are the two square

---

[†]Square root of a number or polynomial is one of two identical factors whose product is the number or polynomial. Additional discussion of square roots will be found in Section 4 of Chapter VII.

roots of Step 1, separated by the sign of the remaining trinomial term.

EXAMPLES

1. Factor $4x^2 + 12x + 9$.

$4x^2 + 12x + 9$ is a perfect square trinomial.

Step 1. $\sqrt{4x^2} = 2x$

$\sqrt{9} = 3$

Step 2.   The remaining trinomial term $(+12x)$ is positive. Therefore, $(2x + 3)(2x + 3)$ is the factored form of $4x^2 + 12x + 9$.

2. Factor $25x^2 - 40x + 16$.

$25x^2 - 40x + 16$ is a perfect square trinomial.

Step 1. $\sqrt{25x^2} = 5x$

$\sqrt{16} = 4$

Step 2.   The remaining trinomial term $(-40x)$ is negative. Therefore, $(5x - 4)(5x - 4)$ is the factored form of $25x^2 - 40x + 16$.

3. Factor $16a^2 - 24ab + 9b^2$.

$16a^2 - 24ab + 9b^2$ is a perfect square trinomial.

Step 1. $\sqrt{16a^2} = 4a$

$\sqrt{9b^2} = 3b$

Step 2. $(4a - 3b)(4a - 3b)$ is the factored form of $16a^2 - 24ab + 9b^2$.

4. Factor $4x^2 + 24xy + 36y^2$.

$4x^2 + 24xy + 36y^2$ is a perfect square trinomial and is also a type covered by Program **5.1** (contains a common factor 4). Remove the factor 4 first:

$$4(x^2 + 6xy + 9y^2)$$

Since $x^2 + 6xy + 9y^2$ is also a perfect square trinomial:

Step 1. $\sqrt{x^2} = x$

$\sqrt{9y^2} = 3y$

Step 2.

$x^2 + 6xy + 9y^2 = (x + 3y)(x + 3y)$. Therefore, the factored form of $4x^2 + 24xy + 36y^2$ is:

$$(4)(x + 3y)(x + 3y)$$

EXERCISE 5-2

*Expand the following, using Program **5.2**:*

1. $(x + y)^2$ **2.** $(a + b)^2$ **3.** $(m + n)^2$

4. $(x - y)^2$ **5.** $(a - b)^2$ **6.** $(m - n)^2$

7. $(x + 3)^2$ **8.** $(b - 4)^2$ **9.** $(3x + 1)^2$

10. $(2y - 1)^2$ **11.** $(3a - 2)^2$ **12.** $(4x - 5)^2$

13. $(3a - 2b)^2$ **14.** $(2x + 3y)^2$ **15.** $(8x + 7y)^2$

16. $(pm + qn)^2$ **17.** $(ax - by)^2$ **18.** $(1.2x - 0.9y)^2$

19. $(\frac{1}{2}x - \frac{3}{4}y)^2$ **20.** $(\frac{5}{3}a - \frac{2}{5}b)^2$ **21.** $(12)^2$

22. $(15)^2$ **23.** $(25)^2$ **24.** $(61)^2$

In the parentheses, supply the coefficient necessary to make the expression a perfect square trinomial:

25. $x^2 - (\ \ )xy + y^2$ **26.** $9x^2 + (\ \ )x + 16$

27. $25a^2 + (\ \ )ab + 81b^2$ **28.** $25x^2 - (\ \ )x + 144$

29. $\frac{1}{4}p^2 + (\ \ )pt + \frac{1}{9}t^2$ **30.** $16a^4 + (\ \ )a^2 + 9$

31. $a^2x^2 + (\ \ )x + c^2$ **32.** $4a^2b^2 - (\ \ )d + 25c^2d^2$

Factor the following trinomials completely when possible:

33. $x^2 + 2xy + y^2$ **34.** $a^2 - 2ab + b^2$

35. $x^2 + 4x + 4$ **36.** $2x^2 - 12x + 18$

37. $16a^2 + 8a - 1$ **38.** $9x^2y^2 + 6xy + 1$

39. $18x^2 + 24xy + 8y^2$ **40.** $4a^2 - 12a + 9$

41. $25x^4 + 80x^2y^2 + 64y^4$ **42.** $16a^2 + 14a + 9$

43. $\frac{1}{25}x^2 + \frac{2}{15}xy + \frac{1}{9}y^2$ **44.** $x^2 + 0.2x + 0.01$

45. $405x^2 + 360xy + 80y^2$ **46.** $\dfrac{a^2}{16} + \dfrac{ab}{6} + \dfrac{b^2}{9}$

47. $a^2x^2 + 2abx + b^2$ **48.** $c^2x^2y^2 + 2cxyz + z^2$

4. PRODUCT OF THE SUM AND DIFFERENCE OF TWO TERMS

An analysis of the expansion of two factors that are the sum and difference of the same two terms—represented by the binomials $(x + y)$ and $(x - y)$—when carried out by Program **2.9** (multiplication of two polynomials) leads to this generalization:

The product of the sum and difference of two terms is equal to the square of the first less the square of the second.

$$(x + y)(x - y) = \begin{cases} \begin{array}{r} x - y \\ x + y \\ \hline (\times) \\ x^2 - xy \\ + xy - y^2 \\ \hline x^2 + 0xy - y^2 = x^2 - y^2 \end{array} \end{cases}$$

5.4

To compute the product of the sum and difference of two terms:

Step 1. Square the two terms.

Step 2. Write the desired product as the square of the first term minus the square of the second term.

EXAMPLES

1. Expand $(3x + 4)(3x - 4)$.

Step 1. $(3x)^2 = 9x^2$

$(4)^2 = 16$

Step 2. $9x^2 - 16$

2. Expand $(4 + 3x)(4 - 3x)$.

Step 1. $(4)^2 = 16$

$(3x)^2 = 9x^2$

Step 2. $16 - 9x^2$

3. Expand $(a - b)(a + b)$.

Step 1. $(a)^2 = a^2$

$(b)^2 = b^2$

Step 2. $a^2 - b^2$

4. Multiply 98×102.

Express 98 as $(100 - 2)$, and 102 as $(100 + 2)$.

$$98 \times 102 = (100 - 2)(100 + 2)$$
$$= (100)^2 - (2)^2$$
$$= 10{,}000 - 4 = 9996$$

Since the product of the sum and difference of two terms is invariably the difference of the squares of the two terms, then inversely, an algebraic expression involving the difference of two squares will have for factors the sum and difference of the square roots of these two terms.

5.5

To factor the difference of two squared terms:

Step 1. Find the square root for each of the terms.

Step 2. Write one factor as the sum of the two square roots of Step 1 and the other factor as the difference of these two square roots.

EXAMPLES

1. Factor $a^2 - 4b^2$.

Step 1. $\sqrt{a^2} = a$

$$\sqrt{4b^2} = 2b$$

Step 2. $(a + 2b)(a - 2b)$.

2. Factor $1 - 16x^2y^2$.

Step 1. $\sqrt{1} = 1$

$$\sqrt{16x^2y^2} = 4xy$$

Step 2. $(1 + 4xy)(1 - 4xy)$

3. Factor $12x^2y - 27y^3$.

$12x^2y - 27y^3$ is an example of the type covered by Program **5.1**. Remove the common factor $3y$:

$$12x^2y - 27y^3 = 3y(4x^2 - 9y^2)$$

Since

$$(4x^2 - 9y^2) = (2x + 3y)(2x - 3y)$$

the complete factorization of $12x^2y - 27y^3$ is $(3y)(2x + 3y)(2x - 3y)$.

EXERCISE 5-3

Expand directly:

1. $(m - n)(m + n)$

2. $(s + t)(s - t)$

3. $(x - 3)(x + 3)$

4. $(y + 7)(y - 7)$

5. $(2x - y)(2x + y)$

6. $(3x - 2y)(3x + 2y)$

7. $\left(\dfrac{m}{2} - n\right)\left(\dfrac{m}{2} + n\right)$

8. $(8x - 9y)(8x + 9y)$

9. $(ab - c)(ab + c)$

10. $(15)(25) = (20 - 5)(20 + 5)$

11. $(37)(43)$

12. $(0.6x - 0.5y)(0.6x + 0.5y)$

Factor completely:

13. $a^2 - b^2$

14. $a^2 - 9b^2$

15. $16a^2 - 1$

16. $9x^2y^2 - 64z^2$

17. $25p^2 - 4q^2$

18. $\frac{1}{4}x^2 - \frac{9}{16}y^2$

19. $0.01x^2 - 0.16y^2$

20. $a^4 - b^4$

21. $12m^2 - 27n^2$

22. $36a^2b^2 - a^2$

23. $a^5b^2 - a^3$

24. $\dfrac{x^2y^2}{27} - \dfrac{x^2}{12}$

5. PRODUCT OF ANY TWO BINOMIALS

The two preceding types of special products, Programs **5.2** and **5.4**, have been singled out because their forms are readily recognized. In this section we develop a general program for computing the product of any two binomials directly (which includes Programs **5.2** and **5.4**). A consideration of the longer multiplication procedure (Program **2.9**) reveals certain consistencies which then allow us to develop a general procedure for arriving at the product by inspection.

$$
\begin{array}{r}
2x + y \\
x - 3y \\
(\times)\ \overline{} \\
2x^2 + xy \\
-6xy - 3y^2 \\
\overline{2x^2 - 5xy - 3y^2}
\end{array}
$$

Note in this computation that: (a) the first term of the product $(2x^2)$ is the product of the first terms of the binomials, (b) the last term of the product $(-3y^2)$ is the product of the second terms of the binomials (including the sign between the terms), and (c) the middle term of the product $(-5xy)$ is the result of adding the two products formed by multiplying the first term of each binomial with the second term of the other binomial. Part (c) is often referred to as computing the **cross-products**. Below at the right is a horizontal method for writing the product directly, with the computation above repeated and lettered to identify corresponding steps. Note that the middle term in the horizontal method is computed mentally.

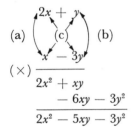

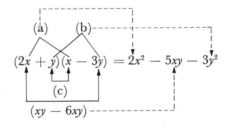

5.6

To multiply any two binomials directly:

Step 1. Compute the product of the first terms of the two binomials.

Step 2. Compute the sum of the two cross-products.

Step 3. Compute the product of the second terms of the two binomials.

Step 4. Express the product of the two binomials as the sum of the results of Steps 1, 2, and 3.

[*Note:* The three steps in sequence given in Program **5.6** do not have to be performed in that order. However, when the binomials are alike except for numerical coefficients, this sequence of steps leads to a trinomial whose terms are arranged in descending powers of one variable and ascending powers of the other. Such an orderly arrangement of the terms is an important aid to factoring.]

EXAMPLES

1. Write the product directly for $(3x - 2y)(x - 4y)$.

Step 1.   Product of the first terms:

$$(3x)(x) = 3x^2$$

Step 2.   Sum of cross-products:

$$\left.\begin{array}{l}(3x)(-4y) = -12xy \\ (-2y)(x) = -2xy\end{array}\right\} -14xy$$

Step 3.   Product of the second terms:

$$(-2y)(-4y) = 8y^2$$

Step 4. $(3x - 2y)(x - 4y) = 3x^2 - 14xy + 8y^2$

2. Write the product directly for $(4x - 5)(x + 6)$.

$$\begin{array}{cc}Step\ 1 & Step\ 3 \\ 4x^2 & -30\end{array}$$

$$Step\ 4$$
$$(4x - 5)(x + 6) = 4x^2 + 19x - 30$$
$$\left.\begin{array}{l}-5x \\ +24x\end{array}\right\} +19x$$
$$Step\ 2$$

3. Expand directly $(3x - 2y)(3x - 2y)$.

$$\begin{array}{cc}9x^2 & 4y^2\end{array}$$

$$(3x - 2y)(3x - 2y) = 9x^2 - 12xy + 4y^2$$
$$\left.\begin{array}{l}-6xy \\ -6xy\end{array}\right\} -12xy$$

[*Note:* $(3x - 2y)(3x - 2y) = (3x - 2y)^2$ and, as can be expected, the expansion is a perfect square trinomial.]

4. Multiply 24×32.

Express 24 as $20 + 4$, or $2t + 4$, and 32 as $30 + 2$, or $3t + 2$ (where $t = 10$). Then:

$$\begin{aligned}24 \times 32 &= (2t + 4)(3t + 2) \\ &= 6t^2 + 16t + 8 \\ &\quad \text{(if } t = 10, \text{ then } t^2 = 100) \\ &= 6(100) + 16(10) + 8 \\ &= 600 + 160 + 8 \\ &= 768\end{aligned}$$

EXERCISE 5-4

Expand, using Program **5.6**:

1. $(x + 3)(x + 2)$ **2.** $(5a + 1)(a + 7)$

3. $(3x + 4)(x + 3)$ **4.** $(2x - 5)(3x - 2)$

5. $(4y - 7)(2y - 5)$ **6.** $(3x + 1)(x - 5)$

7. $(2x - 7)(x + 4)$ **8.** $(5a - 4)(3a + 1)$

9. $(2m + 3)(2m + 5)$ **10.** $(8p - 3)(5p + 2)$

11. $(3x - 7)(3x - 7)$ **12.** $(3x - 7)(5x + 2)$

13. $(2a - 3)(2a + 3)$ **14.** $\left(\dfrac{x}{2} + 4\right)\left(\dfrac{x}{2} + 6\right)$

15. $(3xy - 7)(2xy + 4)$ **16.** $(2a^2 - 3b^2)(2a^2 - 5b^2)$

17. $(0.1y - 0.2)(0.6y - 0.5)$ **18.** $(7x + 2y)(7x + 2y)$

19. $(ax + b)(cx + d)$ **20.** $(kx - c)(mx + d)$

21. $(\frac{3}{4}x - \frac{1}{2})(\frac{1}{3}x + \frac{1}{2})$ **22.** $(2px + 3)(x - p)$

23. $(12)(23) = (t + 2)(2t + 3)$ **24.** $(11)(14)$

6. FACTORING TRINOMIALS OF THE FORM $ax^2 + bxy + cy^2$

    Finding two binomial factors whose product is a certain trinomial of the form $ax^2 + bxy + cy^2$—by no means always possible—usually involves a systematic trial-and-error approach. A few generalizations, however, coupled with experience in the inverse process of expanding factors, will tend to cut down the number of necessary trials.

    From our programs for finding the products of two binomials directly, discussed in Section 5, we know that the first term of the trinomial to be factored must contain as its factors the first terms of the desired binomials; that the last term of the trinomial must contain as its factors the second terms of the desired binomials; and that the middle term of the trinomial must be the sum of the cross-products of the first and second terms of the binomial factors.

    Another aid to factoring trinomials is an awareness of the distribution of signs in the trinomial. In the products of the following pairs of binomials, note how the signs are distributed among the terms of their expansions to the right:

(a) $(3x + 2y)(x + y) = 3x^2 + 5xy + 2y^2$

(b) $(3x - 2y)(x - y) = 3x^2 - 5xy + 2y^2$

(c) $(3x + 2y)(x - y) = 3x^2 - xy - 2y^2$

(d) $(3x - 2y)(x + y) = 3x^2 + xy - 2y^2$

We may generalize with the following sign analysis:

    1. When the sign of the last term of a factorable trinomial is positive, the signs of the second terms in both binomial factors agree with that of the middle term of the trinomial.

2. When the sign of the last term of a factorable trinomial is negative, the signs of the second terms of the two binomial factors differ, and the prevailing sign of the sum of the resulting cross-products agrees with that of the middle terms of the trinomial.

5.7

To factor a trinomial of the form $ax^2 + bxy + cy^2$ that is the product of two binomials:

Step 1. Determine two factors of the first term of the trinomial (ax^2).

Step 2. Determine two factors of the last term of the trinomial (cy^2).

Step 3. Use these factors to form two binomials so that each binomial contains one factor from each of Steps 1 and 2. Insert signs according to the sign analysis of the trinomial.

Step 4. Check the sum of the resulting cross-products in Step 3. If it agrees with the middle term of the trinomial (bxy), the binomials of Step 3 are the desired factors. If it does not agree, repeat the steps with new factor combinations and with new sign arrangements if the binomials are to differ in sign.

EXAMPLES

1. Factor $x^2 - 5x + 6$.

Step 1.   Factors of the first term of the trinomial (x^2) are x and x.

Step 2.   Factors of the last term of the trinomial (6) are 6 and 1; also 3 and 2.

Step 3.   Coupling the x and x with the 6 and 1, respectively, we get $(x \quad 6)(x \quad 1)$. From analysis of the signs of the trinomial, the signs of the second terms of the binomials will be negative. Thus: $(x - 6)(x - 1)$.

Step 4.   The sum of the cross-products for $(x - 6)(x - 1)$ is $-7x$. This does not agree with the middle term of the trinomial $(-5x)$; therefore, $(x - 6)$ and $(x - 1)$ are not the factors of $x^2 - 5x + 6$.

Step 3.   Couple the x and x factors of the first term with the factors 3 and 2 of the last term for $(x - 3)(x - 2)$.

Step 4.   The sum of the cross-products for $(x - 3)(x - 2)$ is $-5x$; consequently, $(x - 3)(x - 2)$ is the factored form of $x^2 - 5x + 6$.

[*Note:* When the numerical coefficient of the first term of the trinomial is 1, as in this example, the correct pair of factors for the numerical coefficient of the last term can be quickly determined by the following:

(a) If the sign of the third term of the trinomial is *positive*, the correct pair of factors for the numerical coefficient of that third term will be those whose *sum* is the numerical coefficient of the middle term of the trinomial.

(b) If the sign of the third term of the trinomial is *negative*, the correct pair of factors for the numerical coefficient of that third term will be those whose *difference* is the numerical coefficient of the middle term of the trinomial.]

2. Factor $x^2 + 2x - 15$.

Step 1. The factors of x^2 are x and x.

Step 2. Factors of 15 are 15 and 1; also 5 and 3. The sign of the third term is negative. The difference between the factors 15 and 1 is 14; the difference between the factors 5 and 3 is 2, which is the numerical coefficient of the middle term. We choose 5 and 3.

Step 3. Couple the x and x factors with the 5 and 3 in two temporarily signless binomials: $(x \quad 5)(x \quad 3)$. Since the sign of the third term in the trinomial is negative, the signs of the second terms of the two binomials will differ; and since the middle term is positive, we place the $(+)$ sign before the 5 and the $(-)$ sign before the 3: $(x + 5)(x - 3)$.

Step 4. Check the sum of the cross-products: $5x - 3x = 2x$. Hence the factored form of $x^2 + 2x - 15$ is $(x + 5)(x - 3)$.

3. Factor $2x^2 + xy - 3y^2$.

Step 1. Factor $2x^2$: $2x \cdot x$.

Step 2. Factor $3y^2$: $3y \cdot y$.

Step 3. Couple these factors into binomial form. Because the last term is negative, the signs of the second terms of the binomials will differ; because the sign of the middle term (xy) is positive, the $(+)$ sign should go with the greater cross-product. Thus: $(2x - y)(x + 3y)$.

Step 4. Check the sum of the cross-products: $6xy - xy \neq xy$.

Step 3. Try a new combination: $(2x + 3y)(x - y)$.

Step 4. Check the sum of the cross-products: $3xy - 2xy = xy$. Hence $(2x + 3y)(x - y)$ is the factored form of $2x^2 + xy - 3y^2$.

4. Factor $6x^2 - x - 12$.

The factors of $6x^2$ are $6x$ and x; $3x$ and $2x$. The factors of 12 are 12 and 1; 6 and 2; 4 and 3. From an analysis of the signs in the trinomial, establish the combination of signs in the binomials:

$$(\cdots + \cdots)(\cdots - \cdots)$$

Test the various factor combinations of the third term, 12, with the $6x$ and x factors of $6x^2$, keeping in mind that the sum of the cross-products must be negative (because the middle term in the trinomial is negative).

Factors (?)	*Middle term check*
$(6x + 1)(x - 12)$	$x - 72x \neq -x$
$(6x + 2)(x - 6)$	$2x - 36x \neq -x$
$(6x + 3)(x - 4)$	$3x - 24x \neq -x$

Factors (?)	*Middle term check*
$(6x + 4)(x - 3)$	$4x - 18x \neq -x$
$(6x + 6)(x - 2)$	$6x - 12x \neq -x$

(The other combination leads to a positive middle term.)

Before abandoning the $6x$ and x combination, interchange them:

$$(x + 1)(6x - 12) \qquad 6x - 12x \neq -x$$

(The other combinations lead to a positive middle term.)

From this it is clear that $6x$ and x are *not* the proper pair of factors for $6x^2$. So try the pair $2x$ and $3x$:

$$(2x + 1)(3x - 12) \qquad 3x - 24x \neq -x$$
$$(2x + 2)(3x - 6) \qquad 6x - 12x \neq -x$$

(The other combinations lead to a positive middle term.)

Finally, interchange the factors $2x$ and $3x$:

$$(3x + 1)(2x - 12) \qquad 2x - 36x \neq -x$$
$$(3x + 2)(2x - 6) \qquad 4x - 18x \neq -x$$
$$(3x + 4)(2x - 3) \qquad 8x - 9x = -x$$

Hence the factored form of $6x^2 - x - 12$ is $(3x + 4)(2x - 3)$.

Example 4 was purposely labored to demonstrate better the technique for factoring a trinomial. Actually, increased experience with factoring tends to narrow the guesswork considerably. But often when the correct combinations are not found readily, a suspicion arises that the trinomial is not factorable. The following is a quick means for checking whether or not a trinomial is indeed factorable:

Trinomials of the form $ax^2 + bxy + cy^2$ (in which the numerical coefficients a, b, and c denote integers) can be factored if the expression $(b^2 - 4ac)$ is a perfect square.†

EXAMPLES

1. Can $3x^2 + 2x - 4$ be factored?

If $3x^2 + 2x - 4 = ax^2 + bxy + cy^2$, then $a = 3$, $b = 2$, $c = -4$, and $y = 1$.

$$(b^2 - 4ac) = (2)^2 - 4(3)(-4) = 4 - (-48) = +52$$

Since $+52$ is not a perfect square, $3x^2 + 2x - 4$ cannot be factored into the product of two binomials with integral coefficients.

2. Can $3x^2 + 10x - 8$ be factored?

Here $a = 3$, $b = 10$, and $c = -8$.

$$b^2 - 4ac = (10)^2 - 4(3)(-8) = 100 + 96 = 196 = (14)^2$$

---

†Table I, page 349, lists squares of numbers through 100.

Hence, $3x^2 + 10x - 8$ can be factored: $(3x - 2)(x + 4)$.

3. Can $2x^2 + 11xy + 12y^2$ be factored?

$$b^2 - 4ac = 121 - 4(2)(12) = 121 - 96 = 25 = (5)^2$$

Hence, $2x^2 + 11xy + 12y^2$ can be factored: $(2x + 3y)(x + 4y)$.

EXERCISE 5-5

Factor completely. For those trinomials that you cannot factor, apply the (b² − 4ac) *test and write its numerical value instead of the factors.*

1. $x^2 + 5x + 6$ **2.** $y^2 - 7y + 12$ **3.** $a^2 - 10a + 16$

4. $a^2 + 12a + 36$ **5.** $y^2 + 4y - 5$ **6.** $x^2 + 2x - 15$

7. $x^2 - 2x + 3$ **8.** $18x^2 + 33x + 14$ **9.** $3x^2 + 29x + 40$

10. $6a^2 + 11a - 10$ **11.** $4x^2 - 16xy + 15y^2$ **12.** $3a^2 + 14a - 5$

13. $2x^2 - 3x - 9$ **14.** $100 - 20y + y^2$ **15.** $4 - 12b + 9b^2$

16. $\dfrac{x^2}{4} + 6x + 36$ **17.** $x^2 - 3x + 2$ **18.** $12a^2 - 35ab + 18b^2$

19. $m^3 + 10m^2n + 24mn^2$ **20.** $9x^2 - 28xy + 3y^2$ **21.** $18a^4 + 21a^2 - 4a$

22. $a^4 - 6a^2 + 8$ **23.** $30x^2 + 4xy - 2y^2$ **24.** $15r^2 + 20r + 8$

25. $0.4c^2 - 1.1c + 0.6$ **26.** $4apx^2 - 17apx + 18ap$ **27.** $0.2x^2 + 0.2x - 6$

28. $12x^4 - 11x^2 + 2$ **29.** $a - 7a^2 - 18$ **30.** $\dfrac{x^2}{8} + \dfrac{5x}{2} + 12$

31. $10x^2 - 3y^2 - xy$ **32.** $-18 + 42x^2 - 24x$

7. FACTORING BY GROUPING

The programs developed in the previous sections of this chapter can be usefully extended to other types of algebraic expressions. For instance, according to Program **5.1**:

$$3x^2 - 12xy = 3x(x - 4y)$$

However, had the common term been a binomial, say $(3 + x)$, instead of $3x$, it too could have been factored out. To start with a simple illustration, suppose we have the following polynomial:

$$a(x + y) + b(x + y)$$

Clearly, $(x + y)$ is a common factor in each major addend of the sum, and as such, it can be factored out:

$$a(x + y) + b(x + y) = (x + y)(a + b)$$

Further

$$a(x + y) + b(x + y) = ax + ay + bx + by$$

Hence

$$ax + ay + bx + by = (x + y)(a + b)$$

Thus we see that in some cases a polynomial such as $ax + ay + bx + by$, containing no universally common term, may still be factored by appropriate grouping.

EXAMPLES

1. Factor $xy + 5y + bx + 5b$.

If we factor y out of the first two terms and b out of the last two terms, we can express the original sum as

$$\underbrace{xy + 5y}_{} + \underbrace{bx + 5b}_{}$$
$$y(x + 5) + b(x + 5)$$

Now if we factor out the common $(x + 5)$, we will have transformed the given polynomial into an equivalent product:

$$y(x + 5) + b(x + 5) = (x + 5)(y + b)$$

2. Factor $3m + 3n - am - an$.

Factor 3 from the first two terms and $-a$ from the last two terms:

$$3m + 3n - am - an = 3(m + n) - a(m + n)$$
$$= (m + n)(3 - a)$$

3. Factor $cx + y - x - cy$.

Rearrange, and factor out c:

$$cx + y - x - cy = cx - cy + y - x$$
$$= c(x - y) + y - x$$

Express $y - x$ as $-1(x - y)$; then

$$c(x - y) + (y - x) = c(x - y) + (-1)(x - y)$$
$$= (x - y)(c - 1)$$

4. Factor $c^2d - 3c^2 - 9d + 27$.

Factor c^2 from the first two terms and -9 from the last two terms:

$$c^2(d - 3) - 9(d - 3) = (d - 3)(c^2 - 9)$$

But $(c^2 - 9)$ can be factored further, by **5.5**, as $(c - 3)(c + 3)$. Hence
$$c^2d - 3c^2 - 9d + 27 = (c - 3)(c + 3)(d - 3)$$

There are many other polynomial sums, whose terms are other than monomials, which are factorable. Generally, the more experience one has with factoring the different basic types, the more likely he is to recognize a prototype in these more complicated sums.

EXAMPLES

1. Factor $(x - b)^2 + 9(x - b) + 14$.

Insight may be obtained if we let $a = (x - b)$, in which case

$$(x - b)^2 + 9(x - b) + 14 = a^2 + 9a + 14$$

The latter readily factors to $(a + 7)(a + 2)$. However, since $a = (x - b)$, then

$$(a + 7) = [(x - b) + 7]$$
$$(a + 2) = [(x - b) + 2]$$

and

$$(x - b)^2 + 9(x - b) + 14 = (x - b + 7)(x - b + 2)$$

2. Factor $(y + 2)^2 - (x + 3)^2$.

If we let $a = (y + 2)$ and $b = (x + 3)$, then

$$(y + 2)^2 - (x + 3)^2 = a^2 - b^2 = (a - b)(a + b)$$

Substituting back for a and b, we get

$$(y + 2)^2 - (x + 3)^2 = [(y + 2) - (x + 3)][(y + 2) + (x + 3)]$$
$$= [y + 2 - x - 3][y + 2 + x + 3]$$
$$= [y - x - 1][y + x + 5]$$

With experience, need for the substitution "crutch" will diminish, and factoring can proceed directly.

3. Factor $2(x - 3)^2 + (x - 3) - 6$.

$$-3(x - 3) + 4(x - 3) = (x - 3)$$

Simplifying:

$$2(x - 3)^2 + (x - 3) - 6 = [2(x - 3) - 3][(x - 3) + 2]$$
$$= [2x - 6 - 3][x - 3 + 2]$$
$$= [2x - 9][x - 1]$$

EXERCISE 5-6

Factor by grouping:

1. $ax + bx + 3a + 3b$ **2.** $cx + cy - 4x - 4y$

3. $dx - c + x - dc$ **4.** $cn - 4c - bn + 4b$

5. $gx - 3gh + ghx - 3g$ **6.** $ab + 6a + 2b + 12$

7. $xy - 5x - 15 + 3y$ **8.** $3s - 2t - 6 + st$

9. $a^2x + a^2d - x - d$

10. $xy - hx + x^2 - hy$

11. $2at + bt - bs - 2as$

12. $2ay - 6ax - 3bx + by$

13. $5ac - ad - 2bd + 10bc$

14. $bc + 2c - 4b - 8$

15. $kx^2 + ky^2 - mx^2 - my^2$

16. $a^2x^2 - 3x^2 - 4a^2 + 12$

Factor:

17. $(x + 1)^2 + 2(x + 1) + 1$

18. $(a - b)^2 + 4(a - b) + 4$

19. $(x + 5)^2 - (x + 2)^2$

20. $4a^2(b + c) - b^2(b + c)$

21. $2(x + y)^2 - (x + y) - 3$

22. $4(m + n)^2 + 20(m + n) + 25$

23. $2(a^2 + 2ab + b^2) - 3(a + b) - 9$

24. $(x - y)(x + y)^2 + 6(x + y)(x - y) + 9(x - y)$

25. $(x^2 + 4x + 4) - (x^2 - 6x + 9)$

26. $(25x^2 + 10x + 1) - 2(x^2 - 5x - 1 - x^2) + 1$

Factor completely:

27. $8a^2 - 2ab - 3b^2$

28. $y^2 - 2xy^2 + 2x - 1$

29. $a + 3b - a^2 + 9b^2$

30. $a^4 - a^2 - 12$

31. $2a^3 + 5a^2 - 4a - 10$

32. $20x^2 + 63xy - 45y^2$

33. $0.01a^2 - 0.0001b^2$

34. $ab - a - b + 1$

35. $\frac{25}{36}x^2y^2 - x^2$

36. $2(x + y)^2 + 12(x + y) + 10$

37. $\dfrac{9s^2}{16} - \dfrac{4t^2}{9}$

38. $a^4 - 4a^2 + 4$

39. $(9a^2 + 30a + 25) - (a - 4)^2$

40. $2x^4 - 10x^2 - 12$

41. $16p^4 - 81q^4$

42. $(m + n)^2 - m^2$

43. $a^2 + 6ab + 9b^2 - 4$

44. $x^6 - x^4 - 16x^2 + 16$

45. $2x^4 + 6x^3 - 8x^2 - 24x$

46. $(x^2 + 4x - 6)^2 - 36$

47. $0.25x^2 - 0.3x + 0.09$

48. $a^2 + 2ab + b^2 + 4a + 4b + 3$

49. $2a^2 - 4ab + 2b^2 + 5a - 5b - 12$

50. $9a^2 - 6ab + b^2 - 25x^2 - y^2 + 10xy$

8. SUMS AND DIFFERENCES OF TWO CUBES

By multiplication, it can be verified that

$$(x + y)(x^2 - xy + y^2) = x^3 + y^3$$

and

$$(x - y)(x^2 + xy + y^2) = x^3 - y^3$$

There is a fairly consistent and easily remembered pattern among the terms of the factors of these two expressions. By using

$$x^3 + y^3 = (x + y)(x^2 - xy + y^2)$$

and

$$x^3 - y^3 = (x - y)(x^2 + xy + y^2)$$

as formulas, it is possible to factor directly any sum or difference of two cubes.

EXAMPLES

1. Factor $a^3 + b^3$.

Using $x^3 + y^3 = (x + y)(x^2 - xy + y^2)$ as a formula, and replacing x with a, and y with b, we get

$$a^3 + b^3 = (a + b)(a^2 - ab + b^2)$$

2. Factor $27p^3 - 8$.

Using $x^3 - y^3 = (x - y)(x^2 + xy + y^2)$ as a formula, and substituting $3p$ for x, and 2 for y (since $(3p)^3 = 27p^3$ and $(2)^3 = 8$), we get

$$27p^3 - 8 = (3p)^3 - (2)^3 = [3p - 2][(3p)^2 + (3p)(2) + (2)^2]$$
$$= (3p - 2)(9p^2 + 6p + 4)$$

[*Note:* We could have achieved the same result by considering $27p^3 - 8$ as $27p^3 + (-8)$ and

$$x^3 + y^3 = (x + y)(x^2 - xy + y^2)$$

as the formula. Thus $x = 3p$ and $y = -2$, and

$$27p^3 - 8 = (3p)^3 + (-2)^3$$
$$= [3p + (-2)][(3p)^2 - (3p)(-2) + (-2)^2]$$
$$= (3p - 2)(9p^2 + 6p + 4)]$$

3. Factor $(a - 3)^3 + 64$.

Using the $x^3 + y^3$ formula, with $x = (a - 3)$ and $y = 4$, we get

$$(a - 3)^3 + 64 = [(a - 3) + 4][(a - 3)^2 - (a - 3)(4) + (4)^2]$$
$$= [a - 3 + 4][a^2 - 6a + 9 - 4a + 12 + 16]$$
$$= (a + 1)(a^2 - 10a + 37)$$

EXERCISE 5-7

Factor:

1. $x^3 - y^3$

2. $p^3 + q^3$

3. $x^3 + 8$

4. $m^3 + 125$

5. $27 - y^3$

6. $8s^3 - 1$

7. $8r^3 + 27t^3$

8. $x^3y^3 + 1$

9. $64a^3b^3 - c^3$

10. $125a^3 + 64b^3$

11. $x^6 - y^3$ **12.** $64x^3 + y^6$

13. $27x^9 - y^6$ **14.** $216b^{12} + 27c^6$

15. $64t^{21} - 8t^3$ **16.** $(a + 1)^3 - y^3$

17. $(a - 1)^3 + 27$ **18.** $(a - 1)^3 - (a - 3)^3$

19. $(x + 3)^3 + (x - 3)^3$ **20.** $(2x + y)^3 - (3x - y)^3$

REVIEW—CHAPTER V

Write the products for each of the following directly:

1. $-4(m + 7)$ **2.** $-5d(c^4 - d^5)$

3. $-4km(k - 3m + 2n)$ **4.** $-d^2(d^2 - 4d + 9)$

5. $0.07ab(-0.3a - 0.4b + 0.5ab)$ **6.** $-10(3x^3 - 5x^2 + x - 15)$

Factor:

7. $14a - 21b$ **8.** $5m^2 - 30m$

9. $54xy - 90y$ **10.** $30d^2pq + 6dpq^2$

11. $7m^3 + 5m^2 - 9m$ **12.** $3ab + ba - 12b$

13. $8a^3b + 24a^2b^2 + 16ab^3$ **14.** $k^6m^5n^4 + k^4m^4n^4 - k^4m^5n^3$

15. $6a^2m^3 + 8bm^3 - 3m^3n^2 - m^3$

Expand, using Program 5.2:

16. $(a + b)^2$ **17.** $(x + 5y)^2$ **18.** $(s - 2t)^2$

19. $(8 + b)^2$ **20.** $(5m + 1)^2$ **21.** $(7x - 5)^2$

22. $(3s - 7t)^2$ **23.** $(4a + 9b)^2$ **24.** $(bm - 2cn)^2$

25. $(\frac{1}{4}c - \frac{3}{8}d)^2$ **26.** $(17)^2$ **27.** $(31)^2$

In the parentheses, supply the coefficient necessary to make the expression a perfect square trinomial:

28. $a^2 - (\ )ab + b^2$ **29.** $9x^2 + (\ )xy + 100y^2$

30. $\frac{25}{81}a^2 - (\ )ab + \frac{1}{25}b^2$ **31.** $a^4z^2 - (\ )z + 9$

Factor those which are perfect square trinomials:

32. $a^2 - 2ab + b^2$ **33.** $x^2 + 10x + 25$

34. $49m^2 + 42m - 7$ **35.** $8m^2 + 10mn + 3n^2$

36. $16s^4 + 40s^2t^2 + 25t^4$ **37.** $\frac{4}{25}x^2 - \frac{3}{5}xy + \frac{9}{16}y^2$

38. $36m^2 + 168m + 196$ **39.** $a^4m^2 + 4a^2my^2 + 4y^4$

Expand:

40. $(c + d)(c - d)$ **41.** $(a - 5)(a + 5)$

42. $(3m + n)(3m - n)$ **43.** $\left(\frac{x}{3} - y\right)\left(\frac{x}{3} + y\right)$

44. $(mn + t)(mn - t)$ **45.** $(18)(22)$

Factor completely:

46. $m^2 - n^2$

47. $36t^2 - 1$

48. $9a^2 - 64b^2$

49. $0.64x^2 - 0.0001y^2$

50. $8a^2 - 32b^2$

51. $m^5 - m^7n^2$

Expand, using Program **5.6** :

52. $(m + 2)(m + 7)$

53. $(3t + 7)(t + 4)$

54. $(5a - 2)(3a - 8)$

55. $(3a - 8)(4a + 11)$

56. $(7t + 4)(7t + 1)$

57. $(8a - 3)(8a - 3)$

58. $(5m - 7)(5m + 7)$

59. $(4mn - 3)(5mn + 3)$

60. $(0.2x - 0.5)(0.4x - 0.7)$

61. $(c + mx)(d + nx)$

62. $(\frac{3}{5}m + \frac{1}{5})(\frac{1}{3}m - \frac{1}{5})$

63. $(55)(12)$

Factor completely. For those trinomials which you cannot factor, apply the $(b^2 - 4ac)$
test and write its numerical value instead of the factors.

64. $a^2 + 10a + 9$

65. $t^2 - 9t + 20$

66. $a^2 + 10a - 11$

67. $x^2 - 4x + 5$

68. $5s^2 + 64s + 48$

69. $15m^2 - 26m + 8$

70. $3t^2 - 14t - 49$

71. $9 - 21a + 6a^2$

72. $b^2 - 13b + 14$

73. $x^3 - 10x^2 + 21x$

74. $4m^4 + 2m^2 - 5m$

75. $40t^2 + 28st - 48s^2$

76. $0.8x^2 - 1.8x + 0.9$

77. $0.4a^2 - 7.7a - 6$

78. $21m^2 - 2n^2 + mn$

79. $b^2 + 4b - 10$

Factor by grouping:

80. $mx + nx + 5m + 5n$

81. $ax - x - y + ay$

82. $adx + 3dx - 5ad - 15d$

83. $mn + 7n - 5m - 35$

84. $a^2x + a^2y - 4x - 4y$

85. $2ab - dy - 2by + ad$

86. $3am - an - 2bn + 6bm$

87. $3as^2 + 3at^2 - bs^2 - bt^2$

Factor:

88. $(a + 3)^2 + 4(a + 3) + 4$

89. $(t + 2)^2 - (t - 3)^2$

90. $6(2m + n)^2 - (2m + n) - 7$

91. $4(9x^2 + 6xy + y^2) - 8(3x + y) - 5$

92. $(a^2 + 6a + 9) - (a^2 - 10a + 25)$

93. $16m^2 - 2mn - 5m^2$

94. $3c^4 + 5c^3 - 15c - 25$

95. $4s + 3t - 16s^2 + 9t^2$

96. $0.01m^4 - 0.0001n^2$

97. $x^4y^2 - \frac{1}{169}z^4$

98. $\frac{16}{81}x^2 - \frac{9}{25}y^2z^2$

99. $(9m^2 + 30m + 25) - (m - 5)^2$

100. $256x^4 - 81y^4$

101. $9x^2 + 12xy + 4y^2 - 16$

102. $5m^4 + 15m^3 - 20m^2 - 60m$

103. $0.64a^2 - 0.8a + 0.25$

104. $3x^2 - 6xy + 3y^2 + 13x - 13y - 30$

105. $m^3 - n^3$ **106.** $m^3 + 64$

107. $125 - x^3$ **108.** $64a^3 + 27b^3$

109. $125s^3t^3 - u^3$ **110.** $a^9 - b^3$

111. $m^{12} - 125n^6$ **112.** $27x^{15} - 64y^9$

113. $(m - 4)^3 + 8$ **114.** $(s + 5)^3 - (s - 4)^3$

VI

FRACTIONS

In keeping with the number-numeral distinction, it may be said that a number has many equivalent "names" (i.e., numerals). For example, the rational number that is midway between 0 and 1 may be "named" equivalently by the fractions

$$\frac{1}{2}, \quad \frac{4}{8}, \quad \frac{7}{14}, \quad \frac{50}{100}, \quad \frac{3+2}{7+3}, \quad \frac{9-1}{15+1}, \quad \text{etc.}$$

One way to test whether or not two fractions are equivalent is to apply the following generalization:

Two fractions, a/b and p/q, are equivalent to one another if and only if $a \times q = b \times p$.

Thus, $\frac{4}{6}$ and $\frac{8}{12}$ are equivalent because $4 \times 12 = 6 \times 8$. On the other hand, $\frac{3}{4}$ and $\frac{7}{10}$ are not equivalent because $3 \times 10 \neq 4 \times 7$. Moreover, if two fractions are equivalent, some appropriate multiplication or division of *both* numerator and denominator of one of the fractions will produce the other fraction. For example, we may show that $\frac{8}{16}$ is equivalent to $\frac{4}{8}$ by dividing numerator and denominator of the former by 2, or equivalent to $\frac{7}{14}$ by multiplying both numerator and denominator of $\frac{8}{16}$ by $\frac{7}{8}$ (or dividing by $\frac{8}{7}$).

In algebra, we extend the fraction notation of arithmetic to include polynomials as possible numerators

1. LOWEST TERMS

117

and denominators. Such fractions are commonly referred to as **algebraic fractions**. Thus the algebraic fractions

$$\frac{a}{x}, \quad \frac{3a}{3x}, \quad \frac{a^2}{ax}, \quad \frac{a(x+y)}{x(x+y)}, \quad \frac{a(p+q)^2}{x(p+q)^2}, \quad \frac{ab^2-ac^2}{xb^2-xc^2}$$

are all equivalent to one another, since any one fraction may be obtained from any of the others by some appropriate nonzero multiplication or division of both numerator and denominator. For instance, we may show that $a(x+y)/x(x+y)$ is equivalent to a/x by dividing both numerator and denominator by $(x+y)$; and we may show that $a(p+q)^2/x(p+q)^2$ is equivalent to a^2/ax by multiplying both terms by $a/(p+q)^2$:

$$\frac{a\cancel{(p+q)^2} \cdot \dfrac{a}{\cancel{(p+q)^2}}}{x\cancel{(p+q)^2} \cdot \dfrac{a}{\cancel{(p+q)^2}}} = \frac{a^2}{ax}$$

A fraction is said to be in **lowest terms** in arithmetic when its numerator and denominator denote integers which are relatively prime to each other, that is, they have no common integra' divisors except 1. Thus, for the set of equivalent fractions displayed in the first paragraph of this section, $\frac{1}{2}$ is the fraction in lowest terms. An algebraic fraction is similarly considered to be in lowest terms when no common factor having integral coefficients exists between the numerator and the denominator, except for the trivial $+1$ and -1.

[*Note:* In this chapter, as in Chapter V, a polynomial having integral coefficients is considered factorable if it can be expressed as a product of two or more factors whose numerical coefficients are integers.]

6.1

To express the equivalent of an algebraic fraction in lowest terms:

Step 1. Factor the numerator completely; factor the denominator completely.

Step 2. Divide out (cancel) all common factors that exist in both the numerator and the denominator.

EXAMPLES

1. Express $\dfrac{x^2+2x-8}{x^2+x-12}$ in lowest terms.

Step 1. $x^2+2x-8 = (x-2)(x+4)$

$x^2+x-12 = (x+4)(x-3)$

Step 2. $\dfrac{x^2+2x-8}{x^2+x-12} = \dfrac{(x-2)\cancel{(x+4)}}{\cancel{(x+4)}(x-3)} = \dfrac{x-2}{x-3}$

2. Express $\dfrac{3x^2 - 9ax + 6a^2}{6x^2 - 6a^2}$ in lowest terms.

Step 1. $\quad 3x^2 - 9ax + 6a^2 = (3)(x - a)(x - 2a)$

$\qquad\qquad 6x^2 - 6a^2 = (2)(3)(x - a)(x + a)$

Step 2. $\quad \dfrac{3x^2 - 9ax + 2a^2}{6x^2 - 6a^2} = \dfrac{(3)(x - a)(x - 2a)}{(2)(3)(x - a)(x + a)} = \dfrac{x - 2a}{2(x + a)}$

3. Express $\dfrac{cp - 2a - ap + 2c}{p^2 + 4p + 4}$ in lowest terms.

Step 1. $\quad cp - 2a - ap + 2c = (c - a)(p + 2)$

$\qquad\qquad p^2 + 4p + 4 = (p + 2)(p + 2)$

Step 2. $\quad \dfrac{cp - 2a - ap + 2c}{p^2 + 4p + 4} = \dfrac{(c - a)(p + 2)}{(p + 2)(p + 2)} = \dfrac{(c - a)}{(p + 2)}$

2. NEGATIVE OF A POLYNOMIAL

In Section 6 of Chapter I, we discussed the difference between a negative number and the negative of a number. Polynomials cannot be classified as positive or negative since the variables in them ordinarily represent numbers which may be positive or negative, depending upon the replacement set. For example, $(c - a)$ may represent either a positive or negative number, according to the relative sizes of the numbers that replace c and a. When $c > a$, the expression $(c - a)$ represents a positive number; when $c < a$, the expression $(c - a)$ represents a negative number.

On the other hand, the negative of $(c - a)$ does exist: it is $-(c - a)$. Recognizing the negatives of some polynomials, such as $(p - q)$, is a fairly simple matter when they are preceded by a minus sign: $-(p - q)$. When they are expressed in expanded form, however, the fact is not so evident: $-(p - q) = (-1)(p - q) = (-1)(p) + (-1)(-q) = -p + q = q - p$. Thus, the negative of $p - q$ is $q - p$; conversely, the negative of $q - p$ is $p - q$.

EXAMPLES

1. Write the negative of $(x - y)$.

The negative of $(x - y)$ is $-(x - y) = -x + y = (y - x)$. Conversely, then, $(x - y) = -(y - x)$.

2. Is $(a - b)$ the negative of $(b - a)$?

The negative of $(a - b)$ is $-(a - b)$, which in expanded form is $-a + b$, which, in turn, can be expressed as $b - a$. Hence $(b - a) = -(a - b)$ and $(a - b) = -(b - a)$.

3. Write the negative of $x^2 - x - 6$.

The negative of $x^2 - x - 6$ is

$$-(x^2 - x - 6) = -x^2 + x + 6 = 6 + x - x^2$$

Note the effect upon the factors of these two expressions:

$$x^2 - x - 6 = (x - 3)(x + 2)$$
$$6 + x - x^2 = (3 - x)(2 + x)$$

Hence

$$(x - 3)(x + 2) = (-1)(3 - x)(2 + x)$$

and

$$(3 - x)(2 + x) = (-1)(x - 3)(x + 2)$$

4. Express $\dfrac{x^2 + x - 12}{15 - 2x - x^2}$ in lowest terms.

According to Program **6.1**:

Step 1. $x^2 + x - 12 = (x - 3)(x + 4)$

$15 - 2x - x^2 = (5 + x)(3 - x)$

Step 2. $\dfrac{\cancel{(x - 3)}(x + 4)}{(5 + x)\cancel{(3 - x)}} = \dfrac{(-1)(x + 4)}{5 + x} = \dfrac{-(x + 4)}{(5 + x)}$

$$\left[\text{or } -\frac{x + 4}{5 + x} \text{ or } \frac{x + 4}{-(5 + x)} \right]$$

[*Note:* Because we shall be dealing frequently with minus signs in fractions, it might be well to review Section 11, Chapter I, and recall that a fraction may be thought of as possessing three signs: the sign of the fraction, the sign of the numerator, the sign of the denominator, and that reversing any two of the three signs yields an equivalent fraction.]

EXERCISE 6-1

Express in lowest terms:

1. $\dfrac{18a^2}{30a^2 - 12a}$

2. $\dfrac{3x - 9}{2x - 6}$

3. $\dfrac{ab - 4b}{5a - 20}$

4. $\dfrac{ab - ac}{b^2 - c^2}$

5. $\dfrac{xy}{x^2y^2 - 3xy}$

6. $\dfrac{x^4 - 16}{x^3 + 4x}$

7. $\dfrac{x^2 - 5x + 6}{x^2 - 4x + 4}$

8. $\dfrac{x^2 + 6x + 9}{x^2 - 9}$

9. $\dfrac{6x^2 - 11x - 10}{3x^2 - 19x - 14}$

10. $\dfrac{2x^2 - x - 10}{x^2 - 2x - 8}$

11. $\dfrac{ax - ay - 2x + 2y}{a^2 - 7a + 10}$

12. $\dfrac{x^3 - 5x^2 + 6x}{2x^3 - 8x}$

13. $\dfrac{4x^3y - 12x^2y + 9xy}{2axy - 3ay - 2bxy + 3by}$

14. $\dfrac{4a^2 - b^2}{(2a - b)(2a^2 + ab)}$

15. $\dfrac{(x + y)^2 - 3(x + y) - 10}{x^2 + xy + 2x}$

16. $\dfrac{(a - b)^2 - 9}{bc - ac - 3c}$

17. $\dfrac{a^2x + b^2y - b^2x - a^2y}{ax^2 - ay^2 - by^2 + bx^2}$

18. $\dfrac{x^2 - 2xy + y^2 - 16}{(x^2 - 8x + 16) - y^2}$

Make the statements true by inserting the proper sign ($+$ or $-$) in the parentheses:

19. $\dfrac{x}{x - y} = (\quad)\dfrac{x}{y - x} = (\quad)\dfrac{-x}{y - x} = (\quad)\dfrac{-x}{x - y}$

20. $-\dfrac{a}{3x - 4} = (\quad)\dfrac{-a}{3x - 4} = (\quad)\dfrac{-a}{4 - 3x} = (\quad)\dfrac{a}{4 - 3x}$

21. $\dfrac{m - n}{a - b} = (\quad)\dfrac{n - m}{b - a} = (\quad)\dfrac{n - m}{a - b} = (\quad)\dfrac{m - n}{b - a}$

22. $\dfrac{1}{(x - y)(s - t)} = (\quad)\dfrac{1}{(y - x)(s - t)} = (\quad)\dfrac{1}{(y - x)(t - s)}$

Express in lowest terms:

23. $\dfrac{a - ax}{x - 1}$

24. $\dfrac{4 - 2t}{3t - 6}$

25. $-\dfrac{(a - 1)}{p - ap}$

26. $-\dfrac{4x^2 - 9}{9x^2 - 4x^4}$

27. $\dfrac{3x^2 + 2x - 8}{4 + x - 3x^2}$

28. $-\dfrac{6x^2 - 23x + 21}{12 - 5x - 2x^2}$

29. $\dfrac{(x - 2)(x - 3)(x - 4)}{(2 - x)(3 - x)(4 - x)}$

30. $\dfrac{12x^2y - 4x^3 - 9xy^2}{4x^3 + 4x^2y - 15xy^2}$

3. MULTIPLICATION WITH ALGEBRAIC FRACTIONS

As with arithmetic fractions, the product of two or more algebraic fractions is computed by multiplying numerators together and denominators together. Because the factors of the numerators will be the factors of the product's numerator, and the factors of the denominators will be the factors of the product's denominator, time and effort can usually be saved by dividing out like factors that occur in *a* numerator and *a* denominator of the fractions to be multiplied. When this is done thoroughly, the resulting product is automatically in lowest terms.

6.2

To compute the product of algebraic fractions:

Step 1. Factor completely the numerators and denominators of the given fractions.

Step 2. Divide out (cancel) any factor common to any numerator and any denominator.

Step 3. Write as the desired product a fraction whose numerator consists of the remaining numerator factors and whose denominator consists of the remaining denominator factors.

EXAMPLES

1. Compute the product: $\dfrac{x^2 + x - 6}{2x^2 - 7x - 15} \cdot \dfrac{2x^2 - 3x - 9}{x^2 - 5x + 6}$.

Step 1. $\left.\begin{aligned} x^2 + x - 6 &= (x + 3)(x - 2) \\ 2x^2 - 3x - 9 &= (2x + 3)(x - 3) \end{aligned}\right\}$ numerators

$\left.\begin{aligned} 2x^2 - 7x - 15 &= (2x + 3)(x - 5) \\ x^2 - 5x + 6 &= (x - 2)(x - 3) \end{aligned}\right\}$ denominators

Step 2. $\dfrac{(x + 3)\cancel{(x - 2)}}{\cancel{(2x + 3)}(x - 5)} \cdot \dfrac{\cancel{(2x + 3)}\cancel{(x - 3)}}{\cancel{(x - 2)}\cancel{(x - 3)}}$

Step 3. $\dfrac{x + 3}{x - 5}$ is the desired product in lowest terms.

2. Compute the product: $\dfrac{2x^2 - 18}{x^2 + 6x - 7} \cdot \dfrac{x^2 - 1}{8x^2 + 4x - 24}$.

$$\dfrac{2x^2 - 18}{x^2 + 6x - 7} \cdot \dfrac{x^2 - 1}{8x^2 + 4x - 24} = \dfrac{2(x + 3)(x - 3)}{(x + 7)(x - 1)} \cdot \dfrac{(x - 1)(x + 1)}{\underset{2}{4}(x + 2)(2x - 3)}$$

$$= \dfrac{(x + 3)(x - 3)(x + 1)}{2(x + 7)(x + 2)(2x - 3)}$$

[*Note:* The product may be expanded or left in factor form; the latter is the usual convention.]

3. Multiply: $\dfrac{ac - bc + ad - bd}{c^2 - d^2} \cdot \dfrac{d^2 + 2cd + c^2}{b^2 - a^2}$.

$$\dfrac{ac - bc + ad - bd}{c^2 - d^2} \cdot \dfrac{d^2 + 2cd + c^2}{b^2 - a^2}$$

$$= \dfrac{\cancel{(a - b)}(c + d)}{(c - d)\cancel{(c + d)}} \cdot \dfrac{\cancel{(d + c)}(d + c)}{\overset{-1}{\cancel{(b - a)}}(b + a)} = -\dfrac{(c + d)^2}{(c - d)(b + a)}$$

EXERCISE 6-2

Multiply; express products in lowest terms:

1. $\dfrac{24}{3x - 6} \cdot \dfrac{x^2 - 2}{4}$

2. $\dfrac{a^2 - b^2}{a^2} \cdot \dfrac{ab}{a + b}$

3. $\dfrac{x(x - y)^3}{y(x^2 - y^2)} \cdot \dfrac{y(x + y)}{x^2(x - y)^2}$

4. $\dfrac{m(m + 2n)^2}{m - 3n} \cdot \dfrac{3n - m}{m^2 + 2mn}$

5. $\dfrac{9xy}{x + 4} \cdot \dfrac{2x^2 + 5x - 12}{3x^2y^2 - 6xy}$

6. $\dfrac{4x^2 - y^2}{y - 2x} \cdot \dfrac{3y}{xy + 2x^2}$

7. $\dfrac{6a - 18}{9a^2 + 6a - 24} \cdot \dfrac{12a - 16}{8a - 12}$

8. $\dfrac{a + b}{2a + b} \cdot \dfrac{4a^2 - b^2}{a^2 - b^2}$

9. $\dfrac{3ay^2 - 9ay}{10y^2 + 5y} \cdot \dfrac{2y^3 + y^2}{a^2y - 3a^2}$

10. $\dfrac{x^2 - 8x + 15}{x^2 + 4x - 21} \cdot \dfrac{x^2 - 6x - 16}{x^2 + 9x + 14}$

11. $\dfrac{x^2 - 2x - 3}{x^2 - 1} \cdot \dfrac{x^2 + 6x + 9}{x^2 - 9}$

12. $\dfrac{xy - ay - bx + ab}{2x - a} \cdot \dfrac{a - 2x}{a - x}$

13. $\dfrac{ab + 3b + 2a + 6}{3b + 6} \cdot \dfrac{a^2 - 5a + 6}{a^2 - 9}$

14. $\dfrac{xy - 2y - 3x + 6}{2y - 4x + 8 - xy} \cdot \dfrac{4x - 2y - 8 + xy}{xy - 2y - 3x + 6}$

15. $\dfrac{x^2 - 2x + 1}{7x^2 - 7x} \cdot \dfrac{42x^2}{x^2 - 4x + 3}$

16. $\dfrac{9a + 9b}{3a^2 + 6ab + 3b^2} \cdot \dfrac{2a - 2b}{4a^2 - 16b^2}$

17. $\dfrac{2x - y}{x + y} \cdot \dfrac{x - y}{y - 2x} \cdot \dfrac{x + y}{y - x}$

18. $\dfrac{x - 2xy + y^2 - 16}{2x - y} \cdot \dfrac{2y - 4x}{2x - 8 - 2y}$

19. $\dfrac{a^2 - b^2}{a^3 - 3a^2b + 3ab^2 - b^3} \cdot \dfrac{b^2 - 2ab + a^2}{3a^2 + 2ab - b^2}$

20. $\dfrac{x - 3y}{x^2y - 9y^3} \cdot \dfrac{x^2 + 2xy - 3y^2}{x - y}$

4. DIVISION WITH ALGEBRAIC FRACTIONS

If we multiply or divide two numbers (which may be a divisor and dividend) by the same nonzero number, we obtain two numbers whose quotient is the same as that of the original pair. Consequently, if we multiply both the divisor and dividend of $6 \div 3 = 2$ by, say 4, we obtain a different divisor and dividend but the same quotient: $24 \div 12 = 2$. Similarly, dividing both terms by 7 results in $\frac{6}{7} \div \frac{3}{7} = 2$; etc. Thus we are able to compute the quotient of two fractions by performing a corresponding multiplication. To compute the quotient for $a/b \div c/d$, for example, we multiply divisor and dividend each with d/c (the reciprocal of c/d):

$$\underset{\substack{\big\uparrow \\ \text{dividend}}}{\dfrac{a}{b}\left(\times \dfrac{d}{c} \right)} \div \underset{\substack{\big\uparrow \\ \text{divisor}}}{\dfrac{c}{d}\left(\times \dfrac{d}{c} \right)}$$

$$\left(\dfrac{a}{b} \times \dfrac{d}{c} \right) \div \left(\dfrac{\cancel{c}}{\cancel{d}} \times \dfrac{\cancel{d}}{\cancel{c}} \right)$$

$$\left(\dfrac{a}{b} \times \dfrac{d}{c} \right) \div 1$$

Since the quotient for any number divided by 1 is itself, it follows that

$$\frac{a}{b} \div \frac{c}{d} = \left(\frac{a}{b} \times \frac{d}{c} \right) \div 1 = \frac{a}{b} \times \frac{d}{c} = \frac{ad}{bc}$$

In effect, then, the quotient of any two fractions—arithmetic as well as algebraic—may be computed by the simple alternative of multiplying the dividend fraction and the reciprocal of the divisor (i.e., the numerator and denominator of the divisor interchanged, or the divisor *inverted*).

6.3

To compute the quotient of two algebraic fractions:

Step 1. Invert the divisor.

Step 2. Multiply the dividend and the inverted divisor of Step 1 for the desired quotient.

EXAMPLES

1. Divide: $\dfrac{6a^5}{25b^4} \div \dfrac{a}{5b^2}$.

Step 1.   The reciprocal of the divisor $\dfrac{a}{5b^2}$ is the divisor inverted: $\dfrac{5b^2}{a}$.

Step 2. $\dfrac{6a^5}{25b^4} \cdot \dfrac{5b^2}{a} = \dfrac{6 \cdot a^{5} \cdot 5 \cdot b^{2}}{25 \cdot b^{4} \cdot a} = \dfrac{6a^4}{5b^2}$

2. Compute: $\dfrac{2a + 6}{a^2 - 6a + 9} \div \dfrac{a + 3}{a^2 - 5a + 6}$.

Step 1.   Invert $\dfrac{a + 3}{a^2 - 5a + 6}$ and factor:

$$\frac{a^2 - 5a + 6}{a + 3} = \frac{(a - 3)(a - 2)}{(a + 3)}$$

Step 2. $\dfrac{2(a + 3)}{(a - 3)(a - 3)} \cdot \dfrac{(a - 3)(a - 2)}{(a + 3)} = \dfrac{2(a - 2)}{a - 3}$

3. Compute: $\dfrac{c^2 - 2cd + d^2}{b^2 + 4} \div \dfrac{db - cb + 2d - 2c}{b^4 - 16}$.

Step 1.   Invert $\dfrac{db - cb + 2d - 2c}{b^4 - 16}$ and factor:

$$\frac{b^4 - 16}{db - cb + 2d - 2c} = \frac{(b^2 + 4)(b - 2)(b + 2)}{(d - c)(b + 2)}$$

Step 2. $\dfrac{(c - d)(c - d)}{b^2 + 4} \cdot \dfrac{(b^2 + 4)(b - 2)(b + 2)}{(d - c)(b + 2)}$

$$= (-1)(c - d)(b - 2) = [(-1)(c - d)][(b - 2)]$$
$$= (d - c)(b - 2)$$

EXERCISE 6-3

Divide; express quotients in lowest terms:

1. $\dfrac{x^2 + x - 6}{x - a} \div \dfrac{x^2 - 4}{x - a}$

2. $\dfrac{x^2 - 6x + 8}{x - 3} \div \dfrac{x - 4}{9 - x^2}$

3. $\dfrac{x - 3y}{x^2 - 6x + 8} \div \dfrac{3x - 9y}{x^2 - 16}$

4. $\dfrac{3c - bc}{5x - ax} \div \dfrac{3a - ab}{5m - am}$

5. $\dfrac{3x + 6}{x^2 + 4x} \div \dfrac{3x^2 - 9}{x^2 + 2x}$

6. $\dfrac{x^2 - 9}{3x - 3y} \div \dfrac{x^2 + 9 - 6x}{y^2 - x^2}$

7. $\dfrac{2m + 4}{m^2 - 7m - 18} \div \dfrac{4m - 16}{m^2 - 81}$

8. $\dfrac{x^2 - 16}{x^2 - 9} \div \dfrac{x - 4x^2}{9x + 27}$

9. $\dfrac{2x^2 - 7x + 6}{2x^2 - 13x + 15} \div \dfrac{4x^2 - 12x + 5}{2x^2 - 11x + 5}$

10. $\dfrac{6x^2 + 5xy - 6y^2}{9x^2 - 21xy + 10y^2} \div \dfrac{6x^2 - 25xy + 14y^2}{6x^2 - 31xy + 35y^2}$

11. $(a^2 - b^2) \div \dfrac{a^2 - 4ab + 3b^2}{a + b}$

12. $\dfrac{as - bs + bt - at}{3m^3 - 3mn^2} \div \dfrac{as - at}{n^2 - 2mn + m^2}$

13. $\dfrac{6a^2 + 11ab - 10b^2}{a^2 + b^2} \div (3a^2 + ab - 2b^2)$

14. $\dfrac{3x^2 - 14x + 8}{2x^2 - 3x - 20} \div \dfrac{6 - 25x + 24x^2}{15 - 34x - 16x^2}$

15. $\dfrac{x^2 - 5x + 6}{x - 2} \div (x - 2)$

16. $\dfrac{ab - 3b + 2a - 6}{ac - 5a - 3c + 15} \div \dfrac{2b - 6d - 12 + bd}{-5d + 2c - 10 + cd}$

17. $\dfrac{a^2 - 2ab + b^2 - 16}{a^2 - 5ba + 6b^2} \div \dfrac{a - b + 4}{6a - 18b}$

18. $\dfrac{a^2 - 18 + 3a}{30 - a^2 + a} \div \dfrac{15 - 2a - a^2}{-(a^4 - 36a^2)}$

19. $\dfrac{x^2 - y^2}{x^3 + 3x^2y + 3xy^2 + y^3} \div \dfrac{3y - 3x}{x^2 + 2xy + y^2}$

20. $\dfrac{27x^3 - 54x^2y + 36xy^2 - 8y^3}{9x^2 - 12xy + 4y^2} \div \dfrac{9x^2 - 3xy - 2y^2}{27x^3 + 27x^2y + 9xy^2 + y^3}$

5. ADDITION AND SUBTRACTION WITH ALGEBRAIC FRACTIONS

In arithmetic we learn that, to add or subtract with fractions having different denominators, a first step is to find equivalents of the two fractions that have the same, or common, denominator. For instance, to compute the sum of $\frac{1}{2}$ and $\frac{1}{3}$, we replace the two original fractions with equivalents having the same denominator. Then we compute the sum for these two equivalents of the original fractions as we would any pair of fractions having a common denominator: Add the numerators and express their sum as a numerator of a fraction whose denominator is the common denominator.

$$\frac{1}{2} + \frac{1}{3}$$
$$\downarrow \quad \downarrow$$
$$\frac{3}{6} + \frac{2}{6} = \frac{3+2}{6} = \frac{5}{6}$$

Although there are an infinite number of possible pairs of fractions having common denominators equivalent to the original fractions, e.g., $\frac{3}{6} + \frac{2}{6}$, $\frac{6}{12} + \frac{4}{12}$, $\frac{9}{18} + \frac{6}{18}$, etc., only one pair can have the lowest or least common denominator (denominators restricted to positive integers).

The same holds for computing sums and differences with algebraic fractions. When the denominators are the same, we can compute the sum (or difference) by adding (or subtracting) the numerators and expressing that result over the common denominator. But when the denominators are different, we need first to find fractions that are equivalent to the originals and have a common denominator, preferably the least common denominator. This involves two steps: (1) finding the least common denominator of the given fractions; then (2) writing an equivalent fraction having that common denominator for each of the given fractions.

Many persons learned to find the least common denominator in elementary school by a trial-and-error method (select the largest denominator, double it, triple, etc., until a common denominator is found). This works well enough with most arithmetic fractions, but not so well with algebraic fractions. For the latter, a more systematic procedure is needed.

6.4

To find the least common denominator (LCD) of several algebraic fractions:

Step 1. Factor each denominator completely.

Step 2. Form a product of the factors of the denominators so that each factor appears in it the number of times it appears in the denominator which contains it most. This is the least common denominator (usually left in factor form).

EXAMPLES

1. Find the LCD for

$$\frac{x+2}{x^2+2x-3}, \quad \frac{7}{2x^2-8x+6}, \quad \frac{x^2-2}{9x^2-54x+81}, \quad \frac{2-x}{3x^2-15x+18}$$

Step 1. $x^2 + 2x - 3 = (x + 3)(x - 1)$

$2x^2 - 8x + 6 = (2)(x - 3)(x - 1)$

$9x^2 - 54x + 81 = (3)(3)(x - 3)(x - 3)$

$3x^2 - 15x + 18 = (3)(x - 2)(x - 3)$

Step 2. $(x + 3)(x - 1)(2)(3)(3)(x - 3)^2(x - 2)$

$\text{LCD} = (2)(3)(3)(x + 3)(x - 1)(x - 3)^2(x - 2)$

or

$$18(x + 3)(x - 1)(x - 3)^2(x - 2)$$

2. Find the LCD for $\dfrac{x + 5}{x^2 - 4}, \dfrac{3}{x^2 + x - 6}, \dfrac{8 - x}{6 - x - x^2}.$

Step 1. $x^2 - 4 = (x - 2)(x + 2)$

$x^2 + x - 6 = (x - 2)(x + 3)$

$6 - x - x^2 = (2 - x)(3 + x)$

Step 2. $(x - 2)(x + 2)(x + 3) = \text{LCD}$

[*Note:* $(2 - x)$ is the negative of $(x - 2)$ and divides that factor exactly (-1) times; therefore, there is no need to introduce the factor $(2 - x)$ among the factors of the LCD.]

Once the least common denominator has been determined in computing sums and differences, the next step is to express each of the fractions as equivalent fractions having this same common denominator. Contained in the LCD is the necessary "appropriate" multiplier for both the numerator and denominator of each algebraic fraction that will produce an equivalent fraction having the desired least common denominator.

6.5

To express an algebraic fraction, having a specified denominator, equivalent to a given fraction:

Step 1. Divide the specified denominator by the denominator of the given fraction.

Step 2. Multiply the numerator and denominator of the given fraction by the quotient of Step 1.

EXAMPLES

1. Express a fraction equivalent to $\dfrac{3x + 2}{x - 7}$, having a denominator of $(x - 7)(x + 2).$

Step 1.   Divide $(x - 7)(x + 2)$ by $(x - 7)$:

$$\frac{(x - 7)(x + 2)}{(x - 7)} = (x + 2)$$

Step 2. Multiply numerator and denominator of the given fraction $\dfrac{3x+2}{x-7}$ by $(x+2)$:

$$\frac{(3x+2)(x+2)}{(x-7)(x+2)}$$

2. Express each of the three fractions:

(1) $\dfrac{x+2}{x^2+x-6}$, (2) $\dfrac{3x}{20-6x-2x^2}$, (3) $\dfrac{x-7}{x^2+8x+15}$

as equivalent fractions, each having as its denominator the LCD of the three fractions.

(a) Factor the three denominators:

   (1) $x^2 + x - 6 = (x-2)(x+3)$

   (2) $20 - 6x - 2x^2 = 2(2-x)(5+x)$

   (3) $x^2 + 8x + 15 = (x+3)(x+5)$

(b) Find the LCD of the denominators:

$$(2)(x-2)(x+3)(x+5)$$

(c) Divide the LCD by each denominator:

   (1) $\dfrac{(2)\cancel{(x-2)}\cancel{(x+3)}(x+5)}{\cancel{(x-2)}\cancel{(x+3)}} = 2(x+5)$

   (2) $\dfrac{\cancel{(2)}\cancel{(x-2)}(x+3)\cancel{(x+5)}}{\overset{-1}{\cancel{(2)}\cancel{(2-x)}\cancel{(5+x)}}} = (-1)(x+3) \text{ or } -(x+3)$

   (3) $\dfrac{(2)(x-2)\cancel{(x+3)}\cancel{(x+5)}}{\cancel{(x+3)}\cancel{(x+5)}} = 2(x-2)$

(d) Multiply the numerator and denominator of each fraction by its respective quotient in (c):

   (1) $\dfrac{(x+2)}{(x-2)(x+3)}\dfrac{(2)(x+5)}{(2)(x+5)} = \dfrac{(2)(x+2)(x+5)}{(2)(x-2)(x+3)(x+5)}$

   (2) $\dfrac{(3x)}{(2)(2-x)(5+x)}\dfrac{(-1)(x+3)}{(-1)(x+3)}$

$$= \frac{(-1)(3x)(x+3)}{(-1)(2)(2-x)(x+3)(5+x)}$$

$$= \frac{-3x(x+3)}{(2)(x-2)(x+3)(x+5)}$$

$$\underset{\underset{(-1)(2-x)}{\Large\uparrow}}{}$$

   (3) $\dfrac{(x-7)}{(x+3)(x+5)}\dfrac{(2)(x-2)}{(2)(x-2)} = \dfrac{(2)(x-7)(x-2)}{(2)(x-2)(x+3)(x+5)}$

We are now ready to organize the foregoing into a single program for adding and subtracting with algebraic fractions.

6.6

To compute sums (differences) of algebraic fractions:

Step 1. Find the LCD of the terms, if the denominators are different.

Step 2. Express each fraction as an equivalent fraction whose denominator is the LCD of Step 1.

Step 3. Write the sum (difference) of the numerators as the numerator of the sum (difference), and write the LCD as its denominator.

Step 4. (Optional, but usual.) Simplify the numerator of Step 3, and express the resulting fraction in lowest terms.

EXAMPLES

1. Add: $\dfrac{x - 23}{x^2 - x - 20} + \dfrac{x - 3}{x - 5}$.

Step 1. $x^2 - x - 20 = (x - 5)(x + 4)$

$$x - 5 = (x - 5)$$

$$\text{LCD} = (x - 5)(x + 4)$$

Step 2. $\dfrac{x - 23}{(x - 5)(x + 4)} + \dfrac{(x - 3)(x + 4)}{(x - 5)(x + 4)}$

Step 3. $\dfrac{(x - 23) + (x - 3)(x + 4)}{(x - 5)(x + 4)}$

Step 4. $\dfrac{x - 23 + x^2 + x - 12}{(x - 5)(x + 4)} = \dfrac{x^2 + 2x - 35}{(x - 5)(x + 4)}$

$$= \dfrac{(x + 7)\cancel{(x - 5)}}{\cancel{(x - 5)}(x + 4)} = \dfrac{(x + 7)}{(x + 4)}$$

Hence

$$\frac{x - 23}{x^2 - x - 20} + \frac{x - 3}{x - 5} = \frac{x + 7}{x + 4}$$

2. Simplify: $\dfrac{7x - 8}{x^2 - 9} - (3x + 2) + \dfrac{3x^2}{x - 3}$.

Step 1. $x^2 - 9 = (x - 3)(x + 3)$

$$x - 3 = (x - 3)$$

$$\text{LCD} = (x - 3)(x + 3) \text{ or } (x^2 - 9)$$

Step 2. $\dfrac{7x - 8}{(x - 3)(x + 3)} - \dfrac{(3x + 2)(x^2 - 9)}{(x - 3)(x + 3)} + \dfrac{3x^2(x + 3)}{(x - 3)(x + 3)}$

Step 3. $\dfrac{(7x - 8) - [(3x + 2)(x^2 - 9)] + 3x^2(x + 3)}{(x - 3)(x + 3)}$

$$= \dfrac{(7x - 8) - (3x^3 + 2x^2 - 27x - 18) + (3x^3 + 9x^2)}{(x - 3)(x + 3)}$$

$$= \dfrac{7x - 8 - 3x^3 - 2x^2 + 27x + 18 + 3x^3 + 9x^2}{(x - 3)(x + 3)}$$

$$= \frac{7x^2 + 34x + 10}{(x - 3)(x + 3)}$$

[*Note:* $7x^2 + 34x + 10$ is not factorable because $(b^2 - 4ac) = 1156 - 280 = 876 \neq$ perfect square; therefore, the result is in lowest terms. cf. p. 108.]

3. Add: $\dfrac{3x}{2x - 1} + \dfrac{2x}{1 - 2x}$.

Step 1.   LCD $= 2x - 1$

[*Note:* $(2x - 1) \div (1 - 2x) = -1$; or $(2x - 1) = -(1 - 2x)$.]

Step 2. $\dfrac{3x}{2x - 1} + \dfrac{-2x}{2x - 1}$

Step 3. $\dfrac{3x - 2x}{2x - 1} = \dfrac{x}{2x - 1}$

EXERCISE 6-4

Simplify:

1. $\dfrac{x + 3y}{18} + \dfrac{x - 2y}{24}$

2. $\dfrac{s - t}{14s} - \dfrac{s - t}{21t}$

3. $\dfrac{3a - 6b}{12a^2b} + \dfrac{3a - 5b}{18ab^2}$

4. $\dfrac{1}{f_1} + \dfrac{1}{f_2}$

5. $\dfrac{5}{r} - \dfrac{6}{t} + \dfrac{8}{r}$

6. $a - 3 + \dfrac{5}{a}$

7. $2x - \dfrac{6x^2}{3x - 2}$

8. $\dfrac{5}{m} - \dfrac{6}{n} + 3$

9. $x - 2 + \dfrac{4x + 3}{x - 7}$

10. $x + 3 - \dfrac{2x - 5}{7 - x}$

11. $\dfrac{3x - 5}{x + 3} + \dfrac{2x - 1}{x - 4}$

12. $\dfrac{1}{p + q} - \dfrac{1}{p - q}$

13. $\dfrac{3}{a - b} - \dfrac{4}{b - a}$

14. $\dfrac{x - 2}{4x} - \dfrac{3x + 5}{6x}$

15. $\dfrac{3x}{x^2 - 4} - \dfrac{2}{x + 2}$

16. $\dfrac{8y}{y^2 - 9} + \dfrac{4}{3 - y}$

17. $\dfrac{3a - 4}{a^2 - a - 20} - \dfrac{3}{a - 5}$

18. $\dfrac{x + 2}{x^2 - 9} + \dfrac{3x - 1}{x^2 + x - 12}$

19. $\dfrac{3c - d}{2c - d} + \dfrac{3c^2}{d^2 - 4c^2}$

20. $\dfrac{x - 4}{x - 2} + \dfrac{3 + 5x}{2 - x}$

21. $\dfrac{1}{x^2 - 5x + 6} - \dfrac{4}{4 - x^2}$

22. $\dfrac{2x}{x^2 + xy} - \dfrac{3}{xy + y^2}$

23. $2 - \dfrac{x}{x - 2} + \dfrac{3(2x - 9)}{x^2 - 5x + 6}$

24. $2 - \dfrac{3 - 2x}{2x - 3} + 2x$

25. $\dfrac{x+2}{2x^2-3x-9} - \dfrac{x-2}{3x^2-11x+6}$

26. $\dfrac{2}{x^2-9} - \dfrac{3}{x^2-1} + \dfrac{1}{x^2+2x-3}$

27. $\dfrac{8}{x^2-x-2} + \dfrac{3+x}{x^2-4} + \dfrac{4-2x}{x^2+3x+2}$

28. $\dfrac{3}{6x^2-11x+3} + \dfrac{3x+1}{12x-4} - \dfrac{2x-3}{9x^2-1}$

6. COMPLEX FRACTIONS

Fractional notation permits various degrees of complexity. Most of the fractions we have seen so far may be classified as **simple fractions**, i.e., the numerators and denominators themselves do not contain fractions. In contrast, **complex fractions** are fractions in which either the numerator or denominator, or both, involve fractions. Complex fractions can be simplified (i.e., expressed by a simpler equivalent) if we apply some of the procedures learned up to this point. There are no uniform patterns by which to proceed, and attacks will vary. One often useful approach is to express the complex fraction as a division of the numerator by the denominator.

EXAMPLES

1. Simplify: $\dfrac{\dfrac{3}{xy^2}}{\dfrac{2}{x^2y}}$.

$$\frac{\dfrac{3}{xy^2}}{\dfrac{2}{x^2y}} = \frac{3}{xy^2} \div \frac{2}{x^2y} = \frac{3}{xy^2} \cdot \frac{x^2y}{2} = \frac{3x}{2y}$$

2. Simplify: $\dfrac{\dfrac{1}{s}+\dfrac{1}{t}}{\dfrac{s}{t}-\dfrac{t}{s}}$.

$$\frac{\dfrac{1}{s}+\dfrac{1}{t}}{\dfrac{s}{t}-\dfrac{t}{s}} = \frac{\dfrac{t+s}{st}}{\dfrac{s^2-t^2}{st}} = \frac{t+s}{st} \cdot \frac{st}{s^2-t^2} = \frac{1}{s-t}$$

3. Simplify: $\dfrac{1-\dfrac{3y}{x+y}}{1-\dfrac{y}{x-y}}$.

First simplify the numerator and denominator separately:

$$1 - \frac{3y}{x+y} = \frac{x+y}{x+y} - \frac{3y}{x+y} = \frac{x-2y}{x+y}$$

$$1 - \frac{y}{x-y} = \frac{x-y}{x-y} - \frac{y}{x-y} = \frac{x-2y}{x-y}$$

Replace the original numerator and denominator by their equivalents:

$$\frac{1 - \dfrac{3y}{x+y}}{1 - \dfrac{y}{x-y}} = \frac{\dfrac{x-2y}{x+y}}{\dfrac{x-2y}{x-y}} = \frac{\cancel{x-2y}}{x+y} \cdot \frac{x-y}{\cancel{x-2y}} = \frac{x-y}{x+y}$$

$$\left[\; Note: \text{ The numerator and denominator of } \dfrac{\dfrac{x-2y}{x+y}}{\dfrac{x-2y}{x-y}} \text{ may be divided by}\right.$$

$(x - 2y)$ directly:

$$\left. \frac{\dfrac{\cancel{x-2y}}{x+y}}{\dfrac{\cancel{x-2y}}{x-y}} = \frac{1}{x+y} \cdot \frac{x-y}{1} = \frac{x-y}{x+y} \;\right]$$

4. Simplify: $1 - \dfrac{2}{3 - \dfrac{1}{2 + \dfrac{3}{x-1}}}$

(a) $1 - \dfrac{2}{3 - \dfrac{1}{2 + \dfrac{3}{x-1}}} \left[= \dfrac{2x-2+3}{x-1} = \dfrac{2x+1}{x-1} \right]$

(b) $1 - \dfrac{2}{3 - \dfrac{1}{\dfrac{2x+1}{x-1}}} \left[= \dfrac{x-1}{2x+1} \right]$

(c) $1 - \dfrac{2}{3 - \dfrac{x-1}{2x+1}} \left[= \dfrac{6x+3-x+1}{2x+1} = \dfrac{5x+4}{2x+1} \right]$

(d) $1 - \dfrac{2}{\dfrac{5x+4}{2x+1}} \left[= \dfrac{2(2x+1)}{5x+4} = \dfrac{4x+2}{5x+4} \right]$

(e) $1 - \dfrac{4x+2}{5x+4} = \dfrac{5x+4-4x-2}{5x+4} = \dfrac{x+2}{5x+4}$

EXERCISE 6-5

Simplify:

1. $\dfrac{\dfrac{x-5}{x^2-25}}{2x+3}$

2. $\dfrac{\dfrac{x^2-y^2}{(x+y)^2}}{\dfrac{3x+3y}{x-y}}$

3. $\dfrac{1 - \dfrac{2x}{3 + x}}{4 - \dfrac{3x}{x + 3}}$

4. $\dfrac{3a + \dfrac{2a}{a - 6}}{a - \dfrac{4a}{a + 3}}$

5. $\dfrac{1 - \dfrac{1}{a}}{1 + \dfrac{1}{1 - \dfrac{1}{a}}}$

6. $\dfrac{2 + \dfrac{1}{x}}{1 - \dfrac{2}{2 - \dfrac{1}{x}}}$

7. $\dfrac{a + \dfrac{1}{a - 2}}{a + \dfrac{3}{a - \dfrac{4}{a}}}$

8. $\dfrac{1 - \dfrac{10a - 5}{3a - 4}}{2a - \dfrac{3a + 1}{1 + \dfrac{a - 3}{2a - 1}}}$

7. FRACTIONAL EQUATIONS

Equations in which the variable appears in the denominator of at least one term are sometimes called **fractional equations**. When both members of the equation are multiplied with a common denominator of the fraction terms, a new equation results, called the **derived equation**. The derived equation will be without a fraction but the equation may or may not be an equivalent equation (i.e., same solution set). When the derived equation is first degree, it can be solved as any other first-degree equation (by Program **3.1**). It is very important, though, that the solution obtained be substituted for the variable in the given equation to assure that it satisfies the given equation.

6.7

To solve a fractional equation whose derived equation is first degree:

Step 1. Find the LCD of the terms of the equation that are in fraction form.

Step 2. Multiply both members of the equation by LCD of Step 1 to obtain a derived equation without fractions.

Step 3. Solve the derived equation of Step 2.

Step 4. Substitute the solution obtained in Step 3 for the variable of the given equation; if it satisfies the given equation, it is the solution of that equation.

EXAMPLES

1. Solve for x: $\dfrac{2}{3x} = \dfrac{3}{4x} - \dfrac{1}{2}$.

Step 1.   LCD $= 12x$

Step 2. $\left(\cancel{12x} \cdot \dfrac{2}{\cancel{3x}}\right) = \left(\cancel{12x} \cdot \dfrac{3}{\cancel{4x}}\right) - \left(\cancel{12x} \cdot \dfrac{1}{\cancel{2}}\right)$

$\qquad\quad 8 \quad = \quad 9 \quad - \quad 6x$

Step 3. $8 = 9 - 6x$

$\qquad 6x = 1$

$\qquad x = \frac{1}{6}$ (possible solution of the given equation)

Step 4. $\dfrac{2}{3x} = \dfrac{3}{4x} - \dfrac{1}{2}$

$x = \frac{1}{6}:\qquad \dfrac{2}{3(\frac{1}{6})} \overset{?}{=} \dfrac{3}{4(\frac{1}{6})} - \dfrac{1}{2}$

$\qquad\qquad \dfrac{2}{\frac{1}{2}} \overset{?}{=} \dfrac{3}{\frac{2}{3}} - \dfrac{1}{2}$

$\qquad\qquad 4 \overset{?}{=} \dfrac{9}{2} - \dfrac{1}{2}$

$\qquad\qquad 4 = 4$

The given equation is satisfied; the solution set for the given equation is $\{\frac{1}{6}\}$.

2. Solve: $\dfrac{8}{3x + 1} + 2 = \dfrac{2x}{x - 1}.$

Step 1. LCD $= (3x + 1)(x - 1)$

Step 2. $\left[\cancel{(3x+1)}(x - 1) \cdot \dfrac{8}{\cancel{3x+1}}\right] + [(3x + 1)(x - 1)\cdot 2]$

$\qquad\qquad\qquad\qquad\qquad\qquad = (3x + 1)\cancel{(x-1)} \cdot \dfrac{2x}{\cancel{(x-1)}}$

$\quad 8(x - 1) + 2(3x + 1)(x - 1) = 2x(3x + 1)$

Step 3. $8x - 8 + 2(3x^2 - 2x - 1) = 6x^2 + 2x$

$\qquad 8x - 8 + \cancel{6x^2} - 4x - 2 = \cancel{6x^2} + 2x$

$\qquad\qquad 8x - 4x - 10 = 2x$

$\qquad\qquad\qquad 2x = 10$

$\qquad\qquad\qquad x = 5$ (possible solution of the given equation)

Step 4. $\dfrac{8}{3x + 1} + 2 = \dfrac{2x}{x - 1}$

$x = 5:\qquad \dfrac{8}{3(5) + 1} + 2 \overset{?}{=} \dfrac{2(5)}{(5) - 1}$

$\qquad\qquad \dfrac{8}{16} + 2 \overset{?}{=} \dfrac{10}{4}$

$\qquad\qquad 2\frac{1}{2} = 2\frac{1}{2}$

3. Solve: $\dfrac{3x}{x - 2} = 4 + \dfrac{6}{x - 2}.$

Step 1. LCD $= (x - 2)$

Step 2. $\left[(x-2) \cdot \dfrac{3x}{(x-2)}\right] = [(x-2)\cdot 4] + \left[(x-2) \cdot \dfrac{6}{x-2}\right]$

$$3x = 4(x-2) + 6$$

Step 3. $3x = 4x - 8 + 6$

$-x = -2$

$x = 2$

Step 4.   Substitute 2 for the variable x in the given equation:

$$\frac{3x}{x-2} = 4 + \frac{6}{x-2}$$

$x = 2:$ $\dfrac{3(2)}{(2)-2} \overset{?}{=} 4 + \dfrac{6}{(2)-2}$

$\dfrac{6}{0} \overset{?}{=} 4 + \dfrac{6}{0}$

Since $\frac{6}{0}$ is undefined, 2 is not a solution of the given equation. Thus the given equation has no solution.

EXERCISE 6-6

Solve for x:

1. $3 = \dfrac{1}{4x}$

2. $\dfrac{7}{3x+1} = \dfrac{3}{2x-1}$

3. $\dfrac{x-5}{x+4} = \dfrac{x-1}{x-4}$

4. $\dfrac{3}{x-2} + \dfrac{2x}{x-1} = 2$

5. $\dfrac{2x}{x+1} + \dfrac{1}{3x-2} = 2$

6. $\dfrac{a+x}{a-x} = \dfrac{a+b}{a-b}$

7. $\dfrac{3x-2}{4x-1} = \dfrac{3x+1}{4x-7}$

8. $\dfrac{3x-2}{2x-3} = \dfrac{6x-6}{4x-7}$

9. $\dfrac{6}{x+2} - \dfrac{5}{x} = \dfrac{5-4x}{x^2+2x}$

10. $\dfrac{3}{x+0.3} - 0.6 = \dfrac{0.3-x}{0.3+x}$

11. $\dfrac{3}{x+2} + \dfrac{2x}{4-x^2} = \dfrac{2}{x-2}$

12. $\dfrac{2}{4-x^2} = \dfrac{x}{4-x^2} + \dfrac{3}{x+2}$

13. $\dfrac{9}{x+1} - \dfrac{2x+3}{x-2} = \dfrac{7x-2x^2}{x^2-x-2}$

14. $\dfrac{x+a}{x-b} - \dfrac{x-a}{x+b} - \dfrac{a^2-b^2}{x^2-b^2} = 0$

15. $\dfrac{5x}{x^2+x-6} - \dfrac{3x}{x^2+2x-8} = \dfrac{2}{x+3}$

16. $\dfrac{1-\dfrac{2}{x}}{3-\dfrac{4}{x}} = \dfrac{2}{7}$

17. $\dfrac{3 - \dfrac{2}{x}}{6 - \dfrac{3}{x}} = \dfrac{x + 4}{2x + 9}$

18. $\dfrac{2x + 4}{x^2 + 2x - 8} - \dfrac{x - 4}{x^2 - 3x + 2} = \dfrac{x + 6}{x^2 + 3x - 4}$

REVIEW—CHAPTER VI

Insert the proper sign $(+ \text{ or } -)$ in the parentheses to make the statements true:

1. $\dfrac{m}{a - b} = (\quad)\dfrac{-m}{b - a} = (\quad)\dfrac{m}{b - a} = (\quad)\dfrac{-m}{a - b}$

2. $\dfrac{a - 5}{(a - 4)(a - 2)} = (\quad)\dfrac{a - 5}{(a - 2)(a - 4)} = (\quad)\dfrac{5 - a}{(2 - a)(4 - a)}$

$= (\quad)\dfrac{5 - a}{(2 - a)(a - 4)} = (\quad)\dfrac{-(a - 5)}{(4 - a)(2 - a)}$

Express in lowest terms:

3. $\dfrac{2x^2 - 17x + 21}{x^2 - 12x + 35}$

4. $\dfrac{6x^2 - 23x + 20}{6x^2 - 5x - 4}$

5. $\dfrac{(3 - a)(a^2 + 9)}{a^4 - 81}$

6. $\dfrac{xy - ay - ab + bx}{ab - bx + ay - xy}$

7. $\dfrac{ab + 6 - 3b - 2a}{4b + 2a - 8 - ab}$

8. $\dfrac{x^2 - 4x + 4}{4a - 4ax + ax^2}$

9. $\dfrac{x^2 - y^2 + x - y}{x^2 + 2xy + y^2 - 1}$

10. $\dfrac{x^3 + 2x^2 - 9x - 18}{6 + x - x^2}$

11. $\dfrac{3(m^2 + 2mn + n^2) + 7m + 7n - 6}{m + n + 3}$

12. $\dfrac{(x^2 - 8x + 16) - 4}{1 - (x^2 - 2x + 1)}$

Simplify:

13. $\dfrac{2x + y}{4x^2 - y^2} \cdot \dfrac{x^3 - 2x^2y}{x^2 - 2xy}$

14. $\dfrac{3 + 2x}{2 + 35x^2 + 19x} \cdot \dfrac{49x^2 - 1}{4x^2 + 12x + 9}$

15. $\dfrac{2a - 3}{a^2 - 1} \cdot \dfrac{2a^2 + a - 3}{9 - 4a^2}$

16. $\dfrac{a^2 + 5a + 6}{a^2 - 1} \cdot \dfrac{a^2 - 2a - 3}{a^2 - 9}$

17. $\dfrac{m^2 - 5m + 6}{m^2 - 4m + 3} \cdot \dfrac{m^2 - 5m + 4}{m^2 - 6m + 8}$

18. $\dfrac{1 - 2x + x^2}{1 - x^2} \cdot \dfrac{x}{x - 1} \cdot \dfrac{1 + x}{1 + x^2}$

19. $\dfrac{ab - b + 2a - 2}{cd - 4d + 3c - 12} \cdot \dfrac{d^2 - d - 12}{b^2 + 4b + 4}$

20. $\dfrac{am - na + bm - bn}{ax + bx - ay - by} \cdot \dfrac{mx - nx - my + ny}{m^2 - n^2}$

21. $\dfrac{4m^2 - 4mn + n^2}{3m^2 + 13mn - 10n^2} \cdot \dfrac{9m^2 - 12mn + 4n^2}{3m^2 + 4mn - 4n^2} \cdot \dfrac{m^2 - 4n^2}{2m^2 - 5mn + 2n^2}$

22. $\dfrac{a^2 - b^2}{a^3 - 3a^2b + 3ab^2 - b^3} \cdot \dfrac{3a^2 - 6ab + 3b^2}{a^2 + 3ab + 2b^2}$

23. $\dfrac{8a^3b^2}{7xy^2} \div \dfrac{a^4b^2 - 3a^2b^3}{2x^3y - x^2y^2}$

24. $\dfrac{4a^2 - 28a + 49}{12a^2 - 17a + 6} \div \dfrac{4a^2 - 49}{12a^2 - a - 6}$

25. $\dfrac{2x^2 - 13x + 15}{3x^2 - 17x + 10} \div \dfrac{4x^2 - 9}{3px - 2p}$

26. $\dfrac{a^2 + 4a + 3}{a^2 + a - 6} \div \dfrac{a^2 + 3a + 2}{pa^2 - 2pa}$

27. $\dfrac{9a^2 + 6ab + b^2 - 4}{2a^2 - 5ab + 3b^2} \div \dfrac{10 + 5b + 15a}{a^2 - 7ab + 6b^2}$

28. $\dfrac{ab + a + b + 1}{mn + 2n + 2m + 4} \div \dfrac{a^2 - 1}{n^2 + 4n + 4}$

29. $\dfrac{6x^2 - 19x + 10}{8x^2 - 14x - 15} \div \dfrac{15x^2 - 16x + 4}{15x^2 - 11x + 2}$

30. $\left(2 - \dfrac{2 + 8m - 3m^2}{9 - m^2}\right) \div \left(6 - \dfrac{14 + 7m}{3 + m}\right)$

31. $\dfrac{x^3 - 1}{x^4 + x^2 + 1} \cdot \dfrac{x^3 + 1}{2x^2 + x - 1} \div \dfrac{2x^2 - x - 1}{2x^2 + x - 1}$

32. $\dfrac{a^2 - 1}{a^2 - a(a + 1)} \cdot \dfrac{a^2 - 3a}{a^2 - 3a + 2} \div \dfrac{a^2 - 3a - 4}{a - 2}$

33. $\dfrac{2}{a - 2} - \dfrac{3}{5a - 10}$

34. $\dfrac{3}{x + y} - \dfrac{5}{xy - y^2} + \dfrac{2}{x^2 - y^2}$

35. $\dfrac{3}{x - 2} - \dfrac{x}{x^2 - 4} - \dfrac{2}{x + 2}$

36. $\dfrac{2m + 6}{m^2 - m - 12} - \dfrac{3(m + 5)}{m^2 + 3m - 10}$

37. $\dfrac{y}{2y - x} + \dfrac{1}{x - 2y} + \dfrac{x}{2y^2 - xy}$

38. $\dfrac{m^2 + 1}{m^3 - m^2 - m - 1} - \dfrac{m^2 - 1}{m^3 + m^2 - m - 1}$

39. $\dfrac{3a^2}{a^4 - 4} + \dfrac{5a^2 - 3}{2a^4 + a^2 - 6}$

40. $\dfrac{1}{ab + a + b + 1} + \dfrac{1}{2 + ab + 2a + b} + \dfrac{1}{2 + 2b + a + ab}$

41. $\dfrac{a - 1}{a + 1} - 1 + \left(\dfrac{a - 1}{a^2 - 1} - \dfrac{5a - 1}{a - 1} + 2 \right)$

42. $\dfrac{2 - 2x}{x^3 - 3x^2 + 3x - 1} + \dfrac{2}{1 - x} + \dfrac{x + 1}{x^2 - 2x + 1}$

43. $\dfrac{\dfrac{4a^2 - 4ab + b^2}{4a^2 - b^2}}{\dfrac{b^2 - 2ab}{b^2 + 2ab}}$

44. $\dfrac{\dfrac{1}{a + 1} - \dfrac{1}{a - 1}}{1 - \dfrac{a - 1}{a + 1}}$

45. $\dfrac{a - \dfrac{a}{a - 2}}{a - \dfrac{a}{a + \dfrac{1}{a + 2}}}$

46. $\dfrac{1 + \dfrac{1}{1 + x(x - 1)}}{3 - \dfrac{3x}{2x + \dfrac{2}{x - 1}}}$

47. $\dfrac{-3 + \dfrac{2}{7 - a}}{5a - \dfrac{4a - 1}{1 - \dfrac{2a + 5}{3a - 2}}}$

Solve:

48. $\dfrac{5}{4x + 2} = \dfrac{1}{x + 1}$

49. $\dfrac{3x - 2}{2x + 3} = \dfrac{9x - 5}{6x + 1}$

50. $\dfrac{1 + 2x}{x - 4} = \dfrac{4x^2 + 5x}{2x^2 - 7x - 4}$

51. $\dfrac{x + 6}{6} - \dfrac{24}{9x + 36} = \dfrac{3x + 24}{18}$

52. $\dfrac{x + 3}{x^2 - 5x + 4} + \dfrac{1}{x - 1} = \dfrac{2}{x - 4}$

53. $\dfrac{7x}{x - 3} - \dfrac{12x^2 - 12}{x^2 - 2x - 3} - \dfrac{5x}{x + 1} = 0$

54. $\dfrac{x}{2x + 1} = \dfrac{x + 1}{2x - 4} + \dfrac{2x - 3}{2x^2 - 3x - 2}$

55. $\dfrac{x - 2}{x + 6} = \dfrac{76 - 14x}{x^2 + 2x - 24} - \dfrac{2 + x}{4 - x}$

56. $\dfrac{x - 3}{x + 1} - 2 = \dfrac{x + 9}{x^2 - x - 2} + \dfrac{x + 4}{2 - x}$

57. $\dfrac{2 - \dfrac{3}{x}}{3 - \dfrac{1}{x}} = \dfrac{2x + 4}{3x + 13}$

VII

EXPONENTS, RADICALS, AND COMPLEX NUMBERS

A **power** was defined in Chapter II as a product of several identical factors, such as $a \times a \times a \times a \times a$. A shorter way to symbolize this product was suggested: a^5, in which the repeated factor (the base) was written once with a numeral (exponent) above and to the right to indicate the number of times that factor was used in the product. Thus:

$$\underbrace{a \times a \times a \times a \times a}_{5 \text{ factors of } a} = a^5 \overset{\longleftarrow \text{exponent}}{\underset{\longleftarrow \text{base}}{\Big\}} \text{power}$$

1. LAWS OF EXPONENTS

Powers, such as a^5, are also referred to as **exponential expressions**. Note that, by the nature of this interpretation, exponents are limited to the counting numbers, i.e., the positive integers.

In Chapter II we also developed a procedure for multiplying powers having the same base (Program **2.2**) and a two-part procedure for dividing powers having the same base (Programs **2.4(a)** and **2.4(b)**). For our present purposes, we consider these as the first two laws of exponents, I and II:

I. $a^m \times a^n = a^{m+n}$ (Program **2.2**)

II. $a^m \div a^n = a^{m-n}$,   if $m > n$ (Program **2.4(a)**)

$\qquad = \dfrac{1}{a^{n-m}}$,   if $m < n$ (Program **2.4(b)**)

and then add three more, making five laws in all.

139

III.　$(a^n)^m = a^{m \times n}$

7.1

To raise a power to a given power, write the base with an exponent equal to the product of the exponents.

EXAMPLES

1. $(b^5)^3 = b^{3 \times 5} = b^{15}$.

　　This can be verified by I (or Program **2.2**): $(b^5)^3 = b^5 \cdot b^5 \cdot b^5 = b^{5+5+5} = b^{15}$.

2. $(x^3)^5 = x^{5 \times 3} = x^{15}$.

3. $(x)^4 = x^{4 \times 1} = x^4$.

IV.　$(ab)^n = a^n b^n$

7.2

To raise a product to a given power, raise each factor of the product to the given power.

EXAMPLES

1. $(bc^2)^3 = b^{3 \times 1} c^{3 \times 2} = b^3 c^6$.

　　This can be verified by I (or Program **2.2**):

$$(bc^2)^3 = bc^2 \cdot bc^2 \cdot bc^2 = b \cdot c^2 \cdot b \cdot c^2 \cdot b \cdot c^2$$
$$= b \cdot b \cdot b \cdot c^2 \cdot c^2 \cdot c^2$$
$$= b^{1+1+1} c^{2+2+2}$$
$$= b^3 c^6$$

2. $(x^2 y^3)^4 = x^{4 \times 2} y^{4 \times 3} = x^8 y^{12}$.

3. $(3xy^2)^3 = 3^{3 \times 1} x^{3 \times 1} y^{3 \times 2} = 3^3 x^3 y^6 = 27 x^3 y^6$.

V.　$\left(\dfrac{a}{b}\right)^n = \dfrac{a^n}{b^n}$　$(b \neq 0)$.

7.3

To raise a number expressed as a fraction to a given power, raise both the numerator and denominator to the given power.

EXAMPLES

1. $\left(\dfrac{p^2}{q^4}\right)^3 = \dfrac{p^{3 \times 2}}{q^{3 \times 4}} = \dfrac{p^6}{q^{12}}$　$(q \neq 0)$.

　　This can be verified by I (or Program **2.2**):

$$\left(\frac{p^2}{q^4}\right)^3 = \left(\frac{p^2}{q^4}\right) \times \left(\frac{p^2}{q^4}\right) \times \left(\frac{p^2}{q^4}\right)$$

$$= \frac{p^2 \times p^2 \times p^2}{q^4 \times q^4 \times q^4} = \frac{p^{2+2+2}}{q^{4+4+4}} = \frac{p^6}{q^{12}}$$

2. $\left(\dfrac{2x}{y^2}\right)^4 = \dfrac{2^{4\times1} \cdot x^{4\times1}}{y^{4\times2}} = \dfrac{2^4 x^4}{y^8} = \dfrac{16x^4}{y^8}$ $(y \neq 0)$.

3. $\left(\dfrac{3x}{5ab^2}\right)^3 \dfrac{3^{3\times1} \cdot x^{3\times1}}{5^{3\times1} \cdot a^{3\times1} \cdot b^{3\times2}} = \dfrac{27x^3}{125a^3 b^6}$ $(a, b \neq 0)$.

EXERCISE 7-1

Simplify; assume that literal exponents denote positive integers:

1. 2^3

2. 2^5

3. -3^2

4. -3^3

5. $(-3)^3$

6. $(-2)^4$

7. -4^3

8. $-(-3)^3$

9. $\left(-\frac{2}{3}\right)^3$

10. $-\left(-\frac{3}{4}\right)^2$

11. $(-0.3)^4$

12. $-(0.2)^5$

13. $a^3 \cdot a^5$

14. $a^3 \cdot a^2 \cdot a$

15. $x^3 \cdot x^2 \cdot x^4$

16. $(a^2 y^3)(a^2 y)$

17. $x^4 \div x$

18. $y^6 \div y^6$

19. $\dfrac{a^2 b^3}{a^3 b}$

20. $\dfrac{x^3 y^3}{x^3 y}$

21. $\dfrac{a^2 b^2 c^4}{a^3 b c^3}$

22. $(x^2)^3$

23. $-(b^2)^2$

24. $(-m^3)^4$

25. $2^2 \cdot 2^4$

26. $(3^2)^3$

27. $(2)^2 (3)^3$

28. $-(a^2 b^3)^3$

29. $(-a^5 b^4)^3$

30. $(2xy^2)^3$

31. $(3a^2 bc)^4$

32. $(-2x^2 yz^3)^3$

33. $(4 - 6)^3$

34. $\left(\dfrac{3+4}{4}\right)^2$

35. $\left(-\dfrac{a^2}{4}\right)^2$

36. $\left(\dfrac{6}{m}\right)^3$

37. $\left(\dfrac{x^3}{y^5}\right)^4$

38. $\left(\dfrac{3a^2}{4b^3}\right)^2$

39. $\left(-\dfrac{3}{v^2}\right)^4$

40. $(a^{2m})^p$

41. $\left(\dfrac{3^n}{2}\right)^3$

42. $\left(-\dfrac{x^2}{y^n}\right)^4$

43. $x^{4n} \div x^n$

44. $x^{5n} \div x$

45. $x^{2n} \div x^{3n}$

46. $a^{2+n} \cdot a^n$

47. $(3a + x)^2$

48. $r^{xy} \cdot r^{x+1}$

49. $3(x^n)^2$

50. $(m^a)^{p+2}$

51. $(a^2 b^3)^r$

52. $\left(-\dfrac{a^2 b^3}{2b}\right)^7$

53. $\left(\dfrac{a^2 b^3}{c}\right)\left(\dfrac{ac^3}{b}\right)$

54. $\left(-\dfrac{x^3 y}{a}\right)^2 \left(\dfrac{x^4 y^2}{a^2}\right)$

55. $\left(\dfrac{x^2 y}{a}\right) \div \left(\dfrac{xy^5}{a^2}\right)$

56. $\left(-\dfrac{ma}{b}\right)^3 \div \left(\dfrac{mc^4}{b^3}\right)$

57. $\left[\left(\dfrac{3a^2}{b}\right)\left(-\dfrac{b^3}{9ac}\right)\right]^4$

58. $\left[\left(-\dfrac{2a^2}{b^2}\right)\left(\dfrac{b^2}{2a^3}\right)\right]^3$

59. $\left[\left(\dfrac{a^{2+n}}{b^3 y}\right) \div \left(-\dfrac{a^n}{by^3}\right)\right]^3$

60. $\left[\left(-\dfrac{x^2}{ay^4}\right)^3 \div \left(\dfrac{x^2}{a^2 y}\right)\right]^3$

2. EXPONENT OF ZERO

The five laws of exponents listed in the previous section have meaning only for exponents that represent positive integers. This is a consequence of defining powers as products of an integral number of identical factors. It is possible to expand the mathematical utility of the laws of exponents, however, by assigning certain meanings to exponents that represent other than positive integers. Such meanings must be internally consistent with the already stated laws of exponents, and consistent otherwise mathematically.

Let us begin by considering an interpretation for the exponent of zero in an expression such as a^0, $(a \neq 0)$. If exponent law I is to hold when we multiply a^m (m denotes any positive integer) with the as yet uninterpreted a^0, then

$$a^m \times a^0 = a^{m+0} = a^m$$

Since the only factor that multiplies with a^m to yield a product of a^m is the number 1, then a^0 may be considered the equivalent of 1. Accordingly, we state:

Definition A: Any power having a nonzero base and an exponent of zero is equal to 1.

EXAMPLES

1. $3^0 = 1$.

2. $b^0 = 1$ $(b \neq 0)$.

3. $3x^0 = 3$ $(x \neq 0)$. The exponent is considered related only to the factor, x, and not to the 3. So, $3x^0 = (3)(x^0) = (3)(1) = 3$.

4. $(3x)^0 = 1$ $(x \neq 0)$. By IV, Program **7.2**: $(3x)^0 = (3)^0(x)^0 = (1)(1) = 1$.

5. $\left(\dfrac{p}{q}\right)^0 = 1$ $(p, q \neq 0)$. By V, Program **7.3**: $\left(\dfrac{p}{q}\right)^0 = \dfrac{p^0}{q^0} = \dfrac{1}{1} = 1$.

6. $(a + 3b)^0 = 1$ $(a, b \neq 0)$.

7. $(-6m)^0 = (-6)^0(m)^0 = (1)(1) = 1$ $(m \neq 0)$.

3. NEGATIVE EXPONENT

Let us now consider an interpretation for exponents that represent negative integers. If we were to multiply an interpretable a^m (m denotes any positive integer, $a \neq 0$), with an as yet uninterpreted a^{-m}, exponent law I and definition A require that:

$$a^m \times a^{-m} = a^{m-m} = a^0 = 1$$

Since the only factor that will multiply with a given factor to yield a product of 1 is the *reciprocal* of the given factor, a consistent interpretation that may be assigned to a^{-m} $(a \neq 0)$ is the reciprocal of a^m, or $1/a^m$. Thus:

> *Definition B:* Any power having a nonzero base a and a negative exponent $-n$ is equivalent to $1/a^n$.

A useful consequence of this definition is the following generalization:

An equivalent fraction results when any nonzero *factor* in the numerator of a fraction is placed in the denominator of that fraction, or any nonzero *factor* in the denominator of a fraction is placed in the numerator of that fraction, *provided the sign of the factor's exponent is reversed at the same time.*

EXAMPLES

1. $a^{-4} = \dfrac{1}{a^4}.$

  [a^{-4} may be read "a to the negative 4 power."]

2. $8^{-1} = \dfrac{1}{8^1} = \dfrac{1}{8}.$

3. $a^3 b^{-2} = a^3 \left(\dfrac{1}{b^2} \right) = \dfrac{a^3}{b^2}.$

4. $\dfrac{1}{a^{-2}} = \dfrac{1}{1/a^2} = 1 \div \dfrac{1}{a^2} = 1 \cdot a^2 = a^2.$

5. $\dfrac{a^{-2}}{b^{-3}} = \dfrac{b^3}{a^2}.$

6. $\left(\dfrac{p}{q} \right)^{-1} = \dfrac{q}{p}.$

7. $\dfrac{(a+b)^{-2}}{(q+p)} = \dfrac{1}{(a+b)^2(p+q)} \left[\text{or } \dfrac{(p+q)^{-1}}{(a+b)^2} \right].$

[*Note:* With the introduction of negative exponents, the two parts of exponent law II may be collapsed into one:

$$a^m \div a^n = a^{m-n}$$

When $m > n$, then $m - n > 0$, and the quotient has a positive exponent; when $m < n$, then $m - n < 0$, and the quotient has a negative exponent; when $m = n$, then $m - n = 0$, and the quotient has 0 as an exponent.]

EXERCISE 7-2

Simplify; assume that literal exponents denote positive integers.

1. $(-3x)^0$ **2.** $-3x^0$ **3.** $m^2 \cdot m^3 \cdot m^0$

4. 3×10^0 **5.** $\left(\dfrac{3p}{q} \right)^0$ **6.** $\left(\dfrac{3p^0}{q} \right)^2$

7. $\left(\dfrac{5+4}{5}\right)^0$ 　　　　**8.** $\dfrac{3x^0y^2z^0}{3xy^0z^0}$ 　　　　**9.** $\left(\dfrac{2xy}{m^2}\right) \div \left(\dfrac{xy}{m}\right)^0$

10. $\left(\dfrac{2xy}{m^2}\right)^0 \div \left(\dfrac{xy}{m}\right)$

Express the following with all exponents negative:

11. a^4 　　　　**12.** a^2b^3 　　　　**13.** $\dfrac{1}{y^2}$

14. $\dfrac{2}{p^3}$ 　　　　**15.** x^0y^3 　　　　**16.** $\dfrac{1}{16}$

17. $\dfrac{x^{-1}}{b}$ 　　　　**18.** $-c^3$ 　　　　**19.** $-(x)^2(yz)^3$

20. $(a-b)^2$ 　　　　**21.** (a^2-b^2) 　　　　**22.** $(a^2-b^2)^{-2}$

Simplify the following to equivalent expressions containing only positive exponents:

23. $(3x)^{-3}$ 　　　　**24.** $\left(\dfrac{2}{x}\right)^{-2}$ 　　　　**25.** $(x^{-2})^{-2}$

26. $a^5 \cdot a^{-3} \cdot a^{-2}$ 　　　　**27.** $(xy^{-1})^{-2}$ 　　　　**28.** $(3x^0y)^{-4}$

29. $(2^3x^{-3})^{-2}$ 　　　　**30.** $\left(\dfrac{a^{-3}}{c^{-4}}\right)^3$ 　　　　**31.** $\left(-\dfrac{a^{-1}}{b}\right)^{-2}$

32. $\left(\dfrac{ax^4}{b^3y}\right)^3\left(\dfrac{by^3}{a^2x^2}\right)^2$ 　　　　**33.** $(x^2y^3)^3\left(\dfrac{a^2x^3}{y^2}\right)^2$ 　　　　**34.** $x^{-1}-y^{-1}$

35. $(x^{-2})^{-2} \cdot (y^{-3})^2$ 　　　　**36.** $2x^{-1}-y^{-2}$ 　　　　**37.** $3x^{-2}+3y^{-2}$

38. $x^{-3} \div y^{-2}$ 　　　　**39.** $a^{-2} \div \left(\dfrac{a}{b}\right)^{-3}$ 　　　　**40.** $\dfrac{a}{b^{-1}}+\dfrac{b}{a^{-1}}+ab$

41. $\left(\dfrac{a}{b}-\dfrac{b}{a}\right)^{-1}$ 　　　　**42.** $(a^{-1}+b^{-1})^{-1}$ 　　　　**43.** $6x^{-2}+y$

44. $(a^{-1}+2b^{-1})^{-2}$ 　　　　**45.** $\dfrac{x^{-1}y}{x^{-1}+y}$

4. ROOTS

The inverse operation of raising to a power is extracting a root. Because these two operations apply to single numbers, they are called **unary** operations—as opposed to *binary* operations which involve two numbers. Thus, in raising to a power:

$$2^2 = 2 \cdot 2 = 4 \quad \text{(or "two squared} = 4\text{")}$$
$$2^3 = 2 \cdot 2 \cdot 2 = 8 \quad \text{(or "two cubed} = 8\text{")}$$
$$2^4 = 2 \cdot 2 \cdot 2 \cdot 2 = 16 \quad \text{(or "two to the fourth} = 16\text{")}$$

When the power operation is reversed and we seek answers to questions like,

"What number squared is 4?" or "What number cubed is 8?" and so on, we seek a root. Thus:

$$2 \text{ is a square root of } 4$$

$$2 \text{ is a cube root of } 8$$

$$2 \text{ is a fourth root of } 16$$

Powers of rational numbers are always rational numbers, but roots of rational numbers are not always rational numbers. For example, the rational number $+9$ has two rational numbers as square roots, $+3$ and -3, since

$$(+3)^2 = +9 \quad \text{and} \quad (-3)^2 = +9$$

On the other hand, the square roots of 5, or 2, or 6, are not to be found among the rational numbers, but in another set, the **irrational numbers**. Included among the irrational numbers are the nonrational roots of rational numbers, except the even roots of negative rational numbers, such as $\sqrt{-4}$, $\sqrt[4]{-8}$, and $\sqrt[10]{-1}$. These latter roots are available only in a still larger set, the complex numbers, to be discussed in Section 12 of this chapter.

Together, the rational and irrational numbers make up the set of numbers called the **real numbers**:

$$\text{real numbers} \begin{cases} \text{rational numbers} \begin{cases} \text{integers} \\ \text{nonintegers} \end{cases} \\ \text{irrational numbers} \end{cases}$$

As we have noted, the rational numbers are closed under the binary operations of addition, subtraction, multiplication, and division (except by zero). They are also closed under the unary operation of raising to a power. But rational numbers are not closed under the inverse unary operation of extracting a root.

Irrational numbers that are the roots of rational numbers are usually expressed with a symbol called a **radical**: $\sqrt{\ }$. A numeral, appearing in the crook of the radical to denote the root, is called the **index**. The numeral appearing within the radical is called the **radicand**. Thus:

$$\text{cube root of } 7 = \overset{\text{index}}{\underset{\text{radical}}{\sqrt[3]{7}}} \text{——radicand}$$

$$\text{square root of } 6 = \sqrt[2]{6} \quad \text{or} \quad \sqrt{6}$$

(The index is normally omitted from the square root radical.)

$$\text{fourth root of } 5 = \sqrt[4]{5}$$

$$n\text{th root of } a = \sqrt[n]{a}$$

In general, the nth root of a given number is one of n equal factors whose product is the given number. For example, $\sqrt{3} \cdot \sqrt{3} = 3$; $\sqrt[3]{4} \cdot \sqrt[3]{4} \cdot$

$\sqrt[3]{4} = 4$. Some numbers have more than one real root, e.g., the square roots of $+4$ are the two numbers $+2$ and -2, since $(+2)(+2) = +4$ and $(-2)(-2) = +4$. This leads us to define the **principal root** of various numbers:

The principal nth root of a positive number is the positive root.
The principal nth root of 0 is 0.
The principal nth root of a negative number is the negative root when n is odd.

(The principal nth root of a negative number when n is even does not exist among the real numbers.)

EXAMPLES

1. The principal square root of 9 is $\sqrt{9}$ or 3; $-\sqrt{9}$ or -3 is also a square root of 9, since $(-3)(-3) = 9$, but -3 is not the principal square root.

2. The principal cube root of 8 is $\sqrt[3]{8}$ or 2.

3. The principal cube root of -8 is $\sqrt[3]{-8}$ or -2.

4. The principal square root of 5 is $\sqrt{5}$; $-\sqrt{5}$ is also a square root, but not the principal square root.

5. The principal square root of -9 does not exist among the real numbers, i.e., $\sqrt{-9}$ is not a real number.

5. FRACTIONAL EXPONENTS

Up to this point, we have definitions for integral exponents, i.e., zero and the positive and negative integers. We now develop a useful and consistent interpretation for exponents in the form of unit fractions, e.g., $a^{1/3}$. If we were to cube this expression (which suggests a power) according to exponent law III, the resulting power would have an exponent of 1:

$$(a^{1/3})^3 = a^{3 \times 1/3} = a^1$$

It is reasonable, then, to interpret $a^{1/3}$ as *cube root of a*, or

$$a^{1/3} = \sqrt[3]{a}$$

Similarly, it can be shown consistent to interpret $a^{1/2}$ as $\sqrt{a}$, $a^{1/7}$ as $\sqrt[7]{a}$, $x^{1/17}$ as $\sqrt[17]{x}$, $8^{1/3}$ as $\sqrt[3]{8}$ or 2, $9^{1/2}$ as $\sqrt{9}$ or 3, etc. In general, we define $a^{1/n} = \sqrt[n]{a}$, where $\sqrt[n]{a}$ represents the *principal nth root of a*, provided a is nonnegative when n denotes an even integer. The latter qualification is intended to rule out expressions such as $\sqrt{-2}$, which have no meaning among the real numbers.

Now that we have an interpretation for exponents of the form $1/n$, all we need now is an interpretation for exponents of the form m/n, and our discussion is complete.

Consider the product:

$$a^p \cdot a^p \cdot a^p = a^{p+p+p} = a^{3p}$$

If we let p represent an exponent of the form $1/n$, say $\frac{1}{4}$, then by exponent law I:

$$a^{1/4} \cdot a^{1/4} \cdot a^{1/4} = a^{1/4+1/4+1/4} = a^{3/4}$$

Expressing the product in another way, as the cube of the factor $a^{1/4}$, we have

$$a^{1/4} \cdot a^{1/4} \cdot a^{1/4} = (a^{1/4})^3 = (\sqrt[4]{a})^3 = a^{3/4}$$

Consider now the related example:

$$\sqrt[4]{a \cdot a \cdot a} = \sqrt[4]{a^3} = (a^3)^{1/4}$$

To be consistent with exponent law III,

$$(a^3)^{1/4} = a^{(1/4) \times 3} = a^{3/4}$$

Thus we see that

$$a^{3/4} = \sqrt[4]{a^3} = (\sqrt[4]{a})^3$$

which is generalized in the following definition.

> *Definition C:* The numerator of a fractional exponent may be interpreted as the power to which the base is to be raised, and the denominator of a fractional exponent as the root to be extracted.

EXAMPLES

1. Simplify $8^{2/3}$.

$$8^{2/3} = \begin{cases} \sqrt[3]{8^2} = \sqrt{64} = 4 \\ (\sqrt[3]{8})^2 = (2)^2 = 4 \end{cases}$$

2. Simplify $9^{3/2}$.

$$9^{3/2} = \begin{cases} \sqrt{9^3} = \sqrt{729} = 27 \\ (\sqrt{9})^3 = (3)^3 = 27 \end{cases}$$

[*Note:* When the radicand yields a rational root, it is usually simpler computationally to extract the root first and then raise to the power.]

6. SUMMARY—LAWS OF EXPONENTS

In the preceding sections of this chapter, we have developed plausible and consistent interpretations of exponents, from the original positive integers to the complete set of rational numbers. Accordingly, we restate the laws of exponents as follows:

If a and b denote real numbers, and m and n denote rational numbers, then:

$$\text{I.} \quad a^m \cdot a^n = a^{m+n}$$

$$\text{II.} \quad a^m \div a^n = a^{m-n} \ (a \neq 0)$$

$$\text{III.} \quad (a^n)^m = a^{mn}$$

$$\text{IV.} \quad (ab)^n = a^n b^n$$

$$\text{V.} \quad \left(\frac{a}{b}\right)^n = \frac{a^n}{b^n} \ (b \neq 0)$$

To *simplify* an exponential expression usually means to carry out all operations possible by the foregoing laws of exponents (I to V), to eliminate all negative exponents from the final expression, and to expand all numerical coefficients having integral exponents.

EXAMPLES

1. $(x^{3/4})^8 = x^{8(3/4)} = x^6$

2. $x^{1/2}x^{2/3} = x^{1/2+2/3} = x^{3/6+4/6} = x^{7/6}$

$$= x^{1+1/6} = x^1 x^{1/6} = x\sqrt[6]{x}$$

3. $(9x^{-3})^{2/3} = 9^{2/3}x^{(2/3)(-3)} = (3^2)^{2/3}(x^{-2})$

$$= 3^{4/3}x^{-2} = \frac{3\sqrt[3]{3}}{x^2}$$

4. $(-\frac{1}{27})^{-2/3} = [(-\frac{1}{3})^3]^{-2/3} = (-\frac{1}{3})^{(-2/3)(3)}$

$$= (-\frac{1}{3})^{-2} = \frac{1}{(-\frac{1}{3})^2} = \frac{1}{\frac{1}{9}} = 9$$

5. $\dfrac{9x^{-2}y^3}{3x^4y^{-5}} = 3^2 \cdot x^{-2} \cdot y^3 \cdot 3^{-1} \cdot x^{-4} \cdot y^5$

$$= 3^{2-1}x^{-2-4}y^{3+5}$$

$$= 3^1 x^{-6}y^8 = \frac{3y^8}{x^6}$$

6. $(a^2 - b^2)\sqrt{(a-b)^{-2}} = (a^2 - b^2)[(a-b)^{-2}]^{1/2}$

$$= (a^2 - b^2)(a-b)^{-1}$$

$$= \frac{(a^2 - b^2)}{(a-b)} = \frac{(a-b)(a+b)}{(a-b)}$$

$$= a + b$$

7. $\dfrac{(3x^2y)^{1/3}}{9x^{-2/3}y^2} = \dfrac{3^{1/3}x^{2/3}y^{1/3}}{3^2 x^{-2/3}y^2}$

$$= 3^{1/3} \cdot x^{2/3} \cdot y^{1/3} \cdot 3^{-2} \cdot x^{2/3} \cdot y^{-2}$$

$$= 3^{-5/3}x^{4/3}y^{-5/3}$$

$$= \frac{x \cdot x^{1/3}}{3 \cdot 3^{2/3} \cdot y \cdot y^{2/3}} = \frac{x\sqrt[3]{x}}{3y\sqrt[3]{9y^2}}$$

EXERCISE 7-3

Write the real-number principal root for each of the following:

1. $\sqrt{4}$ **2.** $\sqrt{36}$ **3.** $\sqrt[3]{8}$

4. $\sqrt[3]{-125}$ **5.** $\sqrt[3]{-8}$ **6.** $\sqrt{0.01}$

7. $\sqrt{\frac{1}{4}}$ **8.** $\sqrt[3]{-\frac{1}{8}}$ **9.** $\sqrt{-36}$

10. $\sqrt{\frac{16}{25}}$ **11.** $\sqrt{0.16}$ **12.** $\sqrt[3]{-0.001}$

13. $\sqrt[3]{-\frac{8}{27}}$ **14.** $\sqrt[6]{-64}$ **15.** $\sqrt[5]{-32}$

16. $\sqrt[3]{0.008}$

Express the following in exponential form:

17. $\sqrt{p}$ **18.** $\sqrt[3]{s}$ **19.** $\sqrt[10]{t}$

20. $\sqrt[7]{-x}$ **21.** $\sqrt[3]{p^2}$ **22.** $\sqrt[5]{x^2 y^2}$

23. $\sqrt[n]{y^2}$ **24.** $\sqrt[7]{s^2 t^3}$ **25.** $\dfrac{1}{\sqrt[7]{p^8}}$

26. $\sqrt[k]{p^{3m}}$ **27.** $\sqrt{c - 6}$ **28.** $\sqrt[3]{(a + b)^2}$

29. $\sqrt[4]{\dfrac{c^3}{b}}$ **30.** $\sqrt[4]{x^2 y^{1/2}}$ **31.** $\dfrac{1}{\sqrt[n]{c + d^2}}$

32. $\sqrt[n]{c^2 + 2cd + d^2}$

Express the following in simplest radical form:

33. $a^{1/5}$ **34.** $b^{1/3} c^{2/3}$ **35.** $a^{5/8}$

36. $3x^{1/2}$ **37.** $(3x)^{1/2}$ **38.** $(8x)^{1/3}$

39. $(2x^2)^0$ **40.** $a^{1/3} b^{2/3} c$ **41.** $(b - c)^{1/2}$

42. $\dfrac{1}{(a + 2)^{-1/3}}$ **43.** $[(x + y)^2]^{1/3}$ **44.** $m^0 n^{-1/6}$

45. $(s^2 t^{1/4})^{1/n}$ **46.** $\dfrac{x^{-1/2}}{y^{1/2}}$ **47.** $\left(p - \dfrac{pr}{r}\right)^{1/3}$

48. $(a^k m^{-p})^{-1/p}$ **49.** $x^{2/3} \div x^{1/2}$ **50.** $m^{2/5} \div m^{2/3}$

51. $\dfrac{a^{7/2}}{a^{1/2}}$ **52.** $\dfrac{b^{5/2}}{\sqrt[3]{b^{10}}}$ **53.** $\dfrac{a^{0.8}}{a^{0.2}}$

54. $\dfrac{a^{0.12}}{a^{0.6}}$ **55.** $\dfrac{m^{2/3} n^{1/3}}{m^0 n^{1/4}}$ **56.** $\left(\dfrac{4x^{-1} y}{x^{2/3} y^{-1}}\right)^{-3}$

57. $\left(-\dfrac{8}{125}\right)^{2/3}$ **58.** $\left(\dfrac{1}{16}\right)^{-3/4}$ **59.** $\sqrt{(x + y)^{-2}}$

60. $\sqrt{x^{-1} y^{-1}}$ **61.** $(-2^{-6})^{2/3}$ **62.** $-(2^6)^{-2/3}$

63. $\dfrac{(2x^2 y^3)^{1/4}}{16 x^{-2} y}$ **64.** $\dfrac{x^{(2m+3)/c}}{x^{(m-4)/c}}$ **65.** $\sqrt[5]{s^{1/2}}$

66. $(2^0 + 3^0 + 4^0)^{-1/2}$ **67.** $\sqrt[4]{a^{-3/4}}$ **68.** $\left(\dfrac{a^{m+2} n^{m-3}}{a^2 n^{-3}}\right)^{-1/2}$

69. $\left(\dfrac{a^{2/3}b^{1/2}}{3a^{3/4}}\right)^4$ **70.** $(x^k y^2)^{-3}(a^{2-2k}b^{2k})^{1/2}$ **71.** $(x^{-3})\sqrt[m]{x^{3m}y^2}$

72. $[(x^3)^{-2/5}(y^2)^{3/4}]^{1/2}$ **73.** $\sqrt[4]{\sqrt[3]{\sqrt{x}}}$ **74.** $\sqrt[a]{\sqrt[b]{\sqrt[c]{x^d}}}$

7. EQUIVALENT RADICAL EXPRESSIONS

In the same way that fractions have equivalents, exponential and radical expressions also have equivalents. For instance, 4^2, 16^1, 2^4, $\sqrt{256}$, $(4096)^{1/3}$, and $8^{4/3}$ are all equivalent "names" for the same number. In computation, it is often useful—sometimes necessary—to replace one expression with an equivalent. The following are four basic programs which yield equivalent radical expressions.

7.4

To produce an equivalent radical expression of lower order (index):

Step 1. Write the terms of the radical expression exponentially with all numerical coefficients in prime factor form.

Step 2. Express fractional exponents in lowest terms.

Step 3. Write the equivalent of the expression of Step 2 in radical form.

EXAMPLES

1. Reduce the order of $\sqrt[4]{9a^2b^2}$.

Step 1. $(3^2a^2b^2)^{1/4} = 3^{2/4}a^{2/4}b^{2/4}$

Step 2. $3^{2/4}a^{2/4}b^{2/4} = 3^{1/2}a^{1/2}b^{1/2} = (3ab)^{1/2}$

Step 3. $\sqrt{3ab}$

2. Reduce to a lower order: $\sqrt[6]{25x^4b^{12}}$.

Step 1. $(5^2x^4b^{12})^{1/6} = 5^{2/6}x^{4/6}b^{12/6}$

Step 2. $5^{2/6}x^{4/6}b^{12/6} = 5^{1/3}x^{2/3}b^2$

Step 3. $b^2\sqrt[3]{5x^2}$

3. Reduce the order: $\sqrt[4]{25x^2y^3}$.

Step 1. $(5^2x^2y^3)^{1/4} = 5^{2/4}x^{2/4}y^{3/4}$

Step 2. $5^{2/4}x^{2/4}y^{3/4} = 5^{1/2}x^{1/2}y^{3/4}$

Step 3. Thus the order of $\sqrt[4]{25x^2y^3}$ is already at its lowest.

7.5

To produce an equivalent radical expression of higher order (index):

Step 1. Write the terms of the radical expression exponentially with all numerical coefficients in prime factor form.

Step 2. Replace each exponent with an equivalent fractional exponent in which the denominator is the desired order.

Step 3. Write the equivalent of the expression of Step 2 in radical form.

EXAMPLES

1. Express the equivalent of $\sqrt[3]{9xy^2}$ with an index of 6.

Step 1. $\sqrt[3]{9xy^2} = (3^2xy^2)^{1/3} = 3^{2/3} \cdot x^{1/3} \cdot y^{2/3}$

Step 2. $3^{2/3}x^{1/3}y^{2/3} = 3^{4/6}x^{2/6}y^{4/6}$

Step 3. $\sqrt[6]{3^4x^2y^4} = \sqrt[6]{81x^2y^4}$

2. Express the equivalent of $\sqrt[4]{7a^2b^3}$ with an index of 12.

Step 1. $\sqrt[4]{7a^2b^3} = 7^{1/4}a^{2/4}b^{3/4}$

Step 2. $7^{1/4}a^{2/4}b^{3/4} = 7^{3/12}a^{6/12}b^{9/12}$

Step 3. $\sqrt[12]{7^3a^6b^9} = \sqrt[12]{343a^6b^9}$

7.6

To produce an equivalent radical expression by removing a factor from the radicand:

Step 1. Factor the radicand so that one or more of its factors are the same power as the index of the radical.

Step 2. Replace any of the power factors of the radicand with its root as a factor outside the radical.

EXAMPLES

1. Simplify $\sqrt{x^3}$.

Step 1. $\sqrt{x^3} = \sqrt{x^2 \cdot x}$

Step 2. $x\sqrt{x}$

2. Simplify $\sqrt{18a^5}$.

Step 1. $\sqrt{18a^5} = \sqrt{2 \cdot 3^2 \cdot a^2 \cdot a^2 \cdot a}$

Step 2. $3a^2\sqrt{2a}$

3. Simplify $\sqrt[3]{24a^5b^8}$.

Step 1. $\sqrt[3]{24a^5b^8} = \sqrt[3]{3 \cdot 2^3 \cdot a^3 \cdot a^2 \cdot b^6 \cdot b^2}$

Step 2. $2ab^2\sqrt[3]{3a^2b^2}$

7.7

To produce an equivalent radical expression by introducing a factor into the radicand:

Step 1. Express the factor to be introduced with a fractional exponent whose denominator is the index of the radical.

Step 2. Insert under the radical the factor raised to the power of the numerator of the fractional exponent of Step 1.

EXAMPLES

1. Express under one radical an equivalent of $b^2\sqrt[6]{a^5}$.

Step 1. $b^2 = b^{12/6}$

Step 2. $\sqrt[6]{a^5 b^{12}}$

2. Write an equivalent of $3\sqrt{5}$ under one radical.

Step 1. $3 = 3^{2/2}$

Step 2. $\sqrt{3^2 \cdot 5} = \sqrt{9 \cdot 5} = \sqrt{45}$

3. Express under one radical: $\sqrt{x} \cdot \sqrt[4]{a}$.

Step 1. $\sqrt{x} = x^{1/2} = x^{2/4}$

Step 2. $\sqrt[4]{ax^2}$

EXERCISE 7-4

Reduce the order of the following:

1. $\sqrt[4]{9a^2}$ **2.** $\sqrt[4]{64a^2b^2}$ **3.** $\sqrt[6]{4m^2}$

4. $\sqrt[6]{-8x^3}$ **5.** $\sqrt[9]{27m^6}$ **6.** $\sqrt[12]{36s^8}$

7. $\sqrt[9]{(x^2 + y^2)^3}$ **8.** $\sqrt[6]{\dfrac{27x^3}{64y^3}}$

Express the following under the given radical:

9. $\sqrt{3} = \sqrt[4]{}$ **10.** $\sqrt{2xy} = \sqrt[6]{}$ **11.** $\sqrt[3]{4xy} = \sqrt[6]{}$

12. $\sqrt[3]{6x^2y} = \sqrt[9]{}$ **13.** $\sqrt{16} = \sqrt[3]{}$ **14.** $\sqrt[4]{13x^2y^3z} = \sqrt[8]{}$

15. $\sqrt{0.2} = \sqrt[6]{}$ **16.** $\sqrt[3]{(a-b)^2} = \sqrt[6]{}$

Write the complete expression under a single radical:

17. $3\sqrt{2}$ **18.** $5\sqrt{3}$ **19.** $2\sqrt[3]{6}$

20. $5\sqrt[3]{4}$ **21.** $m\sqrt{3}$ **22.** $a\sqrt{5a}$

23. $4a\sqrt[3]{2a}$ **24.** $2x^2\sqrt[3]{2x^2}$ **25.** $4x^2y\sqrt{2xy}$

26. $3mn^2\sqrt[3]{2m^2n}$ **27.** $(a-b)\sqrt[6]{(a-b)}$ **28.** $3x^2y\sqrt[3]{2x^2y^2}$

Simplify:

29. $\sqrt{75}$ **30.** $\sqrt[3]{32}$ **31.** $\sqrt[3]{81}$

32. $\sqrt{108}$ **33.** $\sqrt{8x}$ **34.** $\sqrt{50x^3}$

35. $\sqrt[3]{27x^4y^2}$ **36.** $\sqrt{45m^2}$ **37.** $\sqrt{18x^2y^3z^4}$

38. $\sqrt[3]{kx^5y^4}$ **39.** $3\sqrt{12x^3}$ **40.** $x\sqrt{3xy^3}$

41. $\sqrt{\dfrac{x}{16}}$ **42.** $\sqrt{\dfrac{5s^5}{t^6}}$ **43.** $4\sqrt[3]{\dfrac{m^2}{8}}$

44. $y\sqrt[3]{(y-a)^6}$ **45.** $3+a\sqrt{45}$ **46.** $a+b\sqrt{(a+b)^3}$

47. $\sqrt[3]{(a-5)^4}$ **48.** $m\sqrt[6]{(m-n)^8}$

8. RATIONALIZING THE DENOMINATOR

To *rationalize a denominator* means to express a fraction equivalently as one having a nonradical denominator. The following program for rationalizing a denominator is based upon the fact that an equivalent fraction results when the numerator and denominator of a fraction are both multiplied by the same nonzero number. A suitable multiplier is one that yields a denominator whose terms have only positive integral exponents and rational coefficients.

[*Note:* The resulting equivalent fraction with rational denominator may or may not have a rational numerator.]

7.8

To rationalize a denominator:

Step 1. Express the denominator exponentially.

Step 2. Determine a factor which, when multiplied with the given denominator, yields a product containing only positive integral exponents and rational coefficients.

Step 3. Multiply the numerator and denominator by the factor of Step 2.

EXAMPLES

1. Express an equivalent of $\dfrac{5}{\sqrt[3]{a}}$ with a rational denominator.

Step 1. $\sqrt[3]{a} = a^{1/3}$

Step 2.   A suitable factor is $a^{2/3}$, since $a^{1/3} \cdot a^{2/3} = a^{1/3+2/3} = a^{3/3} = a^1$

Step 3. $\dfrac{5 \cdot a^{2/3}}{a^{1/3} \cdot a^{2/3}} = \dfrac{5\,a^{2/3}}{a^1} = \dfrac{5\sqrt[3]{a^2}}{a}$

2. Rationalize the denominator: $\dfrac{p}{\sqrt[3]{x^2y}}$.

Step 1. $\sqrt[3]{x^2y} = x^{2/3}y^{1/3}$

Step 2.   A suitable factor is $x^{1/3}y^{2/3}$ since

$$(x^{2/3}y^{1/3})(x^{1/3}y^{2/3}) = x^{1/3+2/3}y^{1/3+2/3} = x^1y^1$$

Step 3. $\dfrac{p \cdot x^{1/3}y^{2/3}}{x^{2/3}y^{1/3} \cdot x^{1/3}y^{2/3}} = \dfrac{p\sqrt[3]{xy^2}}{xy}$

3. Express an equivalent of $\sqrt{\tfrac{3}{5}}$ with a rational denominator.

$$\left[\text{Note:}\ \sqrt{\frac{3}{5}} = \frac{\sqrt{3}}{\sqrt{5}} = \frac{3^{1/2}}{5^{1/2}}. \right]$$

Step 1.   Express the denominator $\sqrt{5}$ exponentially: $5^{1/2}$.

Step 2.   A suitable factor is $5^{1/2}$ since $5^{1/2} \cdot 5^{1/2} = 5^{1/2+1/2} = 5^1$.

Step 3. $\dfrac{3^{1/2} \cdot 5^{1/2}}{5^{1/2} \cdot 5^{1/2}} = \dfrac{3^{1/2} \cdot 5^{1/2}}{5^1} = \dfrac{(3 \cdot 5)^{1/2}}{5} = \dfrac{(15)^{1/2}}{5} = \dfrac{\sqrt{15}}{5}.$

When the denominator of a radical expression is a binomial containing one or two square-root radicals, it can be rationalized by a method based upon the fact that the product of the sum and difference of two numbers is the difference of their squares (Program **5.4**).

7.9

To rationalize a binomial denominator involving square roots:

Step 1. Multiply both numerator and denominator by the denominator with the sign of the second term reversed.

Step. 2 (Optional, but usual.) Simplify if possible.

[*Note:* This program, as stated, is valid only for square roots; it is not valid when the index of the root is other than 2.]

EXAMPLES

1. Express equivalently with a rational denominator: $\dfrac{1}{\sqrt{5} - \sqrt{3}}.$

Step 1. $\dfrac{(1) \cdot (\sqrt{5} + \sqrt{3})}{(\sqrt{5} - \sqrt{3}) \cdot (\sqrt{5} + \sqrt{3})} = \dfrac{\sqrt{5} + \sqrt{3}}{(\sqrt{5})^2 - (\sqrt{3})^2}$

Step 2. $\dfrac{\sqrt{5} + \sqrt{3}}{(\sqrt{5})^2 - (\sqrt{3})^2} = \dfrac{\sqrt{5} + \sqrt{3}}{5 - 3} = \dfrac{\sqrt{5} + \sqrt{3}}{2}$

2. Rationalize the denominator: $\dfrac{8}{\sqrt{7} + \sqrt{3}}.$

$$\dfrac{(8) \cdot (\sqrt{7} - \sqrt{3})}{(\sqrt{7} + \sqrt{3}) \cdot (\sqrt{7} - \sqrt{3})} = \dfrac{8(\sqrt{7} - \sqrt{3})}{7 - 3}$$

$$= \dfrac{\overset{2}{\cancel{8}}(\sqrt{7} - \sqrt{3})}{\cancel{4}}$$

$$= 2(\sqrt{7} - \sqrt{3}) = 2\sqrt{7} - 2\sqrt{3}$$

3. Rationalize the denominator: $\dfrac{y}{1 - \sqrt{y}}.$

$$\frac{(y)\cdot(1+\sqrt{y})}{(1-\sqrt{y})\cdot(1+\sqrt{y})} = \frac{y(1+\sqrt{y})}{1-y}$$

EXERCISE 7-5

Write equivalents for each of the following in simplest terms with rational denominators:

1. $\dfrac{1}{\sqrt{3}}$ **2.** $\dfrac{2a}{\sqrt{2a}}$ **3.** $\dfrac{3x}{\sqrt{5}}$

4. $\dfrac{7a}{2\sqrt{3}}$ **5.** $\dfrac{3}{\sqrt[3]{4}}$ **6.** $\dfrac{2p}{\sqrt[3]{p^2}}$

7. $\dfrac{12x}{\sqrt[3]{x}}$ **8.** $\sqrt{\dfrac{6a}{5b}}$ **9.** $\sqrt{7m^{-1}}$

10. $\dfrac{\sqrt[3]{9}}{\sqrt[3]{3}}$ **11.** $\sqrt[3]{\dfrac{-3a}{4b^2c^7}}$ **12.** $\sqrt{\dfrac{a-b}{a+b}}$

13. $\dfrac{1}{\sqrt{3}-\sqrt{2}}$ **14.** $\dfrac{2}{\sqrt{3}+\sqrt{5}}$ **15.** $\dfrac{\sqrt{2}}{\sqrt{2}-\sqrt{3}}$

16. $\dfrac{\sqrt{3}}{3-\sqrt{3}}$ **17.** $\dfrac{x}{1-\sqrt{x}}$ **18.** $\dfrac{\sqrt{a}}{\sqrt{a}-\sqrt{b}}$

19. $\dfrac{\sqrt{2}+\sqrt{3}}{\sqrt{2}-\sqrt{3}}$ **20.** $\dfrac{\sqrt{5}-\sqrt{3}}{\sqrt{5}+\sqrt{3}}$ **21.** $\dfrac{\sqrt{3}}{2\sqrt{3}-3\sqrt{2}}$

22. $\dfrac{\sqrt{a}-\sqrt{b}}{\sqrt{a}+\sqrt{b}}$ **23.** $\dfrac{2\sqrt{2}-\sqrt{3}}{3\sqrt{2}+\sqrt{3}}$ **24.** $\dfrac{3\sqrt{x}+2\sqrt{y}}{\sqrt{x}-\sqrt{y}}$

25. $\sqrt{\dfrac{a}{3}-\dfrac{b}{4}}$ **26.** $\sqrt{\dfrac{a}{x}-\dfrac{b}{y}}$

9. ADDITION AND SUBTRACTION WITH RADICAL EXPRESSIONS

It is not really necessary to state separate procedures for computing with radical expressions. Usually it is sufficient to know the laws of exponents, how to convert from exponential to radical form and from radical to exponential form, and how to add, subtract, multiply, and divide with algebraic expressions. However, the programs provided in this and the next two sections can be convenient in those situations where the data are given in radical symbolism.

Radical expressions are said to be *like* when they have the same index and radicand. Thus $3\sqrt[3]{4}$, $6\sqrt[3]{4}$, and $x\sqrt[3]{4}$ are like because each possesses the same index (3) and the same radicand (4); on the other hand, $2\sqrt{6}$

and $2\sqrt{5}$ are not like because they differ in radicand; $3\sqrt[3]{2}$ and $3\sqrt{2}$ are not like because they differ in index.

Computing sums and differences of like radical expressions amounts to addition or subtraction of their coefficients, in accordance with the distributive property:

$$3\sqrt{5} + 4\sqrt{5} = (3 + 4)\sqrt{5} = 7\sqrt{5}$$

For unlike radical expressions, sums and differences are expressed simply as a polynomial:

$$(+)\ \frac{\begin{array}{c}2\sqrt{6}\\3\sqrt[3]{6}\end{array}}{2\sqrt{6} + 3\sqrt[3]{6}} \qquad (-)\ \frac{\begin{array}{c}4\sqrt{7}\\3\sqrt{5}\end{array}}{4\sqrt{7} - 3\sqrt{5}}$$

7.10

To compute the sum (difference) of radical expressions:

Step 1. Simplify each term involving a radical expression.

Step 2. Add (subtract) the like radical expressions by adding (subtracting) their coefficients; express the sum (difference) of the unlike radical expressions as a polynomial.

EXAMPLES

1. Add: $5\sqrt{27} + 7\sqrt{12} + 3\sqrt{3}$.

Step 1. $5\sqrt{27} = 5\sqrt{9\cdot3} = 5\cdot\sqrt{9}\cdot\sqrt{3} = 5\cdot3\cdot\sqrt{3} = 15\sqrt{3}$

$7\sqrt{12} = 7\sqrt{4\cdot3} = 7\cdot\sqrt{4}\cdot\sqrt{3} = 7\cdot2\cdot\sqrt{3} = 14\sqrt{3}$

$3\sqrt{3} = 3\sqrt{3}$

Step 2. $15\sqrt{3} + 14\sqrt{3} + 3\sqrt{3} = (15 + 14 + 3)\sqrt{3} = 32\sqrt{3}$

2. Subtract: $9\sqrt{2} - \dfrac{6}{\sqrt{2}}$.

Step 1. $9\sqrt{2} = 9\sqrt{2}$

$\dfrac{6}{\sqrt{2}} = \dfrac{6\cdot\sqrt{2}}{\sqrt{2}\cdot\sqrt{2}} = \dfrac{6\sqrt{2}}{2} = 3\sqrt{2}$

Step 2. $9\sqrt{2} - 3\sqrt{2} = (9 - 3)\sqrt{2} = 6\sqrt{2}$

3. Simplify: $3\sqrt{125} - 4\sqrt[3]{5} - 15\sqrt{\tfrac{1}{5}} + 3\sqrt[3]{625}$.

Step 1. $3\sqrt{125} = 3\sqrt{25\cdot5} = 3\cdot5\cdot\sqrt{5} = 15\sqrt{5}$

$4\sqrt[3]{5} = 4\sqrt[3]{5}$

$15\sqrt{\tfrac{1}{5}} = 15\cdot\dfrac{\sqrt{1}}{\sqrt{5}} = \dfrac{15}{\sqrt{5}} = \dfrac{15\sqrt{5}}{\sqrt{5}\sqrt{5}} = \dfrac{15\sqrt{5}}{5} = 3\sqrt{5}$

$3\sqrt[3]{625} = 3\sqrt[3]{125\cdot5} = 3\cdot5\cdot\sqrt[3]{5} = 15\sqrt[3]{5}$

Step 2. $15\sqrt{5} - 4\sqrt[3]{5} - 3\sqrt{5} + 15\sqrt[3]{5}$
$$= (15\sqrt{5} - 3\sqrt{5}) + (15\sqrt[3]{5} - 4\sqrt[3]{5})$$
$$= 12\sqrt{5} + 11\sqrt[3]{5}$$

EXERCISE 7-6

Simplify:

1. $4\sqrt{a} + 3\sqrt{a} - 2\sqrt{a}$

2. $\sqrt{18} + \sqrt{50} - \sqrt{72}$

3. $3\sqrt{5} + \sqrt{20} - \sqrt{45} + \sqrt{5}$

4. $\sqrt{4a} + \sqrt{16a} - \sqrt{36a}$

5. $\sqrt[3]{81} - 2\sqrt[3]{3} + 3\sqrt[3]{24}$

6. $3\sqrt{45} - 2\sqrt{25} + \sqrt{5}$

7. $\sqrt{\frac{1}{2}} + 2\sqrt{\frac{1}{8}}$

8. $\sqrt[3]{2} + \sqrt[3]{-54} + \sqrt[3]{16}$

9. $\sqrt[3]{6} - 3\sqrt[3]{48} + 2\sqrt[3]{\frac{1}{36}}$

10. $\sqrt{8} - \sqrt[3]{2} + \sqrt{18}$

11. $\sqrt{\frac{25}{12}} - \sqrt{\frac{9}{8}}$

12. $2\sqrt{24} + 3\sqrt{150} - 3\sqrt{96}$

13. $3\sqrt{72} - 7\sqrt{18} + 2\sqrt[3]{54}$

14. $\sqrt[3]{a^4} + \sqrt[3]{27a^4} - \sqrt[6]{a^8}$

15. $8\sqrt{3} - \dfrac{2}{\sqrt{3}} + \dfrac{7}{2\sqrt{3}}$

16. $\sqrt{\frac{3}{5}} - \sqrt{\frac{16}{15}} + \sqrt{\frac{20}{3}}$

17. $\sqrt{6x^2} - \sqrt{24x^2} + \sqrt{9x^0}$

18. $\sqrt[m]{a^m x} + \sqrt[m]{b^m x} - \sqrt[2m]{a^{2m} x^2}$

19. $\sqrt[3]{(a-b)^2} + \sqrt[3]{(a-b)^5}$

20. $\sqrt{a^2 b} - \sqrt{9b} - a\sqrt{b}$

21. $\sqrt{0.75} + \sqrt{0.12} - \sqrt{0.27}$

22. $\frac{1}{2}\sqrt{x^3 y} - \frac{1}{3}\sqrt{xy^3} - \frac{1}{2}\sqrt{xy}$

23. $\sqrt{3.6} - \sqrt{8.1} + 3\sqrt{0.9}$

24. $\sqrt{\dfrac{2a}{b}} - 3\sqrt{\dfrac{b}{2a^3}} + 4\sqrt{\dfrac{a^3 b}{8}}$

10. MULTIPLICATION WITH RADICALS

Recall that the product of powers may be written by adding the exponents when the bases are the same and by multiplying the bases when the exponents are the same. This carries over into the following program for computing the product of radical expressions.

7.11

To compute the product of radical expressions:

Step 1. Express each of the radical factors as an equivalent radical expression of a common order.

Step 2. Write the product of the coefficients as the coefficient of the product, and the product of the radicands as the radicand of the product (expressed under the radical of the common order).

Step 3. (Optional, but usual.) Simplify if possible.

EXAMPLES

1. Multiply: $2\sqrt{3x}\cdot4\sqrt{6xy}\cdot\sqrt{2py}$.

Step 1.   All factors are of a common order.

Step 2.   Multiply the coefficients:

$$2\cdot4\cdot1 = 8$$

Multiply the radicands:

$$3x\cdot6xy\cdot2py = 36x^2y^2p$$

Express as a product:

$$8\sqrt{36x^2y^2p}$$

Step 3.   Simplify:

$$8\sqrt{36x^2y^2p} = 8\cdot6\cdot x\cdot y\cdot\sqrt{p} = 48xy\sqrt{p}$$

2. Multiply: $4x\sqrt[3]{2x^2y}\cdot y\sqrt{2xy}$.

Step 1.   A common order is 6.

$$\sqrt[3]{2x^2y} = \sqrt[6]{(2x^2y)^2} = \sqrt[6]{4x^4y^2}$$
$$\sqrt{2xy} = \sqrt[6]{(2xy)^3} = \sqrt[6]{8x^3y^3}$$

Step 2. $4x\cdot y = 4xy$

$$\sqrt[6]{4x^4y^2}\cdot\sqrt[6]{8x^3y^3} = \sqrt[6]{4x^4y^2\cdot8x^3y^3} = \sqrt[6]{4\cdot8\cdot x^{4+3}y^{2+3}} = \sqrt[6]{32x^7y^5}$$

Step 3. $4xy\sqrt[6]{32x^7y^5} = 4x^2y\sqrt[6]{32\,xy^5}$

3. Multiply: $x\sqrt[3]{5xy^2}\cdot y\sqrt{2y}\cdot3\sqrt[4]{8x^2y}$.

Step 1.   A common index is 12.

$$\sqrt[3]{5xy^2} = \sqrt[12]{(5xy^2)^4} = \sqrt[12]{5^4x^4y^8}$$
$$\sqrt{2y} = \sqrt[12]{(2y)^6} = \sqrt[12]{2^6y^6}$$
$$\sqrt[4]{8x^2y} = \sqrt[12]{(2^3x^2y)^3} = \sqrt[12]{2^9x^6y^3}$$

Step 2. $x\cdot y\cdot3 = 3xy$

$$\sqrt[12]{5^4x^4y^8}\cdot\sqrt[12]{2^6y^6}\cdot\sqrt[12]{2^9x^6y^3} = \sqrt[12]{5^4\cdot x^4\cdot y^8\cdot 2^6\cdot y^6\cdot 2^9\cdot x^6\cdot y^3}$$
$$= \sqrt[12]{5^4\cdot2^{6+9}x^{4+6}y^{8+6+3}} = \sqrt[12]{5^4\cdot2^{15}x^{10}y^{17}}$$

Step 3. $3xy\sqrt[12]{5^4\cdot2^{15}\cdot x^{10}\cdot y^{17}} = 3xy\cdot2\cdot y\cdot\sqrt[12]{5^4\cdot2^3\cdot x^{10}\cdot y^5}$
$$= 6xy^2\sqrt[12]{5000x^{10}y^5}$$

4. Multiply: $(3\sqrt{x} + 2\sqrt{y})(2\sqrt{x} - 5\sqrt{y})$.

Compute the product either by long multiplication:

$$
\begin{array}{r}
3\sqrt{x} + 2\sqrt{y} \\
2\sqrt{x} - 5\sqrt{y} \\
\hline
6\sqrt{x^2} + 4\sqrt{xy} \\
\cdot - 15\sqrt{xy} - 10\sqrt{y^2} \\
\hline
6\sqrt{x^2} - 11\sqrt{xy} - 10\sqrt{y^2} = 6x - 11\sqrt{xy} - 10y
\end{array}
$$

$(\times)$

or by the horizontal method:

$$6\sqrt{x^2} \qquad -10\sqrt{y^2}$$

$$(3\sqrt{x} + 2\sqrt{y})(2\sqrt{x} - 5\sqrt{y})$$

$$\left. \begin{array}{l} -(4\sqrt{xy}) \\ (-15\sqrt{xy}) \end{array} \right\} 4\sqrt{xy} - 15\sqrt{xy} = -11\sqrt{xy}$$

$$6\sqrt{x^2} - 11\sqrt{xy} - 10\sqrt{y^2} = 6x - 11\sqrt{xy} - 10y$$

EXERCISE 7-7

Multiply, and simplify:

1. $\sqrt[3]{2a} \cdot \sqrt[3]{4a^2}$
2. $(4x\sqrt{3x})^2$
3. $\sqrt{3a^3} \cdot \sqrt{6}$
4. $\sqrt[3]{4} \cdot \sqrt{6}$
5. $\sqrt[3]{24} \cdot \sqrt{6}$
6. $\sqrt{6}(\sqrt{2} - 3\sqrt{3} + 2\sqrt{6})$
7. $18(2\sqrt{12x} - \sqrt{10x} + 2\sqrt{98x})$
8. $(3x\sqrt[3]{2xy^2})(x\sqrt{2x})$
9. $(2a\sqrt[3]{3a^2b})(a\sqrt{2a})(\sqrt[3]{b^2})$
10. $(2\sqrt{6} + 3\sqrt{2})^2$
11. $(3\sqrt{5} - 2\sqrt{3})^2$
12. $(\sqrt{a} - \sqrt{b})^2$
13. $(\sqrt{a} - \sqrt{b})(\sqrt{a} + \sqrt{b})$
14. $(\sqrt{2} - \sqrt{x})(3\sqrt{2} + 4\sqrt{x})$
15. $(\sqrt{3} - 2\sqrt{5})(2\sqrt{3} + \sqrt{5})$
16. $(3 + \sqrt{x - 3})^2$
17. $(\sqrt{a + 3} + 4)^2$
18. $(\sqrt{x - 2} + \sqrt{2})^2$
19. $(\sqrt{a + b} - \sqrt{a - b})^2$
20. $(\sqrt[3]{a^2b} - \sqrt{ab})^2$

Find the value of:

21. $x^2 + x - 1$ when $x = 2 - \sqrt{2}$
22. $2x^2 - x + 1$ when $x = \sqrt{3} - 1$
23. $3x^2 + 2x - 4$ when $x = \sqrt{3} + 2$
24. $3x^2 - 3x - 5$ when $x = 2\sqrt{3} - \sqrt{2}$

11. DIVISION WITH RADICALS

The quotient of radical expressions is most readily found by expressing the two terms in fraction form: the dividend as the numerator and the divisor as the denominator. If the divisor-denominator contains a radical expression, it is ordinarily rationalized; the resulting equivalent fraction (preferably simplified) is the desired quotient.

7.12

To compute the quotient of two radical expressions:

Step 1. Express the dividend as the numerator and the divisor as the denominator of a fraction.

Step 2. Rationalize the denominator.

Step 3. Simplify the rationalized fraction.

EXAMPLES

1. Divide $12\sqrt{5}$ by $\sqrt{3}$.

Step 1. $12\sqrt{5} \div \sqrt{3} = \dfrac{12\sqrt{5}}{\sqrt{3}}$

Step 2. $\dfrac{12\sqrt{5} \cdot \sqrt{3}}{\sqrt{3} \cdot \sqrt{3}} = \dfrac{12\sqrt{15}}{3}$

Step 3. $\dfrac{12\sqrt{15}}{3} = 4\sqrt{15}$ (quotient)

2. Divide $(4x + \sqrt{6})$ by $3\sqrt{2}$.

Step 1. $(4x + \sqrt{6}) \div 3\sqrt{2} = \dfrac{4x + \sqrt{6}}{3\sqrt{2}}$

Step 2. $\dfrac{(4x + \sqrt{6}) \cdot (\sqrt{2})}{(3\sqrt{2}) \cdot (\sqrt{2})} = \dfrac{4x\sqrt{2} + \sqrt{12}}{3 \cdot 2}$

Step 3. $\dfrac{4x\sqrt{2} + \sqrt{12}}{6} = \dfrac{4x\sqrt{2}}{6} + \dfrac{2\sqrt{3}}{6}$

$= \dfrac{2x\sqrt{2}}{3} + \dfrac{\sqrt{3}}{3}$ (or $\tfrac{2}{3}x\sqrt{2} + \tfrac{1}{3}\sqrt{3}$)

3. Divide $(4\sqrt{2} + 6\sqrt[3]{3})$ by $\sqrt[3]{2}$.

Step 1. $(4\sqrt{2} + 6\sqrt[3]{3}) \div \sqrt[3]{2} = \dfrac{4\sqrt{2} + 6\sqrt[3]{3}}{\sqrt[3]{2}}$

Step 2. $\dfrac{(4\sqrt{2} + 6\sqrt[3]{3}) \cdot (\sqrt[3]{2^2})}{(\sqrt[3]{2}) \cdot (\sqrt[3]{2^2})} = \dfrac{4\sqrt{2}\sqrt[3]{2^2} + 6\sqrt[3]{3 \cdot 2^2}}{\sqrt[3]{2^3}}$

Step 3. $\dfrac{4\sqrt{2}\sqrt[3]{2^2} + 6\sqrt[3]{3 \cdot 2^2}}{\sqrt[3]{2^3}} = \dfrac{4\sqrt[6]{2^3}\sqrt[6]{2^4} + 6\sqrt[3]{12}}{2}$

$= \dfrac{4\sqrt[6]{2^7} + 6\sqrt[3]{12}}{2} = \dfrac{4 \cdot 2\sqrt[6]{2} + 6\sqrt[3]{12}}{2}$

$= 4\sqrt[6]{2} + 3\sqrt[3]{12}$

4. Divide $2\sqrt{3} - 3\sqrt{2}$ by $\sqrt{2} - \sqrt{3}$.

Step 1. $(2\sqrt{3} - 3\sqrt{2}) \div (\sqrt{2} - \sqrt{3}) = \dfrac{2\sqrt{3} - 3\sqrt{2}}{\sqrt{2} - \sqrt{3}}$

$= -\dfrac{2\sqrt{3} - 3\sqrt{2}}{\sqrt{3} - \sqrt{2}}$ [†]

Step 2. $-\dfrac{(2\sqrt{3} - 3\sqrt{2}) \cdot (\sqrt{3} + \sqrt{2})}{(\sqrt{3} - \sqrt{2}) \cdot (\sqrt{3} + \sqrt{2})} = -\dfrac{2\sqrt{9} - \sqrt{6} - 3\sqrt{4}}{\sqrt{9} - \sqrt{4}}$

[†] Expressing the fraction negatively is not essential, but it does facilitate the multiplication of Step 2.

Step 3. $-\dfrac{2\sqrt{9} - \sqrt{6} - 3\sqrt{4}}{\sqrt{9} - \sqrt{4}} = -\dfrac{6 - \sqrt{6} - 6}{3 - 2} = -\dfrac{-\sqrt{6}}{1} = \sqrt{6}$

5. Divide $4\sqrt{5} - 2$ by $3\sqrt{2} - 2\sqrt{5}$.

Step 1. $(4\sqrt{5} - 2) \div (3\sqrt{2} - 2\sqrt{5}) = \dfrac{4\sqrt{5} - 2}{3\sqrt{2} - 2\sqrt{5}}$

Step 2. $\dfrac{(4\sqrt{5} - 2) \cdot (3\sqrt{2} + 2\sqrt{5})}{(3\sqrt{2} - 2\sqrt{5}) \cdot (3\sqrt{2} + 2\sqrt{5})}$

$= \dfrac{12\sqrt{10} - 6\sqrt{2} + 8\sqrt{25} - 4\sqrt{5}}{(3\sqrt{2})^2 - (2\sqrt{5})^2}$

Step 3. $= \dfrac{12\sqrt{10} - 6\sqrt{2} + (8 \cdot 5) - 4\sqrt{5}}{18 - 20}$

$= \dfrac{12\sqrt{10} - 6\sqrt{2} + 40 - 4\sqrt{5}}{-2}$

$= -6\sqrt{10} + 3\sqrt{2} - 20 + 2\sqrt{5}$

EXERCISE 7-8

Divide, and simplify quotients:

1. $8 \div 2\sqrt{2}$

2. $3 \div \sqrt[3]{9}$

3. $(\sqrt{6} - 2\sqrt{15}) \div \sqrt{3}$

4. $(3\sqrt{2} - \sqrt{3} + 2\sqrt{6}) \div \sqrt{6}$

5. $(2\sqrt{6} - 3\sqrt{7} + 2\sqrt{3}) \div 3\sqrt{42}$

6. $(4\sqrt{3} + 3\sqrt{2} + \sqrt{8} - 3\sqrt{6}) \div 2\sqrt{6}$

7. $(2\sqrt{3} + 3\sqrt[3]{2}) \div \sqrt[3]{3}$

8. $(3\sqrt[3]{4} - 2\sqrt[3]{3} + 2\sqrt{6}) \div \sqrt[3]{6}$

9. $3 \div (1 + \sqrt{10})$

10. $a \div (1 + \sqrt{a})$

11. $(2\sqrt{5} - 7\sqrt{3}) \div (\sqrt{5} - 2\sqrt{3})$

12. $(3\sqrt{6} - 2\sqrt{3}) \div (\sqrt{6} + 2\sqrt{3})$

13. $(5 + 3\sqrt{2}) \div (\sqrt{6} - \sqrt{3})$

14. $(\sqrt{6} - 3\sqrt{2}) \div (4\sqrt{2} - 3\sqrt{6})$

15. $(\sqrt{7} - 3\sqrt{2}) \div (2\sqrt{2} - 3\sqrt{7})$

16. $(7 - \sqrt{x}) \div (\sqrt{x} + 4)$

17. $(\sqrt{x} - y) \div (y - \sqrt{x})$

18. $(2\sqrt{x} + 3\sqrt{y}) \div (\sqrt{x} - 2\sqrt{y})$

19. $\dfrac{\sqrt{x} - \sqrt{a}}{\sqrt{x} + \sqrt{y}}$

20. $\dfrac{3\sqrt{2} - 2\sqrt{3}}{3\sqrt{2} + \sqrt{3}} \div \dfrac{3\sqrt{2} + 2\sqrt{3}}{4\sqrt{2} + 3\sqrt{3}}$

12. COMPLEX NUMBERS

The real numbers, we have noted, are closed under the operations of addition, subtraction, multiplication, division (except by zero), raising to a power, and extracting most roots. The exception to the latter is extracting even roots of negative numbers. In this case, closure is possible only within

a larger set of numbers—a set that includes a number called the **imaginary unit**, which is symbolized by i and defined to have the property, $i^2 = -1$.†
Thus:

$$\sqrt{-1} = \sqrt{i^2} = i$$

$$\sqrt{-9} = \sqrt{(9)(-1)} = \sqrt{9i^2} = \sqrt{9} \cdot \sqrt{i^2} = 3 \cdot i = 3i$$

$$\sqrt{-5} = \sqrt{(5)(-1)} = \sqrt{5i^2} = \sqrt{5} \cdot \sqrt{i^2} = i\sqrt{5}$$

Numbers such as these, of the form bi, where b denotes a real number and i the imaginary unit, are called **pure imaginary numbers**. They are part of a larger set of numbers called the **imaginary numbers**, which are sums and differences of real and pure imaginary numbers. For example:

$$3 + 2i, \quad 7 - i, \quad 21 + i\sqrt{3}, \quad \sqrt{5} + i, \quad \sqrt{7} - i\sqrt{3}$$

Together, the real and the imaginary numbers make up a larger set called the **complex numbers**. Within the set of complex numbers, there is closure under all of the operations of the algebra of polynomials: addition, subtraction, multiplication, division (except by zero), raising to a power, and extracting roots.

Standard numerals for complex numbers are expressed in binomial form, $a + bi$, in which a and b denote real numbers, and i is the imaginary unit. The first term, a, is referred to as the *real part* of the complex number, and the second term, bi, is referred to as the *imaginary part*. Note that real numbers are, in effect, complex numbers in which the coefficient of the imaginary part, b, is 0: $a + 0i = a$; and pure imaginary numbers are, in effect, complex numbers in which the real part, a, is 0: $0 + bi = bi$.

This complete number system for the algebra of polynomials may be diagrammed as follows:

$$\text{complex numbers } a + bi \begin{cases} \text{imaginary numbers } b \neq 0 \begin{cases} \text{pure } a = 0 \\ \text{nonpure } a \neq 0 \end{cases} \\ \text{real numbers } b = 0 \begin{cases} \text{rational numbers} \begin{cases} \text{integers} \\ \text{nonintegers} \end{cases} \\ \text{irrational numbers} \end{cases} \end{cases}$$

The number whose numeral is i has an important cyclic property:

$$i = i$$

$$i^2 = -1$$

$$i^3 = i^2 \times i = (-1)(i) = -i$$

$$i^4 = i^2 \times i^2 = (-1)(-1) = +1$$

$$i^5 = i^4 \times i = (+1)(i) = i$$

$$i^6 = i^4 \times i^2 = (+1)(-1) = -1$$

---

†The term "imaginary" reflects the frustration experienced by early mathematicians in trying to explain the square root of negative numbers; unfortunately, the term has endured.

$$i^7 = i^4 \times i^3 = (+1)(i^3) = .i^3 = -i$$
$$i^8 = i^4 \times i^4 = (+1)(+1) = +1$$
$$i^9 = i^8 \times i = (+1)(i) = i$$

etc.

In other words:

$$i^1 = i^5 = i^9 = i^{13} = \cdots = i$$
$$i^2 = i^6 = i^{10} = i^{14} = \cdots = -1$$
$$i^3 = i^7 = i^{11} = \cdots = -i$$
$$i^4 = i^8 = i^{12} = \cdots = +1$$

Thus, every integral power of i can be expressed equivalently by either i, -1, $-i$, or 1. For example:

$$i^{15} = i^4 \cdot i^4 \cdot i^4 \cdot i^4 \cdot i^3 = (1)(1)(1)(1)(-i) = -i$$
$$i^{26} = (i^4)^6 \cdot (i^2) = (1)^6(-1) = -1$$
$$i^{100} = (i^4)^{25} = (1)^{25} = 1$$

13. COMPUTING WITH COMPLEX NUMBERS

As noted in the previous section, the complex numbers are closed under the binary operations of addition, subtraction, multiplication, and division (except by zero), and under the unary operations of raising to a power and extracting a root. Also, the complex numbers are commutative and associative under addition and multiplication, and distributive for multiplication over addition.

A complex number, $a + bi$, may be treated under the various operations as though it were an algebraic binomial. The equivalencies among the various powers of i, noted in the previous section, simplify matters of expression and computation.

7.13

To add, subtract, or multiply complex numbers:

Step 1. Replace the i-expression in each term by its equivalent, i, -1, $-i$ or $+1$.

Step 2. Treat i as though it were an algebraic variable and compute the desired sum, difference, or product.

Step 3. Express the result of Step 2 in the form $a + bi$.

EXAMPLES

1. Add: $(6 + 3i^3) + (2 - i) + (4 - 3i^7)$.

Step 1. $6 + 3i^3 = 6 - 3i$

$$2 - i = 2 - i$$

$$4 - 3i^7 = 4 + 3i$$

Steps 2 and 3.

$$
\begin{aligned}
(6 + 3i^3) + (2 - i) + (4 - 3i^7) &= (6 - 3i) + (2 - i) + (4 + 3i) \\
&= 6 - 3i + 2 - i + 4 + 3i \\
&= (6 + 2 + 4) + (-3i - i + 3i) \\
&= 12 - i
\end{aligned}
$$

2. Subtract: $(4 - 8i^3) - (6 - 2i^{12})$.

Step 1. $4 - 8i^3 = 4 + 8i$

$$6 - 2i^{12} = 6 - (2)(+1) = 4$$

Steps 2 and 3.

$$
\begin{aligned}
(4 - 8i^3) - (6 - 2i^{12}) &= (4 + 8i) - 4 \\
&= (4 - 4) + 8i \\
&= 0 + 8i
\end{aligned}
$$

3. Multiply: $(4 - 3i^3)(6 + 7i^7)$.

Step 1. $4 - 3i^3 = 4 + 3i$

$$6 + 7i^7 = 6 - 7i$$

Step 2. $(4 - 3i^3)(6 + 7i^7) = (4 + 3i)(6 - 7i)$

$$= 24 - 10i - 21i^2$$

Step 3. $24 - 10i - 21i^2 = 24 - 10i + 21 = 45 - 10i$

4. Multiply: $(-6 - 2i^2)(3 - 5i^3)(1 + 3i^{13})$

Step 1. $-6 - 2i^2 = -6 + 2 = -4$

$$3 - 5i^3 = 3 + 5i$$

$$1 + 3i^{13} = 1 + 3i$$

Step 2. $(-4)(3 + 5i)(1 + 3i) = (-4)(3 + 14i + 15i^2)$

$$= -12 - 56i - 60i^2$$

Step 3. $-12 - 56i - 60i^2 = -12 - 56i + 60$

$$= 48 - 56i$$

The **conjugate of a complex number**, $a + bi$, is the number $a - bi$. That is, two complex numbers are conjugates of one another if they have the same real part, a, and if their imaginary parts, bi, are negatives of one another. For example, $3 - 2i$ and $3 + 2i$ are conjugate complex numbers; so are $5 + \frac{1}{2}i$ and $5 - \frac{1}{2}i$. Note, however, that $-3 + 5i$ and $3 - 5i$ are not conjugates, but the negatives of one another, since their sum is zero:

$$(-3 + 5i) + (3 - 5i) = -3 + 5i + 3 - 5i = 0$$

Computing the quotient of two complex numbers involves a procedure similar to Program **7.12** and the use of the conjugate of the denominator as multiplier.

7.14

To divide one complex number by another:

Step 1. Replace the i expression in each term by its equivalent, i, -1, $-i$, or $+1$.

Step 2. Express the terms of Step 1—dividend and divisor—as numerator and denominator of a fraction.

Step 3. Multiply numerator and denominator of the fraction of Step 2 by the conjugate of the denominator.

Step 4. Simplify and express the result of Step 3 in the form $a + bi$.

EXAMPLES

1. Divide $3 + i^5$ by $1 - i^3$.

Step 1. $3 + i^5 = 3 + i$

$1 - i^3 = 1 + i$

Step 2. $(3 + i^5) \div (1 - i^3) = (3 + i) \div (1 + i) = \dfrac{3 + i}{1 + i}$

Step 3. The conjugate of $(1 + i)$ is $(1 - i)$:

$$\frac{(3 + i) \cdot (1 - i)}{(1 + i) \cdot (1 - i)} = \frac{3 - 2i - i^2}{1 - i^2}$$

Step 4. $\qquad = \dfrac{3 - 2i - (-1)}{1 - (-1)} = \dfrac{4 - 2i}{2} = 2 - i$

2. Divide $3 - 2i$ by $4 - 3i^7$.

Step 1. $3 - 2i = 3 - 2i$

$4 - 3i^7 = 4 + 3i$

Step 2. $(3 - 2i) \div (4 - 3i^7) = \dfrac{3 - 2i}{4 + 3i}$

Step 3. The conjugate of $(4 + 3i)$ is $(4 - 3i)$:

$$\frac{(3 - 2i) \cdot (4 - 3i)}{(4 + 3i) \cdot (4 - 3i)} = \frac{12 - 17i + 6i^2}{16 - 9i^2}$$

Step 4. $\qquad = \dfrac{12 - 17i - 6}{16 + 9}$

$\qquad = \dfrac{6 - 17i}{25} = \dfrac{6}{25} - \dfrac{17}{25}i$

3. Divide: $(4 - 2i^6) \div (i^5 - 3)$.

$$4 - 2i^6 = 4 - (2)(-1) = 4 + 2 = 6$$

$$i^5 - 3 = -3 + i^5 = -3 + i$$

$$(4 - 2i^6) \div (i^5 - 3) = \frac{6}{-3 + i} = \frac{(6) \cdot (-3 - i)}{(-3 + i) \cdot (-3 - i)}$$

$$= \frac{-18 - 6i}{9 - i^2}$$

$$= \frac{-18 - 6i}{9 - (-1)} = \frac{-18 - 6i}{10} = -\frac{9}{5} - \frac{3}{5} i$$

EXERCISE 7-9

Replace each of the following by its equivalent: i, −1, −i, *or* +1:

1. i^4 **2.** i^6 **3.** i^{12} **4.** i^9

5. i^{13} **6.** i^{19} **7.** i^{72} **8.** i^{103}

9. i^{-2} **10.** i^{-3} **11.** i^{-5} **12.** i^{-8}

Simplify the following sums:

13. $2i + 3i^2 + 2i^3$ **14.** $i^4 - 3i + 2i^2$

15. $i^5 - 2i^2 - 3$ **16.** $3i^5 + 2i^6 + 3i^8 - 3$

17. $i\sqrt{3} + i - 2i\sqrt{4} + 3i^2 + i^{17}$ **18.** $i^5 - 3i^6 + 2i^7 + 8i^2$

19. $i^6 + i^{-2} + i^{-3} - 3i^{-5}$ **20.** $i^{-8} + i^{-17} + i^{-18} + i^{-23}$

Express in the form $a + bi$:

21. $(3 + i) + (2 - 3i)$ **22.** $(2 + 3i) - (4 + 6i)$

23. $(1 - 6i) + (3 + 2i) + (2 - i)$ **24.** $(4 - i\sqrt{3}) + (2 - i\sqrt{8})$

25. $(3 + 6i) + (5 - 3i) + (4 + 2i^3)$ **26.** $(4 + 3i - 2i^2) + (3 + 5i^2 - 4i^5)$

27. $\left(\frac{1}{2} + \frac{1}{\sqrt{3}} i\right) + \left(\frac{3}{4} - i\sqrt{3}\right) - \left(\frac{1}{8} - \frac{2}{3} i\sqrt{3}\right)$

28. $[(3 + i^2) + (4 - i^3)] + [4 - i^8]$

29. $(2 + 3i)(3 - 4i)$ **30.** $(3 - 5i)(2 + 3i)$

31. $(5 - 3i)(2 - 3i)$ **32.** $(6 - 3i^3)(2 - i^5)$

33. $(3 - 2i^5)(4 - 3i^4)$ **34.** $(2 - 3i)^2$

35. $(-3 - 2i)^3$ **36.** $(3 - 2i)(-4 - 3i)(2 + 2i)$

37. $(3 - 2i^3)(4 - 3i^2)(5 + i^5)$ **38.** $(3 - 2i^{-2})(3 + 4i^3)^2(-1 + 3i^{-8})$

39. $(3 + 2i) \div (4 - i)$ **40.** $(2 - 5i) \div (3 - 5i)$

41. $(6 - i^4) \div (2 - 3i^2)$ **42.** $(3 - 2i^3) \div (-2 - i^7)$

43. $(-3 + 2i^{-3}) \div (3 + 2i^{-4})$ **44.** $(2 + 3i)^2 \div (3 - 2i)$

45. $(2 + 3i - 6i^2) \div (3 - 2i - i^5)$ **46.** $(\sqrt{3} - i\sqrt{2}) \div (\sqrt{3} + 2i\sqrt{2})$

REVIEW—CHAPTER VII

Simplify; assume literal exponents represent positive integers:

1. $(-3)^4$

2. $-(5)^4$

3. $(-0.3)^3$

4. $-(3g)^2$

5. $\left(-\dfrac{2}{5}\right)^3$

6. $a^4 \cdot a^2 \cdot a$

7. $m^8 \div m^{10}$

8. $\dfrac{a^3 b^2 c^4}{a^2 b^5 c}$

9. $(-b)^{23}$

10. $(-a^6 b^4 c)^3$

11. $\left(\dfrac{6+3}{6}\right)^3$

12. $\left(\dfrac{x^7 y^2}{xy^3}\right)^4$

13. $\left(-\dfrac{3p}{q^4}\right)^5$

14. $-\left(\dfrac{2m}{y^2}\right)^2$

15. $a^{5n} \div a^4$

16. $p^{st} \cdot p^{s+2}$

17. $(a^2 b^m)^a$

18. $\left(-\dfrac{3a^3}{b^2}\right)^3 \div \left(\dfrac{2a}{b}\right)^2$

19. $\left[-\left(\dfrac{xy}{b^2}\right)\left(\dfrac{b^2 c}{xy}\right)^3\right]^2$

20. $\left[-\left(\dfrac{3xy}{2cd}\right)^3 \div \left(\dfrac{3x^2 y}{2cd^2}\right)^2\right]$

21. $(-3a^2 b^0)^0$

22. $3 \times 10^0 \times 2^2 \times 5^0$

23. $\left(\dfrac{5+4}{4}\right)^0$

24. $\dfrac{2x^0 y^2 z}{(3xy^0)^0}$

25. $\left(\dfrac{3xy}{2m}\right)^0 \div \left(\dfrac{xy}{m^2}\right)$

Express with all exponents negative:

26. $\dfrac{2}{p^3 q}$

27. $2^{-3} x^0 y^4 z^2$

28. $\frac{1}{3^2}$

29. $-a^2 b^3$

30. $\dfrac{3x^m}{s}$

31. $\dfrac{6xy^{-3} z^2}{a^{-2} b^{-3}}$

Simplify each to an expression involving only positive exponents:

32. $\left(\dfrac{2}{xy}\right)^{-3}$

33. $a^4 \cdot a^{-3} \cdot a^m$

34. $(-2x^0 y^{-3})^{-3}$

35. $\left(\dfrac{a^{-4} b^{-2}}{b^{-3} c^2}\right)^3$

36. $\left(\dfrac{ax^5}{m^2}\right)^{-2}\left(\dfrac{ab}{m^2 n}\right)^4$

37. $a^{-1} - y^{-1} + z^{-1}$

38. $3x^{-2} + y^{-2}$

39. $(s^{-2} t^{-2}) \div (st)^3$

40. $\left(\dfrac{m}{n} - \dfrac{n}{m}\right)^{-1}$

41. $(x^{-2} - y^{-2})^{-2}$

42. $\dfrac{mn^{-2}}{n^{-2} + m^2}$

Write the principal root for each of the following:

43. $\sqrt{10,000}$

44. $\sqrt[3]{-1000}$

45. $\sqrt{-0.04}$

46. $\sqrt[5]{243}$ **47.** $\sqrt[3]{-\frac{8}{27}}$ **48.** $\sqrt[4]{0.0016}$

Express each in exponential form:

49. $\sqrt[5]{-y^3}$ **50.** $\sqrt[7]{a^2b^{1/2}}$ **51.** $\sqrt[r]{s^2t^4}$

52. $\sqrt[k]{p^{4k}}$ **53.** $\sqrt[5]{\dfrac{a^p}{b^q}}$ **54.** $\sqrt[4]{m^3n^2s^{1/2}}$

Express each in radical form:

55. $(27a)^{1/3}$ **56.** $(2x^3)^0$ **57.** $(a^2b^{1/2})^{1/p}$

58. $\left(\dfrac{1}{p+q^{-1}}\right)^{-1/2}$ **59.** $\left(a - \dfrac{ab}{b}\right)^{-1/5}$ **60.** $(a^r m^{-b}s)^{-1/b}$

Simplify:

61. $a^{1/2} \div a^{2/3}$ **62.** $a^{0.4} \div a^{0.16}$

63. $\dfrac{a^{3/4}b^0}{a^{1/3}b^{1/2}}$ **64.** $\left(-\frac{32}{243}\right)^{1/5}$

65. $[(-3)^{-4}]^{-1/2}$ **66.** $\dfrac{2x^3y^2z^{1/2}}{(2xy)^3}$

67. $\sqrt[c]{[(a^b)^{1/c}]^a}$ **68.** $\left(\dfrac{x^{m+3}y^{m-3}}{x^{m-4}y^{m+2}}\right)^0$

69. $\left(\dfrac{a^{2/5}b^m}{a^{3/4}b^{2m-1}}\right)^{-3}$ **70.** $\sqrt[3]{\sqrt[4]{\sqrt{\sqrt[3]{a}}}}$

Reduce the order of each:

71. $\sqrt[4]{64a^6b^2}$ **72.** $\sqrt[6]{144x^4y^2}$

73. $\sqrt[2m]{36a^4b^2}$ **74.** $\sqrt[6]{\dfrac{16x^4y^2}{24a^4b^2}}$

Express each under the given radical:

75. $\sqrt{3} = \sqrt[6]{}$ **76.** $\sqrt[3]{3xy^2} = \sqrt[9]{}$

77. $\sqrt{0.04} = \sqrt[3]{}$ **78.** $\sqrt{4x^2y^4} = \sqrt[3]{}$

Write the coefficient before each of the radicals as a part of the radicand:

79. $2\sqrt[3]{5}$ **80.** $3a^2\sqrt[3]{2b}$

81. $3x^2y^2\sqrt[4]{xy^2}$ **82.** $(a+b)\sqrt[m]{a+b}$

Simplify:

83. $\sqrt[3]{-16}$ **84.** $x\sqrt[3]{3x^5y^4z^0}$

85. $\sqrt[3]{ax^4m^{-2}}$ **86.** $x\sqrt[6]{81x^8y^2}$

87. $\sqrt[m]{\dfrac{3^m x^{2m}}{4^{m-1}}}$ **88.** $m\sqrt[6]{(m-n)^0}$

89. $a + b\sqrt[3]{(a+b)^3}$ **90.** $x\sqrt[4]{(x+y)^6}$

Express each in simplest terms with rational denominators:

91. $\dfrac{3x}{\sqrt{p}}$ **92.** $\dfrac{3a}{\sqrt[4]{p^3}}$

93. $\sqrt[3]{8b^{-2}}$

94. $\sqrt{\dfrac{a^2 - 2ab + b^2}{a^2 - b^2}}$

95. $\dfrac{3}{\sqrt{2} - \sqrt{7}}$

96. $\dfrac{\sqrt{7}}{2 + \sqrt{3}}$

97. $\dfrac{\sqrt{x}}{1 - \sqrt{x}}$

98. $\dfrac{\sqrt{5} - \sqrt{3}}{2\sqrt{3} + \sqrt{5}}$

99. $\dfrac{\sqrt[3]{a}}{\sqrt{a} - \sqrt{b}}$

100. $\dfrac{\sqrt{76} - \sqrt{102}}{\sqrt{102} - \sqrt{76}}$

101. $\dfrac{3\sqrt{a} + 4\sqrt{b}}{2\sqrt{b} - \sqrt{a}}$

102. $\sqrt{\dfrac{a}{b} + 1}$

Simplify:

103. $3\sqrt{12} + \sqrt{3} - \sqrt{12} + \sqrt{48}$

104. $\sqrt{45} - \sqrt{60} + \sqrt{15} - \sqrt{135} + \sqrt{20}$

105. $\sqrt[3]{3a^3} + \sqrt[3]{48} - \sqrt[3]{24a^3} - a\sqrt[3]{-3}$

106. $\sqrt{2} - \sqrt[4]{324} + \sqrt[6]{512} + \sqrt[8]{16}$

107. $\sqrt{\tfrac{5}{27}} - \sqrt{\tfrac{12}{125}}$

108. $4\sqrt{5} - \dfrac{2}{\sqrt{5}} + \dfrac{7}{3\sqrt{5}} + \dfrac{\sqrt{5}}{2}$

109. $\sqrt{12x} - \sqrt{48x} - \sqrt{3x^3} + \sqrt{12x^3}$

110. $\sqrt{(a + b)^3} - \sqrt{a + b}$

111. $\dfrac{\sqrt{x^3y}}{3} - x\sqrt{\dfrac{xy}{25}} + \dfrac{\sqrt{4x^3y}}{5}$

112. $\sqrt[3]{0.54} - \sqrt[3]{0.16} + \sqrt[3]{-0.02}$

113. $\dfrac{3}{\sqrt{2}} + \dfrac{5\sqrt{2}}{2}$

114. $\dfrac{2}{\sqrt[n]{a}} + \dfrac{\sqrt[n]{a}}{a}$

Multiply, and simplify:

115. $\sqrt[4]{27x^3} \cdot \sqrt{3x}$

116. $\sqrt[4]{36}(\sqrt{3} - 2\sqrt[3]{2} + \sqrt{6})$

117. $(3m\sqrt[3]{4m^2n})(m\sqrt{n})(\sqrt[6]{mn^4})$

118. $(8 - \sqrt{3})(2 + \sqrt[3]{9})$

119. $(2\sqrt{10} - 3\sqrt{5})(\sqrt{10} + 4\sqrt{5})$

120. $(\sqrt{316} - \sqrt{214})(\sqrt{316} + \sqrt{214})$

121. $(3 - 2\sqrt{5})(\sqrt{3} - \sqrt{5})$

122. $(\sqrt{x + y} - \sqrt{y})^2$

123. $(\sqrt{a - 2} + \sqrt{a + 3})^2$

124. $(\sqrt[3]{9x^2y} - \sqrt{6xy})^2$

125. If $x = 2\sqrt{5} - 2$, then $x^2 - 4x + 7 = ?$

126. If $x = 3\sqrt{2} - \sqrt{3}$, then $2x^2 - 3x + 1 = ?$

Divide, and simplify:

127. $(4\sqrt{3} - 2\sqrt{6} + 3\sqrt{2}) \div \sqrt{18}$

128. $(3\sqrt{110} - 4\sqrt{15} + 2\sqrt{21} - 2\sqrt{11}) \div 2\sqrt{30}$

129. $(3\sqrt[3]{2} - 2\sqrt{3}) \div \sqrt[3]{3}$

130. $(\sqrt{3} - \sqrt{7}) \div (\sqrt{3} - 4\sqrt{7})$

131. $(3\sqrt{2} - 2\sqrt{3}) \div (\sqrt{2} + 3\sqrt{3})$

132. $(2\sqrt{3} - 3\sqrt{2}) \div (3 - \sqrt{2})$

133. $(4 - \sqrt{6}) \div (2\sqrt{3} + 3\sqrt{6})$

134. $(\sqrt{172} - \sqrt{326}) \div (\sqrt{326} - \sqrt{172})$

135. $(\sqrt{3x} - \sqrt{2y}) \div (\sqrt{2x} + \sqrt{3y})$

136. $\dfrac{\sqrt{3} - \sqrt{2}}{\sqrt{3} + 2\sqrt{2}} \div \dfrac{3\sqrt{3} - 2\sqrt{2}}{2\sqrt{3} - \sqrt{2}}$

Simplify:

137. $3i - 4i^2 + 3i^3 - 7$

138. $4i^4 - 3i^3 + 2i^7 + i^8$

139. $4i^{17} + 3i^{12} + 2i^{13} - 2i^{19}$

140. $i\sqrt{3} + 2i - 3\sqrt{3} + 4i^5$

141. $i^{-4} + i^{-3} + 2i^{-8} + i^{-6}$

142. $i^{-8} - i^{-18} + i^{-13} - 2i^{-9} + i^{-38}$

Express in the form $a + bi$:

143. $(1 - 3i) + (2 - 3i) - (4 + 2i)$

144. $(3 + 2i - 6i^2) + (3i - 2 + 4i^5)$

145. $\left(\dfrac{1}{3} + \dfrac{i}{\sqrt{2}}\right) + \left(\dfrac{5}{6} - i\sqrt{2}\right) - \left(\dfrac{2}{3} + \dfrac{3i}{\sqrt{2}}\right)$

146. $(3 - 2i)(4 - 3i)$

147. $(3 - 2i)(4 - 3i)(1 - i)$

148. $(3 - 2i + 5i^6 - 3i^7)^2$

149. $(4 - 3i) \div (-2 + 6i)$

150. $(4 - i^3) \div (2 - 3i^{-6} + i^{-5})$

151. $(\sqrt{3} - i) \div (2\sqrt{3} + 3i)$

152. $(3 - 2i + 4i^2 - 3i^3)^2 \div (2 - 3i + 5i^3)$

153. $\dfrac{(3 - 2i)}{(4 - 3i)} \div \dfrac{(2 - i)}{(2 + 3i)}$

VIII

QUADRATIC EQUATIONS AND INEQUALITIES

A polynomial equation in a single variable is one in which the variable carries only positive integral exponents. The **degree** of a polynomial equation is that of the variable's greatest exponent. When the greatest exponent present is 1, the degree is the *first*, and the equation is called **linear**. When the greatest exponent present is 2, the degree is the *second*, and the equation is called **quadratic**. When the greatest exponents are 3, 4, 5, the degrees are *third*, *fourth*, and *fifth*, respectively, and the equations are referred to as **cubic, quartic** (or **biquadratic**), and **quintic**, respectively.

Polynomial equations in a single variable are said to be in **standard form** when one member is 0 and the other has its terms arranged in descending powers of the variable. The following are examples of polynomial equations in a single variable, in standard form:

1. POLYNOMIAL EQUATIONS

$$x - 2 = 0 \quad \text{(linear)}$$

$$\left.\begin{array}{r} x^2 - 3x + 5 = 0 \\ 7x^2 - 4 = 0 \end{array}\right\} \quad \text{(quadratic)}$$

$$x^3 - 4x - 1 = 0 \quad \text{(cubic)}$$

$$4x^4 - 6x^2 - 2x = 0 \quad \text{(quartic, or biquadratic)}$$

$$x^5 - 3x^4 + 2x^3 - 7x^2 + 4x + 3 = 0 \quad \text{(quintic)}$$

2. GENERAL QUADRATIC EQUATION IN A SINGLE VARIABLE

In the next several sections, we shall be concerned with solving second-degree or quadratic equations in a single variable. Unless otherwise indicated, the replacement set, or domain of the variable, is assumed to be the set of complex numbers. The *general quadratic equation in standard form* is given as

$$ax^2 + bx + c = 0$$

where x is the variable and a, b, c denote real numbers, $a \neq 0$.

Recall that an equation is an "open" sentence which is neither true nor false as it stands. The variable in an equation is a placeholder for numbers. When specific replacements are made for the variable, the sentence may be judged true or false. Those numbers that make the sentence true (i.e., satisfy the equation) are called the solutions of the equation. Provided the domain of the variable is sufficiently extensive, linear equations normally have single solutions, and quadratic equations have two solutions. In general, a polynomial equation in a single variable may have as many solutions (collectively, its **solution set**) as its degree. Thus, a third-degree or cubic equation may have three solutions, a quartic equation may have four, and so on. Duplications may appear among the solutions, however. For example, solving the quadratic equation $x^2 + 2x + 1 = 0$ by Program **8.1** (next section) results in two equal solutions, -1 and -1, which means there is actually but one number in the solution set: $\{-1\}$.

3. SOLVING QUADRATIC EQUATIONS BY FACTORING

Eventually we will develop a general procedure for solving any quadratic equation (Section 5 following), but first we consider a special factoring method that is much faster, when applicable. The left number of the quadratic equation in standard form, $ax^2 + bx + c = 0$, suggests the trinomial product of a pair of binomials (see Section 6 of Chapter V). When this trinomial can be expressed as the product of two binomial factors, we may solve the equation directly. The mathematical basis for the procedure is this: If the product of two factors is zero, then one or both of the factors must be zero; and conversely. That is, if $M \times N = 0$, then either $M = 0$, or $N = 0$, or both $M = 0$ and $N = 0$.

Consequently, if we can factor the nonzero member of a quadratic equation in standard form, we can determine which numbers, as variable replacements, make each factor equal to zero, and thereby establish the solution set for the original equation. For instance, if the quadratic equation is $x^2 - 7x + 10 = 0$, the left member is a trinomial which may be factored as the product of a pair of binomials:

$$x^2 - 7x + 10 = (x - 2)(x - 5)$$

By inspection we can see that, when either 2 or 5 replaces x, the product of $(x - 2)(x - 5)$ must be zero, because in each case one of the binomial factors is equal to zero:

$$(x - 2)(x - 5)$$
$$x = 2: \quad (2 - 2)(2 - 5) = (0)(-3) = 0$$
$$x = 5: \quad (5 - 2)(5 - 5) = (3)(0) \quad = 0$$

If 2 and 5 are the replacements for x that make the product $(x - 2)(x - 5)$ equal zero, then they are also the variable replacements that make the trinomial equivalent, $x^2 - 7x + 10$, equal zero. So, 2 and 5 must be the solutions (solution set) of the quadratic equation $x^2 - 7x + 10 = 0$.

8.1

To solve a quadratic equation by factoring:

Step 1. Express the quadratic equation equivalently in standard form.

Step 2. Factor the nonzero member.

Step 3. Set each factor of Step 2 equal to zero.

Step 4. Solve the linear equations of Step 3 separately.

Step 5. Verify by replacing the variable in the given equation with the solutions found in Step 4; those that satisfy the given equation are the solutions of that equation.

EXAMPLES

1. Solve: $x^2 = 5x - 6$.

Step 1. $\quad x^2 = 5x - 6$

$\qquad\quad x^2 - 5x + 6 = 0$

Step 2. $\quad (x - 3)(x - 2) = 0$

Step 3. $\quad (x - 3) = 0 \qquad (x - 2) = 0$

Step 4. $\qquad x - 3 = 0 \qquad\quad x - 2 = 0$

$\qquad\qquad\quad x = 3 \qquad\qquad\quad x = 2$

Step 5.

	$x^2 = 5x - 6$		$x^2 = 5x - 6$
$x = 3:$	$(3)^2 \overset{?}{=} 5(3) - 6$	$x = 2:$	$(2)^2 \overset{?}{=} 5(2) - 6$
	$9 \overset{?}{=} 15 - 6$		$4 \overset{?}{=} 10 - 6$
	$9 = 9$		$4 = 4$

Both 3 and 2 satisfy the given equation, $x^2 = 5x - 6$; so they are the solutions of that equation.

2. Solve: $x^2 + x = 20$.

Step 1. $x^2 + x = 20$

$x^2 + x - 20 = 0$

Step 2. $(x - 4)(x + 5) = 0$

Step 3. $(x - 4) = 0$ $(x + 5) = 0$

Step 4. $x - 4 = 0$ $x + 5 = 0$

$x = 4$ $x = -5$

Step 5.

$$x^2 + x = 20 \qquad\qquad x^2 + x = 20$$

$$x = 4: \quad (4)^2 + (4) \overset{?}{=} 20 \quad x = -5: \quad (-5)^2 + (-5) \overset{?}{=} 20$$

$$16 + (4) \overset{?}{=} 20 \qquad\qquad 25 - 5 \overset{?}{=} 20$$

$$20 = 20 \qquad\qquad 20 = 20$$

Solution set for $x^2 + x = 20$ is $\{4, -5\}$.

3. Solve: $5x^2 = 3x$.

Step 1. $5x^2 = 3x$

$5x^2 - 3x = 0$

Step 2. $(x)(5x - 3) = 0$

Step 3. $(x) = 0$ $(5x - 3) = 0$

Step 4. $x = 0$ $5x - 3 = 0$

$5x = 3$

$x = \frac{3}{5}$

Step 5.

$$5x^2 = 3x \qquad\qquad 5x^2 = 3x$$

$$x = 0: \quad 5(0)^2 \overset{?}{=} 3(0) \qquad x = \tfrac{3}{5}: \quad 5(\tfrac{3}{5})^2 \overset{?}{=} 3(\tfrac{3}{5})$$

$$0 = 0 \qquad\qquad 5(\tfrac{9}{25}) \overset{?}{=} 3(\tfrac{3}{5})$$

$$\tfrac{9}{5} = \tfrac{9}{5}$$

Solution set for $5x^2 = 3x$ is $\{0, \frac{3}{5}\}$

4. Solve: $x^2 = 16$.

Step 1. $x^2 = 16$

$x^2 - 16 = 0$

Step 2. $(x - 4)(x + 4) = 0$

Step 3. $(x - 4) = 0$ $(x + 4) = 0$

Step 4. $x - 4 = 0$ $x + 4 = 0$

$x = 4$ $x = -4$

Step 5.

$$x^2 = 16 \qquad\qquad\qquad x^2 = 16$$

$$x = 4: \quad (4)^2 \overset{?}{=} 16 \qquad x = -4: \quad (-4)^2 \overset{?}{=} 16$$

$$16 = 16 \qquad\qquad\qquad 16 = 16$$

Thus, $\{x \mid x^2 = 16\} = \{4, -4\}$

5. Solve: $\dfrac{1}{x+3} + \dfrac{1}{x-2} = \dfrac{1}{6}$.

$$\frac{1}{x+3} + \frac{1}{x-2} = \frac{1}{6}$$

$$LCD = (6)(x+3)(x-2)$$

$$\frac{(6)(x-2)}{LCD} + \frac{(6)(x+3)}{LCD} = \frac{(x+3)(x-2)}{LCD}$$

$$(6)(x-2) + (6)(x+3) = (x+3)(x-2)$$

$$6x - 12 + 6x + 18 = x^2 + x - 6$$

$$-x^2 + 11x + 12 = 0$$

$$x^2 - 11x - 12 = 0$$

$$(x-12)(x+1) = 0$$

$$(x-12) = 0 \qquad\qquad (x+1) = 0$$

$$x - 12 = 0 \qquad\qquad x + 1 = 0$$

$$x = 12 \qquad\qquad\qquad x = -1$$

$$\frac{1}{x+3} + \frac{1}{x-2} = \frac{1}{6} \qquad\qquad\qquad \frac{1}{x+3} + \frac{1}{x-2} = \frac{1}{6}$$

$$x = 12: \quad \frac{1}{12+3} + \frac{1}{12-2} \overset{?}{=} \frac{1}{6} \qquad x = -1: \quad \frac{1}{-1+3} + \frac{1}{-1-2} \overset{?}{=} \frac{1}{6}$$

$$\frac{1}{15} + \frac{1}{10} \overset{?}{=} \frac{1}{6} \qquad\qquad\qquad \frac{1}{2} + \frac{1}{-3} \overset{?}{=} \frac{1}{6}$$

$$\frac{2}{30} + \frac{3}{30} \overset{?}{=} \frac{1}{6} \qquad\qquad\qquad \frac{1}{2} - \frac{1}{3} \overset{?}{=} \frac{1}{6}$$

$$\frac{5}{30} \overset{?}{=} \frac{1}{6} \qquad\qquad\qquad\qquad \frac{3}{6} - \frac{2}{6} \overset{?}{=} \frac{1}{6}$$

$$\frac{1}{6} = \frac{1}{6} \qquad\qquad\qquad\qquad \frac{1}{6} = \frac{1}{6}$$

Thus, $\left\{ x \,\middle|\, \dfrac{1}{x+3} + \dfrac{1}{x-2} = \dfrac{1}{6} \right\} = \{12, -1\}.$

EXERCISE 8-1

Solve for x and verify:

1. $x^2 - 6x + 8 = 0$

2. $x^2 + 2x - 15 = 0$

3. $x^2 + 10x + 24 = 0$

4. $x^2 - x = 6$

5. $x^2 = 6x + 7$ **6.** $6x^2 - 5x + 1 = 0$

7. $3x^2 + 22x + 35 = 0$ **8.** $9x^2 + 1 = 6x$

9. $15x^2 = 6 - x$ **10.** $x^2 - 4ax + 4a^2 = 0$

11. $x^2 - 2ax - 15a^2 = 0$ **12.** $25x^2 + 4 = 20x$

13. $6x^2 - 5ax + a^2 = 0$ **14.** $a^2x^2 - ax - 6 = 0$

15. $3a^2x^2 + 7ax = 6$ **16.** $4x^2 - 9x = 0$

17. $2x^2 - 8ax = 0$ **18.** $\dfrac{5}{x+4} = \dfrac{3}{x-2} + 4$

19. $\dfrac{x}{3-x} - \dfrac{2}{x-3} = 0$

20. $(x + 1)(3x + 2) = (3x - 2)(5x - 4)$

21. $\dfrac{x^2}{5} - \dfrac{x}{30} = \dfrac{1}{2}$

22. A tennis court covers 312 sq yds and is 14 yds longer than it is wide. What are the length and width dimensions of the tennis court?

23. The sum of the reciprocals of two consecutive even integers is $\frac{5}{12}$. Find the integers.

24. A contractor estimates that, to cover a certain floor with square asphalt tiles, it will take 528 standard tiles or 297 larger tiles (which are 3 in. longer and wider than the standard tile). What are the dimensions of the larger tile?

25. Within a rectangular garden plot 32 ft by 46 ft, there is to be a uniform border around its outer edge for flowers and a rectangular center grass plot of 1040 sq ft. Find the width of the flower border.

4. PURE QUADRATIC EQUATIONS

Quadratic equations that have zero as numerical coefficient(s) for the linear term(s) can be expressed equivalently in the form $ax^2 + c = 0$, $(a \neq 0)$. Equations of this type are called *pure quadratic equations*. One procedure for solving pure quadratic equations is very similar to that for solving linear equations (Program **3.1**):

$$ax^2 + c = 0$$

$$ax^2 = -c$$

$$x^2 = -\frac{c}{a}$$

The last equation, $x^2 = -c/a$, may be interpreted as "the square of x is $-c/a$." Therefore:

$$x = \pm\sqrt{-\frac{c}{a}}$$

which is a condensed way of stating:

$$x = +\sqrt{-\frac{c}{a}} \quad \text{or} \quad x = -\sqrt{-\frac{c}{a}}$$

The symbol $\pm$ is read "plus or minus."

[*Note:* In solving equations, we are interested in all roots, not just the principal root.]

8.2

To solve a pure quadratic equation:

Step 1. Write an equation equivalent to the given equation so that all terms containing the variable are in one member and all other terms are in the other member; then simplify.

Step 2. Divide both members of the equation of Step 1 by the coefficient of the variable term.

Step 3. Extract the square roots of the nonvariable member of the resulting equation of Step 3 as possible solutions.

Step 4. Verify by replacing the variable in the given equation with the solutions obtained in Step 3; those that satisfy the given equation are the solutions of that equation.

EXAMPLES

1. Solve: $4x^2 - 12 = x^2$.

Step 1. $\quad 4x^2 - 12 = x^2$

$$4x^2 - x^2 = 12$$

$$3x^2 = 12$$

Step 2. $\quad \dfrac{3x^2}{3} = \dfrac{12}{3}$

$$x^2 = 4$$

Step 3. $\quad x = \pm\sqrt{4} = 2, -2 \quad \text{(possible solutions)}$

Step 4.

$4x^2 - 12 = x^2$	$4x^2 - 12 = x^2$
$x = 2: \quad 4(2)^2 - 12 \overset{?}{=} (2)^2$	$x = -2: \quad 4(-2)^2 - 12 \overset{?}{=} (-2)^2$
$4(4) - 12 \overset{?}{=} 4$	$4(4) - 12 \overset{?}{=} 4$
$16 - 12 \overset{?}{=} 4$	$16 - 12 \overset{?}{=} 4$
$4 = 4$	$4 = 4$

The solution set for $4x^2 - 12 = x^2$ is $\{2, -2\}$.

2. Solve: $12x^2 - 5 = 3x^2 + 2$.

Step 1. $12x^2 - 5 = 3x^2 + 2$

$$12x^2 - 3x^2 = 2 + 5$$

$$9x^2 = 7$$

Step 2. $$\frac{9x^2}{9} = \frac{7}{9}$$

$$x^2 = \frac{7}{9}$$

Step 3. $x = \pm\sqrt{\dfrac{7}{9}} = \dfrac{\sqrt{7}}{3}, \ -\dfrac{\sqrt{7}}{3}$ (possible solutions)

Step 4.

$12x^2 - 5 = 3x^2 + 2$ $12x^2 - 5 = 3x^2 + 2$

$x = \dfrac{\sqrt{7}}{3}$: $x = -\dfrac{\sqrt{7}}{3}$:

$12\left(\dfrac{\sqrt{7}}{3}\right)^2 - 5 \overset{?}{=} 3\left(\dfrac{\sqrt{7}}{3}\right)^2 + 2$ $12\left(-\dfrac{\sqrt{7}}{3}\right)^2 - 5 \overset{?}{=} 3\left(-\dfrac{\sqrt{7}}{3}\right)^2 + 2$

$12(\tfrac{7}{9}) - 5 \overset{?}{=} 3(\tfrac{7}{9}) + 2$ $12(+\tfrac{7}{9}) - 5 \overset{?}{=} 3(+\tfrac{7}{9}) + 2$

$\tfrac{28}{3} - 5 \overset{?}{=} \tfrac{7}{3} + 2$ $\tfrac{28}{3} - 5 \overset{?}{=} \tfrac{7}{3} + 2$

$9\tfrac{1}{3} - 5 \overset{?}{=} 2\tfrac{1}{3} + 2$ $9\tfrac{1}{3} - 5 \overset{?}{=} 2\tfrac{1}{3} + 2$

$4\tfrac{1}{3} = 4\tfrac{1}{3}$ $4\tfrac{1}{3} = 4\tfrac{1}{3}$

Solutions of $12x^2 - 5 = 3x^2 + 2$ are

$$\pm\frac{\sqrt{7}}{3}, \quad \text{or} \quad \left\{\frac{\sqrt{7}}{3}, \ -\frac{\sqrt{7}}{3}\right\}$$

3. Solve: $3x^2 + 21 = 0$.

$$3x^2 + 21 = 0$$

$$3x^2 = -21$$

$$x^2 = -7$$

$$x = \pm i\sqrt{7} = i\sqrt{7}, \ -i\sqrt{7} \quad \text{(possible solutions)}$$

$3x^2 + 21 = 0$ $3x^2 + 21 = 0$

$x = i\sqrt{7}:\ 3(i\sqrt{7})^2 + 21 \overset{?}{=} 0$ $x = -i\sqrt{7}:\ 3(-i\sqrt{7})^2 + 21 \overset{?}{=} 0$

$3(i^2)(\sqrt{7})^2 + 21 \overset{?}{=} 0$ $3(-i)^2(\sqrt{7})^2 + 21 \overset{?}{=} 0$

$3(-1)(7) + 21 \overset{?}{=} 0$ $3(i^2)(7) + 21 \overset{?}{=} 0$

$-21 + 21 \overset{?}{=} 0$ $3(-1)(7) + 21 \overset{?}{=} 0$

$0 = 0$ $-21 + 21 \overset{?}{=} 0$

$0 = 0$

Thus, $\{x \mid 3x^2 + 21 = 0\} = \{i\sqrt{7}, \ -i\sqrt{7}\}$

EXERCISE 8-2

Solve for x and verify:

1. $9x^2 = 81$

2. $3x^2 + 8 = 32 - x^2$

3. $4x = \dfrac{36}{x}$

4. $6x^2 - 5 = 2x^2 + 20$

5. $\dfrac{x}{3} + \dfrac{3}{x} = 4x$

6. $\dfrac{3x^2}{2} - 2 = \dfrac{x^2}{3}$

7. $ax^2 + bx^2 = z$

8. $\dfrac{5x}{3} - \dfrac{3}{4x} - \dfrac{x}{2} = 0$

9. $\dfrac{3}{x^2 - 2} = \dfrac{4}{x^2 - 1}$

10. $(x - 2)^2 - 16 = -4x$

11. $\dfrac{3b^2}{2} - x^2 = \dfrac{2x^2}{3}$

12. $\dfrac{1}{x^2} + \dfrac{1}{a^2} = \dfrac{1}{c^2}$

13. $x^2 + 3 = 2$

14. $3x^2 + 4 = x^2 - 2$

15. $5x^2 + 16 = x^2$

16. $4x^2 + 49 = 0$

17. $5x + \dfrac{25}{x} = 0$

18. $\dfrac{x}{3} - \dfrac{4}{x} = x$

19. The product of a positive number subtracted from 12, times that number added to 12, is 128. Find the number.

20. Find the dimensions of a square if increasing its edge by 2 yields an area numerically equal to its new perimeter.

5. SOLVING QUADRATIC EQUATIONS BY FORMULA

Except for the special cases discussed in the two preceding sections, most quadratic equations are best solved by the quadratic formula. This is a general procedure, which means that it applies to *any* quadratic equation. To derive the quadratic formula, a technique known as *completing the square* is important.

8.3

To complete the square of an algebraic expression of the form $x^2 + bx$:

Step 1. Halve the coefficient (b) of the linear or first-degree term and square it.

Step 2. Add the result of Step 1 to the original expression to form a perfect square trinomial.

EXAMPLES

1. Complete the square for $x^2 + 10x$.

Step 1. $\frac{1}{2} \times 10 = 5; 5^2 = 25$

Step 2. $x^2 + 10x + 25 = (x + 5)^2$

2. Complete the square for $x^2 - 6x$.

Step 1. $\frac{1}{2} \times (-6) = -3; (-3)^2 = 9$

Step 2. $x^2 - 6x + 9 = (x - 3)^2$

3. Complete the square for $x^2 + 3x$.

Step 1. $\frac{1}{2} \times 3 = \frac{3}{2}; (\frac{3}{2})^2 = \frac{9}{4}$

Step 2. $x^2 + 3x + \frac{9}{4} = (x + \frac{3}{2})^2$

4. Complete the square for $x^2 - \frac{3}{5}x$.

Step 1. $\frac{1}{2} \times (-\frac{3}{5}) = -\frac{3}{10}; (-\frac{3}{10})^2 = \frac{9}{100}$

Step 2. $x^2 - \frac{3}{5}x + \frac{9}{100} = (x - \frac{3}{10})^2$

We may derive the quadratic formula as follows:

1. Produce an equation equivalent to the general quadratic equation, $ax^2 + bx + c = 0$, in which all terms involving the variable are alone in one member:

$$ax^2 + bx = -c$$

2. Divide each term by the coefficient of the x^2 (quadratic) term:

$$x^2 + \frac{b}{a}x = -\frac{c}{a}$$

3. Add to both members a number that will complete the square of the member involving the variable, and simplify. By **8.3**, that would be $(b/2a)^2$:

$$x^2 + \frac{b}{a}x + \left(\frac{b}{2a}\right)^2 = -\frac{c}{a} + \left(\frac{b}{2a}\right)^2$$

$$x^2 + \frac{b}{a}x + \frac{b^2}{4a^2} = -\frac{4ac}{4a^2} + \frac{b^2}{4a^2}$$

$$\left(x + \frac{b}{2a}\right)^2 = \frac{b^2 - 4ac}{4a^2}$$

4. Reduce the order of both members by means of square root:

$$x + \frac{b}{2a} = \pm\sqrt{\frac{b^2 - 4ac}{4a^2}}$$

$$= \pm\frac{\sqrt{b^2 - 4ac}}{2a}$$

and solve for x:

$$x = \frac{-b + \sqrt{b^2 - 4ac}}{2a} \qquad x = \frac{-b - \sqrt{b^2 - 4ac}}{2a}$$

These two solutions of the general quadratic are frequently combined into a single expression:

$$x = \frac{-b \pm \sqrt{b^2 - 4ac}}{2a}$$

called the **quadratic formula**. Upon replacement of a, b, and c by the corresponding numerical coefficients of a quadratic equation, the right member of the formula yields two numbers which are the possible solutions of the equation.

8.4

To solve a quadratic equation by the quadratic formula:

Step 1. Express the quadratic equation equivalently in the form $ax^2 + bx + c = 0$.

Step 2. Identify the numerical coefficients that correspond to a, b, and c in the equation of Step 1.

Step 3. Substitute the corresponding coefficients for a, b, and c in the quadratic formula:

$$x = \frac{-b \pm \sqrt{b^2 - 4ac}}{2a}$$

to obtain possible solutions.

Step 4. Verify by replacing the variable in the given equation with the solutions obtained in Step 3; those that satisfy the given equation are the solutions of that equation.

EXAMPLES

1. Solve by the quadratic formula: $3x^2 - 10x + 8 = 0$.

Step 1. $(ax^2 + bx + c = 0)$

$3x^2 - 10x + 8 = 0$

Step 2. $a = 3, b = -10, c = 8$

Step 3. $x = \dfrac{-b \pm \sqrt{b^2 - 4ac}}{2a} = \dfrac{-(-10) \pm \sqrt{(-10)^2 - 4(3)(8)}}{2(3)}$

$= \dfrac{10 \pm \sqrt{100 - 96}}{6} = \dfrac{10 \pm \sqrt{4}}{6} = \dfrac{10 \pm 2}{6}$

$= \dfrac{12}{6}, \dfrac{8}{6} = 2, \dfrac{4}{3}$

Step 4. $3x^2 - 10x + 8 = 0$ $3x^2 - 10x + 8 = 0$

$x = 2:\ 3(2)^2 - 10(2) + 8 \overset{?}{=} 0$ $x = \dfrac{4}{3}:\ 3\left(\dfrac{4}{3}\right)^2 - 10\left(\dfrac{4}{3}\right) + 8 \overset{?}{=} 0$

$12 - 20 + 8 \overset{?}{=} 0$ $\dfrac{16}{3} - \dfrac{40}{3} + \dfrac{24}{3} \overset{?}{=} 0$

$0 = 0$ $0 = 0$

Solution set for $3x^2 - 10x + 8 = 0$: $\{2, \frac{4}{3}\}$.

2. Solve: $2x^2 + x = 1$.

Step 1. $2x^2 + x = 1$

$2x^2 + x - 1 = 0$

Step 2. $a = 2, b = 1, c = -1$

Step 3. $x = \dfrac{-b \pm \sqrt{b^2 - 4ac}}{2a} = \dfrac{-1 \pm \sqrt{(1)^2 - 4(2)(-1)}}{2(2)}$

$= \dfrac{-1 \pm \sqrt{1 + 8}}{4} = \dfrac{-1 \pm \sqrt{9}}{4} = \dfrac{-1 \pm 3}{4}$

$= \dfrac{2}{4}, \dfrac{-4}{4} = \dfrac{1}{2}, -1.$

Step 4. $2x^2 + x = 1$ $2x^2 + x = 1$

$x = \frac{1}{2}:\ 2(\frac{1}{2})^2 + (\frac{1}{2}) \overset{?}{=} 1$ $x = -1:\ 2(-1)^2 + (-1) \overset{?}{=} 1$

$2(\frac{1}{4}) + \frac{1}{2} \overset{?}{=} 1$ $2 - 1 \overset{?}{=} 1$

$\frac{1}{2} + \frac{1}{2} \overset{?}{=} 1$ $1 = 1$

$1 = 1$

$\{x \mid 2x^2 + x = 1\} = \{\frac{1}{2}, -1\}$

3. Solve: $x^2 - 3 = -4x$.

Step 1. $x^2 + 4x - 3 = 0$

Step 2. $a = 1, b = 4, c = -3$

Step 3. $x = \dfrac{-b \pm \sqrt{b^2 - 4ac}}{2a} = \dfrac{-4 \pm \sqrt{16 - 4(1)(-3)}}{2(1)}$

$= \dfrac{-4 \pm \sqrt{28}}{2} = \dfrac{-4 \pm 2\sqrt{7}}{2} = -2 \pm \sqrt{7}$

Step 4. $x^2 - 3 = -4x$ $x^2 - 3 = -4x$

$x = -2 + \sqrt{7}:$ $x = -2 - \sqrt{7}:$

$(-2 + \sqrt{7})^2 - 3 \overset{?}{=} -4(-2 + \sqrt{7})$ $(-2 - \sqrt{7})^2 - 3 \overset{?}{=} -4(-2 - \sqrt{7})$

$4 - 4\sqrt{7} + 7 - 3 \overset{?}{=} 8 - 4\sqrt{7}$ $4 + 4\sqrt{7} + 7 - 3 \overset{?}{=} 8 + 4\sqrt{7}$

$8 - 4\sqrt{7} = 8 - 4\sqrt{7}$ $8 + 4\sqrt{7} = 8 + 4\sqrt{7}$

$\{x \mid x^2 - 3 = -4x\} = \{-2 + \sqrt{7}, -2 - \sqrt{7}\}$

4. Solve: $x^2 + 3 = 2x$.

Step 1. $x^2 + 3 = 2x$

$x^2 - 2x + 3 = 0$

Step 2. $a = 1, b = -2, c = 3$

Step 3. $x = \dfrac{-b \pm \sqrt{b^2 - 4ac}}{2a}$

$= \dfrac{-(-2) \pm \sqrt{(-2)^2 - 4(1)(3)}}{2(1)}$

$= \dfrac{2 \pm \sqrt{4 - 12}}{2}$

$= \dfrac{2 \pm \sqrt{-8}}{2} = \dfrac{2 \pm 2i\sqrt{2}}{2} = 1 \pm i\sqrt{2}$

Step 4. $x^2 + 3 = 2x$

$x = 1 \pm i\sqrt{2}:\quad (1 \pm i\sqrt{2})^2 + 3 \overset{?}{=} 2(1 \pm i\sqrt{2})$

$(1 \pm 2i\sqrt{2} + 2i^2) + 3 \overset{?}{=} 2 \pm 2i\sqrt{2}$

$1 \pm 2i\sqrt{2} - 2 + 3 \overset{?}{=} 2 \pm 2i\sqrt{2}$

$2 \pm 2i\sqrt{2} = 2 \pm 2i\sqrt{2}$

Solutions of $x^2 + 3 = 2x$ are $1 + i\sqrt{2}$ and its conjugate, $1 - i\sqrt{2}$.

EXERCISE 8-3

Solve for x by means of the quadratic formula. Express solutions in simplest radical form.

1. $x^2 - 5x + 6 = 0$

2. $x^2 + 8x + 15 = 0$

3. $x^2 - 6x + 9 = 0$

4. $6x^2 - x - 15 = 0$

5. $6x^2 - 17x + 12 = 0$

6. $3x^2 - 19x = 14$

7. $4x^2 + 6x + 1 = 0$

8. $5x^2 - 12x - 12 = 0$

9. $x^2 - 2x + 5 = 0$

10. $x^2 + 13 = 4x$

11. $3x^2 + 2x = \frac{2}{3}$

12. $2x^2 - 3x + 5 = 0$

13. $9x^2 = 3x + 1$

14. $x^2 - 16a = 0$

15. $5 + 8x^2 = 16x$

16. $(x - 2)^2 + (x + 2)^2 = 8c$

17. $x^2 + 1.4x - 1.2 = 0$

18. $x^2 - 2cx + 3 = 0$

19. $\dfrac{x}{4} + \dfrac{1}{2x} = 1$

20. $x^2 - 6ax + 8a^2 = 0$

21. $x = 2a + \dfrac{c}{x}$

22. $\dfrac{2}{x - 3} - \dfrac{4}{x - 1} = 2$

23. $\dfrac{3}{x - 1} - \dfrac{2}{x - 4} = 3$

24. $3x^2 + ax + 1 + 3x + a = 0$

[*Hint:* Write as $3x^2 + (a + 3)x + (a + 1) = 0$.]

25. $2x^2 + ax - 2 - x + a = 0$

26. $ax^2 - 3x + x^2 - a + 2 = 0$

27. The sum S of the first n counting numbers $1, 2, 3, \ldots, n$ is given by the formula $S = \frac{1}{2}n(n + 1)$. Find n for $S = 105$.

28. An open box is formed by cutting a square from each corner of a 5-in. by 7-in. rectangular piece of metal and bending up the edges. If the box is to have a base area of 19 sq in., how long should the corner cuts be?

29. What must be the dimensions of a rectangular field of 40 sq rd such that it can be enclosed with 24 lineal rods of fencing?

30. The edges of two cubes differ by 2 in. and their volumes differ by 296 cu in. What are the dimensions of each?

6. NATURE OF THE SOLUTIONS OF A QUADRATIC EQUATION

The expression, $b^2 - 4ac$, which occurs in the quadratic formula, is called the **discriminant**. It can be used to predict the nature of the solutions, x_1 and x_2, of a quadratic equation in which the coefficients a, b, and c are rational.

1. When $b^2 - 4ac = 0$, the solutions by the quadratic formula are:

$$x_1 = \frac{-b + \sqrt{0}}{2a} \quad \text{and} \quad x_2 = \frac{-b - \sqrt{0}}{2a}.$$

Note that

$$x_1 = x_2 = \frac{-b}{2a}.$$

So when the discriminant is zero, the solutions of the quadratic equation are equal rational numbers.

2. When the discriminant $b^2 - 4ac > 0$ and $b^2 - 4ac = p^2$, where p is a nonzero integer, then

$$x_1 = \frac{-b + p}{2a} \quad \text{and} \quad x_2 = \frac{-b - p}{2a}.$$

Under these circumstances, the solutions of the quadratic equation are unequal rational numbers.

3. When the discriminant $b^2 - 4ac > 0$ and $b^2 - 4ac = n$, where n is not the square of an integer, then

$$x_1 = \frac{-b + \sqrt{n}}{2a} \quad \text{and} \quad x_2 = \frac{-b - \sqrt{n}}{2a}.$$

Under these circumstances, the solutions of the quadratic equation are unequal irrational numbers.

4. When the discriminant $b^2 - 4ac < 0$, then $b^2 - 4ac$ represents an imaginary number, and the solutions of the quadratic equation are unequal imaginary numbers, that is, the equation has no real solutions.

In summary:

Discriminant: $b^2 - 4ac$	Solution numbers:
Zero	Rational and equal
Positive; perfect square	Rational and unequal
Positive; not a perfect square	Irrational and unequal
Negative	Imaginary

EXAMPLES

1. Predict the nature of the solutions of each of the following quadratic equations:

(a) $x^2 - 5x - 6 = 0$; (b) $x^2 - 4x + 1 = 0$; (c) $9x^2 + 6x + 1 = 0$;
(d) $x^2 - 2x + 5 = 0$.

(a) $x^2 - 5x - 6 = 0$; $a = 1, b = -5, c = -6$

$$b^2 - 4ac = (-5)^2 - 4(1)(-6) = 25 + 24 = 49$$

Since the value of the discriminant is positive and a perfect square ($7^2 = 49$), the solutions are predicted to be rational and unequal. (By Program **8.1** or **8.5**, the solutions can be shown to be 2 and 3.)

(b) $x^2 - 4x + 1 = 0$; $a = 1, b = -4, c = 1$

$$b^2 - 4ac = (-4)^2 - 4(1)(1) = 16 - 4 = 12$$

Since the value of the discriminant (12) is positive but not a perfect square, the solutions are predicted to be irrational and unequal. (By Program **8.5**, the solutions can be shown to be $2 + \sqrt{3}$ and $2 - \sqrt{3}$.)

(c) $9x^2 + 6x + 1 = 0$; $a = 9, b = 6, c = 1$

$$b^2 - 4ac = (6)^2 - 4(9)(1) = 36 - 36 = 0$$

Since the value of the discriminant is 0, the solutions are predicted to be rational and equal. (By Program **8.1** or **8.5**, the solutions can be shown to be $-\frac{1}{3}$ and $-\frac{1}{3}$.)

(d) $x^2 - 2x + 5 = 0$; $a = 1, b = -2, c = 5$

$$b^2 - 4ac = (-2)^2 - 4(1)(5) = 4 - 20 = -16$$

Since the value of the discriminant is -16, the solutions are predicted to be imaginary numbers. (By Program **8.5**, the solutions can be shown to be $1 + 2i$ and $1 - 2i$.)

2. For what value of k will the solutions of the equation $3x^2 + kx + 3 = 0$ be (a) equal; (b) imaginary; (c) real but unequal?

$$3x^2 + kx + 3 = 0;\ a = 3,\ b = k,\ c = 3$$
$$b^2 - 4ac = (k)^2 - 4(3)(3) = k^2 - 36$$

(a) For the solutions to be equal, the discriminant, $b^2 - 4ac$, must equal 0. The value for k that will make the discriminant for the given equation 0 can be found by setting:

$$k^2 - 36 = 0$$

and solving for k:

$$k = 6,\ -6$$

Thus the discriminant for the given equation will be 0 when the coefficient of the linear term (k) is either 6 or -6, which is to say, pairs of equal solutions will be had for each of the two equations: $3x^2 + 6x + 3 = 0$ and $3x^2 - 6x + 3 = 0$.

(b) For any value of k that lies between -6 and 6, its square (k^2) will be less than 36. Therefore, for that range of values the discriminant, $k^2 - 36$, will be negative, which is to say that, when any number between -6 and 6 is used as the coefficient of the linear term (i.e., replaces k in the given equation), the solutions of the resulting equation will be imaginary numbers.

(c) For any value of k that is greater than 6 or less than -6, the square will be greater than 36; hence, the discriminant, $k^2 - 36$, will always be positive. Thus, when any number greater than 6 or less than -6 is used as the coefficient of the linear term of the given equation, solutions are certain to be real and unequal.

7. SUM AND PRODUCT OF SOLUTIONS OF A QUADRATIC EQUATION

If we add the solutions of a quadratic equation as expressed by the quadratic formula, we obtain the following:

$$x_1 + x_2 = \frac{-b + \sqrt{b^2 - 4ac}}{2a} + \frac{-b - \sqrt{b^2 - 4ac}}{2a} = \frac{-2b}{2a} = -\frac{b}{a}$$

If we multiply the two solutions, we obtain this result:

$$x_1 \cdot x_2 = \left(\frac{-b + \sqrt{b^2 - 4ac}}{2a}\right)\left(\frac{-b - \sqrt{b^2 - 4ac}}{2a}\right)$$
$$= \frac{(-b)^2 - (\sqrt{b^2 - 4ac})^2}{(2a)(2a)} = \frac{b^2 - b^2 + 4ac}{4a^2}$$

$$= \frac{4ac}{4a^2} = \frac{c}{a}$$

Now, if we divide each term of the general quadratic equation in standard form, $ax^2 + bx + c = 0$, by the coefficient of the quadratic term, a, we derive the following equation:

$$x^2 + \frac{b}{a}x + \frac{c}{a} = 0$$

Note that the coefficient of the linear term in this derived equation is the negative of the sum of the solutions $(x_1 + x_2 = -b/a)$, and that the constant term is the same as the product of the two solutions $(x_1 \cdot x_2 = c/a)$. Thus, the following are equivalent forms of the same equation:

$$ax^2 + bx + c = 0$$

$$x^2 + \frac{b}{a}x + \frac{c}{a} = 0$$

$$x^2 - (x_1 + x_2)x + (x_1 \cdot x_2) = 0$$

This correspondence is particularly useful in formulating quadratic equations which have specified solutions.

EXAMPLES

1. Produce quadratic equations that have for solutions: (a) 3, -1; (b) $2 + \sqrt{5}$, $2 - \sqrt{5}$; (c) $\frac{2}{3}$, $\frac{2}{3}$; (d) $1 + i$, $1 - i$.

(a) 3, -1:

$$x_1 + x_2 = 3 + (-1) = +2$$
$$x_1 \cdot x_2 = (3)(-1) = -3$$

$x^2 - (x_1 + x_2)x + (x_1 \cdot x_2) = 0$

$x^2 - 2x - 3 = 0$ is a quadratic equation that has 3, -1 as solutions.

(b) $2 + \sqrt{5}$, $2 - \sqrt{5}$:

$$x_1 + x_2 = (2 + \sqrt{5}) + (2 - \sqrt{5}) = 4$$
$$x_1 \cdot x_2 = (2 + \sqrt{5})(2 - \sqrt{5}) = (2)^2 - (\sqrt{5})^2$$
$$= 4 - 5 = -1$$

$x^2 - (x_1 + x_2)x + (x_1 \cdot x_2) = 0$

$x^2 - 4x - 1 = 0$ is a quadratic equation that has $2 + \sqrt{5}$ and $2 - \sqrt{5}$ as solutions.

(c) $\frac{2}{3}$, $\frac{2}{3}$:

$$x_1 + x_2 = \frac{2}{3} + \frac{2}{3} = \frac{4}{3}$$
$$x_1 x_2 = \left(\frac{2}{3}\right)\left(\frac{2}{3}\right) = \frac{4}{9}$$

$x^2 - (x_1 + x_2)x + (x_1 \cdot x_2) = 0$

$x^2 - \frac{4}{3}x + \frac{4}{9} = 0$ and $9x^2 - 12x + 4 = 0$ are equivalent quadratic equations that have $\frac{2}{3}$ and $\frac{2}{3}$ as solutions.

(d) $1 + i, 1 - i$:

$$x_1 + x_2 = (1 + i) + (1 - i) = 2$$

$$x_1 \cdot x_2 = (1 + i)(1 - i) = (1)^2 - (i)^2 = 1 + 1 = 2$$

$$x^2 - (x_1 + x_2)x + (x_1 \cdot x_2) = 0$$

$x^2 - 2x + 2 = 0$ is a quadratic equation that has $1 + i$ and $1 - i$ as solutions.

EXERCISE 8-4

Predict the nature of the solutions of each of the following:

1. $x^2 - 2x + 1 = 0$

2. $x^2 - 7x + 12 = 0$

3. $2x^2 - 3x + 2 = 0$

4. $2x^2 - 3x - 2 = 0$

5. $3x^2 - 4x + 7 = 0$

6. $2x^2 + 4x - 3 = 0$

7. $3x^2 + 1 = 5x$

8. $x^2 - \frac{1}{2}x + \frac{1}{16} = 0$

9. $3x^2 - 5 = 2x$

10. $3x^2 = 2x - 5$

For what real value of k will the solutions be equal?

11. $x^2 + kx + 16 = 0$

12. $4x^2 - kx + 1 = 0$

13. $9x^2 + 2kx + 4 = 0$

14. $3x^2 - kx + 2 = 0$

15. $3x^2 + 4x + k = 0$

16. $2x^2 - 3x - k = 0$

17. $kx^2 + 4x - 2 = 0$

18. $3x^2 - kx - 5 = 0$

For each of the following pairs of numbers, write a quadratic equation that will have those numbers as solutions:

19. $4, 3$

20. $3, -1$

21. $2, \frac{1}{2}$

22. $\frac{3}{4}, -\frac{1}{2}$

23. $\frac{2}{7}, \frac{2}{7}$

24. $6, -6$

25. $2 + \sqrt{3}, 2 - \sqrt{3}$

26. $3 + i, 3 - i$

27. $3 + 2\sqrt{2}, 3 - 2\sqrt{2}$

28. $-3 + 2i, -3 - 2i$

29. $-a, b$

30. $2 + 3i\sqrt{2}, 2 - 3i\sqrt{2}$

8. RADICAL EQUATIONS

An equation in which the variable occurs as a radicand is called a **radical equation**. For example:

$$\sqrt[3]{x} = 2$$

By cubing both members:

$$(\sqrt[3]{x})^3 = (2)^3$$

$$x = 8$$

we derive an equation for which the solution is obvious.

However, raising both members of an equation to a power does not always produce an equivalent equation. When the power is an even number, the derived equation may have more solutions than the original equation. Solutions of the derived equation which are not solutions of the original equation are called **extraneous solutions**. Consequently, it is important that all solutions of the derived equation be verified as solutions of the original equation.

EXAMPLES

1. Solve: $x - 2 - \sqrt{x + 4} = 0$.

Arrange the equation so that a single radical expression appears in one member:

$$x - 2 - \sqrt{x + 4} = 0$$

$$x - 2 = \sqrt{x + 4}$$

Square both members: $\qquad (x - 2)^2 = (\sqrt{x + 4})^2$

Solve, using Program **8.1**: $\quad x^2 - 4x + 4 = x + 4$

$$x^2 - 5x = 0$$

$$(x)(x - 5) = 0$$

$$(x) = 0 \qquad (x - 5) = 0$$

$$x = 0 \qquad\qquad x = 5$$

Possible solutions are 0 and 5. Verify by replacing the variable in the given equation by each to see if a true sentence results:

$x - 2 - \sqrt{x + 4} = 0$	$x - 2 - \sqrt{x + 4} = 0$
$x = 0:\;\; (0) - 2 - \sqrt{0 + 4} \overset{?}{=} 0$	$x = 5:\;\; (5) - 2 - \sqrt{5 + 4} \overset{?}{=} 0$
$-2 - \sqrt{4} \overset{?}{=} 0$	$5 - 2 - \sqrt{9} \overset{?}{=} 0$
$-2 - 2 \overset{?}{=} 0$	$5 - 2 - 3 \overset{?}{=} 0$
$-4 \neq 0$	$0 = 0$

Hence the solution set for $x - 2 - \sqrt{x + 4} = 0$ contains but a single number: 5.

2. Solve: $\sqrt{2x + 3} + \sqrt{4x - 3} = 6$.

To solve this equation it will be necessary to square the members twice:

$$\sqrt{2x + 3} + \sqrt{4x - 3} = 6$$

$$\sqrt{2x + 3} = 6 - \sqrt{4x - 3}$$

$$(\sqrt{2x+3})^2 = (6 - \sqrt{4x-3})^2$$
$$2x + 3 = 36 - 12\sqrt{4x-3} + (4x-3)$$
$$12\sqrt{4x-3} = 2x + 30$$
$$6\sqrt{4x-3} = x + 15$$
$$(6\sqrt{4x-3})^2 = (x+15)^2$$
$$36(4x-3) = x^2 + 30x + 225$$
$$144x - 108 = x^2 + 30x + 225$$
$$x^2 - 114x + 333 = 0$$
$$(x-3)(x-111) = 0$$
$$(x-3) = 0 \qquad (x-111) = 0$$
$$x - 3 = 0 \qquad x - 111 = 0$$
$$x = 3 \qquad\qquad x = 111 \quad \text{(possible solutions)}$$

Verification:

$$\sqrt{2x+3} + \sqrt{4x-3} = 6 \qquad\qquad \sqrt{2x+3} + \sqrt{4x-3} = 6$$

$x = 3$: $\qquad\qquad\qquad\qquad\qquad\qquad$ $x = 111$:

$$\sqrt{2(3)+3} + \sqrt{4(3)-3} \overset{?}{=} 6 \qquad \sqrt{2(111)+3} + \sqrt{4(111)-3} \overset{?}{=} 6$$
$$\sqrt{9} + \sqrt{9} \overset{?}{=} 6 \qquad\qquad \sqrt{225} + \sqrt{441} \overset{?}{=} 6$$
$$3 + 3 \overset{?}{=} 6 \qquad\qquad\qquad 15 + 21 \overset{?}{=} 6$$
$$6 = 6 \qquad\qquad\qquad\qquad 36 \neq 6$$

Hence, 3 is a solution of the equation, but 111 is not.

3. Solve: $\sqrt{x-2} - \sqrt{2x+3} = 2$

$$\sqrt{x-2} - \sqrt{2x+3} = 2$$
$$\sqrt{x-2} = 2 + \sqrt{2x+3}$$
$$(\sqrt{x-2})^2 = (2 + \sqrt{2x+3})^2$$
$$x - 2 = 4 + 4\sqrt{2x+3} + (2x+3)$$
$$-4\sqrt{2x+3} = 4 + 2x + 3 - x + 2$$
$$-4\sqrt{2x+3} = x + 9$$
$$(-4\sqrt{2x+3})^2 = (x+9)^2$$
$$16(2x+3) = x^2 + 18x + 81$$
$$32x + 48 = x^2 + 18x + 81$$
$$0 = x^2 - 14x + 33$$
$$0 = (x-11)(x-3)$$
$$x - 11 = 0 \qquad x - 3 = 0$$
$$x = 11 \qquad\qquad x = 3$$

Verification:

$$\sqrt{x-2} - \sqrt{2x+3} = 2$$
$$x = 11: \quad \sqrt{11-2} - \sqrt{2(11)+3} \overset{?}{=} 2$$
$$\sqrt{9} - \sqrt{25} \overset{?}{=} 2$$
$$3 - 5 \overset{?}{=} 2$$
$$-2 \neq 2$$

$$\sqrt{x-2} - \sqrt{2x+3} = 2$$
$$x = 3: \quad \sqrt{3-2} - \sqrt{2(3)+3} \overset{?}{=} 2$$
$$\sqrt{1} - \sqrt{9} \overset{?}{=} 2$$
$$1 - 3 \overset{?}{=} 2$$
$$-2 \neq 2$$

Hence, neither 11 nor 3 is a solution of the given equation. In other words, the equation has *no solutions;* its solution set is the *empty set* (usually symbolized by $\varnothing$ or $\{\ \ \}$).

EXERCISE 8-5

Solve and verify all solutions:

1. $\sqrt{x+3} = 6$ 　　　　　　　**2.** $\sqrt{2x-4} + 2 = 0$

3. $2 + \sqrt{x-2} = 0$ 　　　　　**4.** $\sqrt{x^2+2} + x = 2$

5. $1 - 2x + \sqrt{4x+1} = 0$ 　　**6.** $\sqrt{4x+3} - \sqrt{6x+2} = 0$

7. $3\sqrt{x-1} = 2\sqrt{2x-1}$ 　　**8.** $3\sqrt{x^2-4} = \sqrt{3x^2+4x+6}$

9. $1 + \sqrt{2x} = \sqrt{2x+5}$ 　　**10.** $\sqrt{x-2} - 2 = \sqrt{2x+3}$

11. $\sqrt{2x-5} = 1 + \sqrt{x-3}$ 　**12.** $\sqrt{x+2} + \sqrt{x-1} = \sqrt{2x+5}$

13. $\sqrt{x+8} = \sqrt{2-x} + \sqrt{5-x}$ 　**14.** $\sqrt{2x + \sqrt{2x-4}} = 2$

15. $\sqrt{\dfrac{3x-2}{9}} = \dfrac{x+2}{6}$ 　　**16.** $\sqrt[3]{5x^2+2x+3} - 3 = 0$

9. QUADRATIC INEQUALITIES

A **quadratic inequality** is an inequality that may be expressed equivalently as

$$ax^2 + bx + c > 0 \quad \text{or} \quad ax^2 + bx + c < 0$$

where x represents the variable, and a, b, and c represent real numbers, $a \neq 0$. The usual procedure for solving a quadratic inequality makes use of the facts that the product of two positive numbers, or of two negative numbers, is a positive number, and that the product of a positive and a negative number is a negative number. If we think of positive numbers as numbers greater than zero (i.e., $n > 0$) and negative numbers as numbers less than zero (i.e., $n < 0$), we can restate this generalization about products as follows:

If M and N are two factors, and

$$(M \cdot N) > 0, \qquad \text{then} \quad \begin{cases} \text{(a)} & M > 0 \quad \text{and} \quad N > 0, \quad \text{or} \\ \text{(b)} & M < 0 \quad \text{and} \quad N < 0 \end{cases}$$

$$(M \cdot N) < 0, \qquad \text{then} \quad \begin{cases} \text{(a)} & M < 0 \quad \text{and} \quad N > 0, \quad \text{or} \\ \text{(b)} & M > 0 \quad \text{and} \quad N < 0 \end{cases}$$

As an example, to solve

$$x^2 - 5x + 6 > 0$$

we may start by factoring the left member:

$$(x - 2)(x - 3) > 0$$

If the product $(x - 2)(x - 3)$ is to be greater than zero when the variable is replaced by a number, then both factors must be greater than zero, or both factors must be less than zero. For both factors greater than zero:

$$x - 2 > 0 \quad \text{and} \quad x - 3 > 0$$
$$x > 2 \qquad\qquad x > 3$$

From this we can see that *any number greater than 3* will make *both* factors greater than zero (i.e., positive) and their product greater than zero (i.e., positive). Therefore, numbers greater than 3 satisfy the given inequality.

On the other hand, setting each factor less than zero:

$$x - 2 < 0 \quad \text{and} \quad x - 3 < 0$$
$$x < 2 \qquad\qquad x < 3$$

indicates that *any number less than 2* will make *both* factors less than zero (i.e., negative) and their product greater than zero (i.e., positive). Consequently, numbers less than 2 also satisfy the inequality.

The complete solution set for the given inequality may then be displayed on a number line as follows.

Solving a "less than" type of quadratic inequality involves essentially the same procedure as that used for the "greater than" type just discussed. For example, to solve

$$x^2 - 7x + 10 < 0$$

we factor:

$$(x - 2)(x - 5) < 0$$

If the product of the two factors, $(x - 2)$ and $(x - 5)$, is to be less than zero (i.e., negative) when the variable is replaced by a number, then one factor must be greater than zero and the other less than zero. The product of the two factors will be less than zero when (a) $x - 2 < 0$ and $x - 5 > 0$, or when (b) $x - 2 > 0$ and $x - 5 < 0$.

(a) $\qquad\qquad x - 2 < 0 \quad$ and $\quad x - 5 > 0$

$\qquad\qquad\qquad\quad x < 2 \qquad\qquad\qquad x > 5$

or (b) $\qquad\quad\ x - 2 > 0 \quad$ and $\quad x - 5 < 0$

$\qquad\qquad\qquad\quad x > 2 \qquad\qquad\qquad x < 5$

With respect to (a), there is no number that is both "less than 2" and at the same time "greater than 5"; therefore, this pair of conditions represents an impossibility. On the other hand, *any number between 2 and 5* (which may be expressed $2 < x < 5$) satisfies the conditions represented in (b). Numbers between 2 and 5, then, constitute the solution set for the given inequality and may be displayed on a number line as follows.

$$x^2 - 7x + 10 < 0:$$

$$2 < x < 5$$

0   1   2   3   4   5   6   7

The number line used to express the solution set in the two examples is an important aid in the following general program for solving any quadratic inequality.

8.6

To solve a quadratic inequality:

Step 1. Transform the given inequality into an equation by replacing the sign of inequality with an $=$ sign.

Step 2. Find the real-number solutions of the quadratic equation of Step 1.

Step 3. Locate the two solutions of Step 2 on the number line; they will separate the number line into three parts, each corresponding to sets of numbers.†

Step 4. Select any one number from each of the three sets of Step 3 as a trial number, and substitute it for the variable in the given inequality. All numbers in the set for which the representative trial number satisfies the given inequality satisfy the inequality.

EXAMPLES

1. Solve the quadratic inequality: $x^2 - 5x + 6 > 0$.

Step 1. $x^2 - 5x + 6 = 0$

---

†An exception occurs when the solutions are equal. See Example 5, p. 195.

Step 2. $(x - 2)(x - 3) = 0$

$$x - 2 = 0 \qquad x - 3 = 0$$

$$x = 2 \qquad x = 3$$

Step 3.

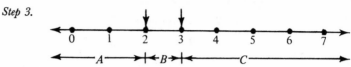

Step 4.

Set	Trial number	$x^2 - 5x + 6 > 0$
A: $(x < 2)$	1	$(1)^2 - 5(1) + 6 > 0$ (true)
B: $(2 < x < 3)$	2.1	$(2.1)^2 - 5(2.1) + 6 > 0$
		$4.41 - 10.5 + 6 > 0$ $\big\}$ (false)
C: $(x > 3)$	4	$(4)^2 - 5(4) + 6 > 0$ (true)

Because trial numbers 1 and 4 satisfy the inequality, all numbers in their respective sets, A and C, satisfy the inequality. Because trial number 2.1 does not satisfy the inequality, no number in set B, which it represents, satisfies the inequality. Thus the solution set for $x^2 - 5x + 6 > 0$ is $\{x | x < 2 \text{ or } x > 3\}$.

2. Solve: $x^2 - 7x + 10 < 0$.

Step 1. $x^2 - 7x + 10 = 0$

Step 2. $(x - 2)(x - 5) = 0$

$$x - 2 = 0 \qquad x - 5 = 0$$

$$x = 2 \qquad x = 5$$

Step 3.

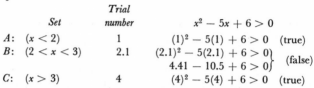

Step 4.

Set	Trial number	$x^2 - 7x + 10 < 0$
A: $(x < 2)$	1	$(1)^2 - 7(1) + 10 < 0$ (false)
B: $(2 < x < 5)$	3	$(3)^2 - 7(3) + 10 < 0$ (true)
C: $(x > 5)$	7	$(7)^2 - 7(7) + 10 < 0$ (false)

Thus, $\{x | x^2 - 7x + 10 < 0\} = \{x | 2 < x < 5\}$.

3. Solve: $2x^2 - x - 6 > 0$.

Step 1. $2x^2 - x - 6 = 0$

Step 2. $(2x + 3)(x - 2) = 0$

$$2x + 3 = 0 \qquad x - 2 = 0$$

$$x = -\tfrac{3}{2} \qquad x = 2$$

Step 3.

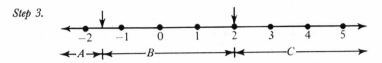

Step 4.

Set	Trial number	$2x^2 - x - 6 > 0$	
A: $(x < -\frac{3}{2})$	-2	$\left.\begin{array}{l} 2(-2)^2 - (-2) - 6 > 0 \\ 8 + \quad 2 - 6 > 0 \end{array}\right\}$	(true)
B: $(-\frac{3}{2} < x < 2)$	0	$2(0)^2 - (0) - 6 > 0$	(false)
C: $(x > 2)$	3	$2(3)^2 - (3) - 6 > 0$	(true)

Solution set for $2x^2 - x - 6 > 0$ is $\{x \,|\, x < -\frac{3}{2} \text{ or } x > 2\}$.

4. Solve: $x^2 - 3 < -4x$.

Step 1. $x^2 - 3 = -4x$

Step 2. Solutions: $-2 + \sqrt{7}$, $-2 - \sqrt{7}$ (approximately, $+0.6$ and -4.6). (See Example 3, p. 182.)

Step 3.

Step 4.

Set	Trial number	$x^2 - 3 < -4x$	
A: $(x < -2 - \sqrt{7})$	-5	$\left.\begin{array}{l} (-5)^2 - 3 < -4(-5) \\ 25 - 3 < 20 \end{array}\right\}$	(false)
B: $(-2 - \sqrt{7} < x < -2 + \sqrt{7})$	0	$\left.\begin{array}{l} (0)^2 - 3 < -4(0) \\ -3 < 0 \end{array}\right\}$	(true)
C: $(x > -2 + \sqrt{7})$	1	$\left.\begin{array}{l} (1)^2 - 3 < -4(1) \\ -2 < -4 \end{array}\right\}$	(false)

Solution set for $x^2 - 3 < -4x$ is $\{x \,|\, -2 - \sqrt{7} < x < -2 + \sqrt{7}\}$.

5. Solve the inequalities:

(a) $x^2 - 2x + 1 > 0$

(b) $x^2 - 2x + 1 < 0$

The corresponding equation for both inequalities is:

$$x^2 - 2x + 1 = 0$$

The solutions are equal:

$$(x - 1)(x - 1) = 0$$

$$x - 1 = 0 \qquad x - 1 = 0$$

$$x = 1 \qquad\quad x = 1$$

Instead of separating the number line into three parts, equal solutions separate it into two:

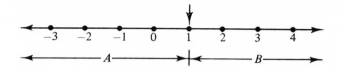

Trial numbers selected from each set, say 0 from A and 2 from B, both satisfy inequality (a):

$$x^2 - 2x + 1 > 0$$

$$x = 0: \quad (0)^2 - 2(0) + 1 > 0 \quad \text{(true)}$$

$$x = 2: \quad (2)^2 - 2(2) + 1 > 0 \quad \text{(true)}$$

The solution set for $x^2 - 2x + 1 > 0$ contains every real number except 1: $\{x \mid x < 1 \text{ or } x > 1\}$. However, the two trial numbers fail to satisfy inequality (b):

$$x^2 - 2x + 1 < 0$$

$$x = 0: \quad (0)^2 - 2(0) + 1 < 0 \quad \text{(false)}$$

$$x = 2: \quad (2)^2 - 2(2) + 1 < 0 \quad \text{(false)}$$

Consequently, there is no solution for this inequality; in other words, its solution set is the empty set:

$$\{x \mid x^2 - 2x + 1 < 0\} = \{ \ \}$$

[*Note:* The reasonableness of these findings can be better appreciated, perhaps, if we express the inequalities in this form:

$$(x - 1)^2 > 0 \quad \text{and} \quad (x - 1)^2 < 0$$

Except when $x = 1$, the expression $x - 1$ represents a nonzero number, and the square of every nonzero number is a positive number, i.e., a number greater than zero. On the other hand, no real number, when squared, is a negative number, i.e., a number less than zero.]

EXERCISE 8-6

Solve:

1. $x^2 - 4 < 0$

2. $x^2 - 9 > 0$

3. $x^2 - 1 > 0$

4. $x^2 - 16 < 0$

5. $m^2 < 3$

6. $n^2 > 3$

7. $x^2 - 6x + 8 < 0$

8. $p^2 - 7p + 12 < 0$

9. $n^2 - 8n + 15 > 0$

10. $t^2 - 10t + 21 > 0$

11. $x^2 + x - 6 < 0$

12. $p^2 + p - 2 < 0$

13. $r^2 + 3r > 4$

14. $y^2 - 2y > 3$

15. $m^2 + 5m + 6 > 0$

16. $n^2 + 6n + 5 > 0$

17. $s^2 + 8s + 12 < 0$ **18.** $x^2 + 8x + 15 < 0$

19. $2x^2 - 7x - 4 < 0$ **20.** $3x^2 + 5x - 2 > 0$

21. $2x^2 + 15 > 13x$ **22.** $3x^2 + 10 < 11x$

23. $6y^2 + y - 2 < 0$ **24.** $8m^2 + 6m - 9 < 0$

25. $6n^2 + 13n + 5 > 0$ **26.** $8n^2 - 30n + 7 > 0$

27. $x^2 + 5 < 5x$ **28.** $y^2 - y < 1$

29. $2m + 6m - 3 > 0$ **30.** $2p - 6p + 1 > 0$

31. $4t^2 + 6t + 1 > 0$ **32.** $9x^2 + 6x - 2 > 0$

33. $4x^2 - 4x > -1$ **34.** $9p^2 + 12p < -4$

35. $12p + 9 < -4p^2$ **36.** $25n^2 - 20n + 4 > 0$

37. $2x^2 - 5x < 0$ **38.** $4t^2 + 2t > 0$

39. $6k^2 + 2k > 0$ **40.** $60y^2 < 140y$

REVIEW — CHAPTER VIII

Solve for all values of x by factoring:

1. $x^2 - 13x + 30 = 0$ **2.** $x^2 + 10x + 16 = 0$

3. $x^2 = 3x + 18$ **4.** $x^2 + 10x + 25 = 0$

5. $12x^2 = 3 + 5x$ **6.** $x^2 + bx = 6b^2$

7. $15x^2 - 13ax + 2a^2 = 0$ **8.** $6 + ax = 15a^2x^2$

9. $2ax = 3x^2$ **10.** $\dfrac{3x + 1}{2x} = \dfrac{2x - 1}{x - 2}$

11. $\dfrac{3}{2x - 1} + \dfrac{x}{1 - 2x} = 0$

12. A department store buyer paid \$96 for a group of similar articles. If she could have gotten the unit price down by \$4, she could have purchased four more articles for the same money. How many articles did she purchase?

13. The hypotenuse of a right triangle is 15 ft and one leg is 3 ft longer than the other. What is the length of the longest leg?

Solve for x:

14. $16x = \dfrac{81}{x}$ **15.** $5x^2 + 4 = x^2 + 13$

16. $\dfrac{x}{2} + \dfrac{2}{x} = 2x$ **17.** $px^2 = s - rx^2$

18. $\dfrac{3}{x^2 - 1} = \dfrac{4}{x^2 + 2}$ **19.** $\dfrac{3m^2}{4} - x^2 = \dfrac{2m^2}{5}$

20. $4x^2 + 9 = x^2$ **21.** $5x^2 + 7 = 2x^2$

22. $\dfrac{x}{4} - \dfrac{3}{x} = x$

23. Twenty-one is the product of a certain negative number increased by 2 and that negative number decreased by 2. What is the certain negative number?

Insert a term in the parentheses to make the following perfect square trinomials:

24. $x^2 - 16x + (\quad)$
  25. $x^2 - 12ax + (\quad)$

26. $x^2 - 7x + (\quad)$
  27. $x^2 + \dfrac{x}{5} + (\quad)$

28. $x^2 - \frac{3}{8}x + (\quad)$
  29. $x^2 - abx + (\quad)$

Solve by means of the quadratic formula:

30. $4x^2 + 11x - 3 = 0$
  31. $12x^2 - 11x + 2 = 0$

32. $x^2 - 7x + 2 = 0$
  33. $2x^2 - 4x + 5 = 0$

34. $4x^2 + x - 14 = 0$
  35. $5x^2 + 22x + 8 = 0$

36. $x^2 + 6x - 13 = 0$
  37. $3x^2 + 10x + 2 = 0$

38. $x^2 - 6x + 10 = 0$
  39. $3x^2 - 4x + 2 = 0$

40. $4x^2 = 4x + 5$
  41. $x^2 - 32c = 3x^2$

42. $x^2 + 2.6x = 1.2$
  43. $\dfrac{1}{x^2} = 1 - \dfrac{2}{x}$

44. $x = \dfrac{a}{x} + b$
  45. $\dfrac{2}{x - 2} + \dfrac{3}{x + 1} = 2$

46. $ax^2 - 2x - 2x^2 - 2 - a = 0$

47. The base of a triangle is 4 in. longer than its altitude. If the area of the triangle is 160 sq in., what must be the length of its base?

48. An object propelled downward through the air moves according to the formula $s = 16t^2 + v_0t$, approximately, in which s represents the distance it travels in feet in t seconds of time, and v_0 represents the initial propelling velocity. How long will it take an object to travel 264 ft if it is propelled downward with an initial velocity of 40 ft/sec?

Predict the nature of the solutions:

49. $x^2 - 3x + 5 = 0$
  50. $3x^2 - 4x - 2 = 0$

51. $12x^2 + 3 = 12x$
  52. $8x^2 - 10x + 3 = 0$

53. $\frac{3}{4}x^2 + \frac{3}{7}x + 2 = 0$

For what real value of k will the solutions be equal?

54. $9x^2 - kx^2 + 16 = 0$
  55. $2x^2 - 8x - k = 0$

56. $2x^2 + kx - 4 = 0$
  57. $kx^2 - 5x + 7 = 0$

Write a quadratic equation that will have each of the following pairs of numbers as solutions.

58. $4, -\frac{3}{4}$
  59. $-\frac{2}{5}, \frac{2}{5}$

60. $-3 + \sqrt{2}, -3 - \sqrt{2}$
  61. $4 + i, 4 - i$

62. $3 + 3\sqrt{3}, 3 - 3\sqrt{3}$
  63. $-1 + 2i, -1 - 2i$

Solve and verify solutions:

64. $\sqrt{x^2 + 3} - 3x = 5$

65. $2\sqrt{x + 1} - 4\sqrt{x - 5} = 0$

66. $2 + \sqrt{2x + 4} = \sqrt{x}$

67. $2\sqrt{x^2 + 3} = \sqrt{3x^2 + 5x + 8}$

68. $2 + \sqrt{x^2 - 4} = x$

69. $\sqrt{x + 1} - \sqrt{2x + 3} + \sqrt{x + 2} = 0$

70. $\sqrt{2x - \sqrt{5x + 1}} = \sqrt{2}$

71. $2^{2/3} = \sqrt[3]{4x^2 + 5x - 5}$

72. $p^2 - 16 > 0$

73. $x^2 + 2x - 15 > 0$

74. $n^2 - 4n - 21 < 0$

75. $2x^2 - 11x + 12 < 0$

76. $6t^2 > 15t$

77. $x^2 + 2x > 5$

78. $2c^2 + 1 < 5c$

79. $4e^2 + 1 < 4e$

80. $9 + k^2 > 6k$

IX

GRAPHING AND SYSTEMS
OF EQUATIONS AND
INEQUALITIES

1. ORDERED PAIRS   In this chapter we expand the concept of mathematical sentences to include those in two variables. The replacement set for both variables, unless otherwise
specified, will be the set of real numbers. Whereas
solutions of equality sentences—equations—in a single
variable are single numbers, solutions of sentences in two
variables are pairs of numbers. For example

$$x + 2y = 7$$

is an equation in two variables, x and y. Among the
pairs of real numbers that make the sentence true are:

$$(x, y): \qquad x + 2y \;\; = 7$$
$$(1, 3): \qquad 1 + 2(3) = 7$$
$$(5, 1): \qquad 5 + 2(1) = 7$$
$$(3, 2): \qquad 3 + 2(2) = 7$$
$$(-1, 4): \qquad -1 + 2(4) = 7$$

Note that the sentence is true for the pair $(1, 3)$, that
is, when 1 replaces x and 3 replaces y. But the sentence
is not true for the reverse of that pair, $(3, 1)$:

$$(3, 1): \qquad 3 + 2(1) = 7 \quad \text{(false)}$$

Consequently, the pair $(1, 3)$ is a solution of the equation
$x + 2y = 7$, but the pair $(3, 1)$ is not a solution.

200

Use of the notation, $(1, 3)$ and $(3, 1)$, involves an arbitrary assumption. The first, or left, component of the pair is assumed to be a replacement for x, and the second, or right, component is assumed to be a replacement for y. Pairs of numbers for which order is important are called **ordered pairs**.

EXAMPLES

1. Decide which of the following ordered pairs are solutions of the equation $2x + y = 12$: $(3, 6)$, $(6, 3)$, $(4, -4)$, $(-5, -2)$, $(7, -2)$, $(0, 12)$.

$$2x + y = 12$$

$(3, 6)$ or $\left.\begin{array}{l}x = 3\\y = 6\end{array}\right\}$	$2(3) + 6 = 12$	true; solution
$(6, 3)$ or $\left.\begin{array}{l}x = 6\\y = 3\end{array}\right\}$	$2(6) + 3 = 12$	false; not a solution
$(4, -4)$ or $\left.\begin{array}{l}x = 4\\y = -4\end{array}\right\}$	$2(4) + (-4) = 12$	false; not a solution
$(-5, -2)$ or $\left.\begin{array}{l}x = -5\\y = -2\end{array}\right\}$	$2(-5) + (-2) = 12$	false; not a solution
$(7, -2)$ or $\left.\begin{array}{l}x = 7\\y = -2\end{array}\right\}$	$2(7) + (-2) = 12$	true; solution
$(0, 12)$ or $\left.\begin{array}{l}x = 0\\y = 12\end{array}\right\}$	$2(0) + (12) = 12$	true; solution

2. Find the ordered pairs that are solutions of the equation $2x - y = 8$ in which the first component of the pair (the x replacement) is -2; 0; 3.

$$2x - y = 8$$

$(-2, ?)$ or $\left.\begin{array}{l}x = -2\\y = ?\end{array}\right\}$
$\begin{aligned}2(-2) - y &= 8\\-y &= 8 + 4\\y &= -12\end{aligned}$

$(0, ?)$ or $\left.\begin{array}{l}x = 0\\y = ?\end{array}\right\}$
$\begin{aligned}2(0) - y &= 8\\y &= -8\end{aligned}$

$(3, ?)$ or $\left.\begin{array}{l}x = 3\\y = ?\end{array}\right\}$
$\begin{aligned}2(3) - y &= 8\\-y &= 8 - 6\\y &= -2\end{aligned}$

Thus, the ordered pairs in question that are solutions of $2x - y = 8$ are: $(-2, -12)$, $(0, -8)$, and $(3, -2)$.

EXERCISE 9-1

Select from among the ordered pairs given for each of the equations those that are solutions of the equation:

1. $2x + y = 4$; $(1, 2)$, $(2, 3)$, $(2, 0)$, $(0, 2)$

2. $x - 3y = -2$; $(1, 1)$, $(3, 2)$, $(2, 3)$, $(4, 2)$

3. $2x - y = 5$; $(1, -3)$, $(2, -1)$, $(3, 2)$, $(4, 3)$

4. $2x - 3y = 4$; $(1, 1)$, $(2, 0)$, $(5, 2)$, $(4, 4)$

5. $3x - 5y + 7 = 0$; $(1, 3)$, $(-2, 4)$, $(-4, -1)$, $(-1, -3)$

6. $3x - 2y = 4$; $(0, 2\frac{1}{2})$, $(-1, -3\frac{1}{2})$, $(4, 8)$, $(1, -\frac{1}{2})$

7. $2x^2 - y = 4$; $(1, -2)$, $(2, 4)$, $(3, 11)$, $(-1, -2)$

8. $2x^2 - 3y^2 = 4$; $(0, 2)$, $(3, 4)$, $(1, 2)$, $(-1, -2)$

Replace the "?" so as to make the ordered pair a solution of the given equation:

9. $x - 3y = 5$; $(?, 1)$, $(-1, ?)$, $(?, -3)$

10. $3x - 2y = 7$; $(3, ?)$, $(-1, ?)$, $(?, 4)$

11. $5x - 7y = 6$; $(?, -1)$, $(4, ?)$, $(?, 0)$

12. $x^2 - 3y = 4$; $(?, -1)$, $(4, ?)$, $(?, 0)$

2. CARTESIAN COORDINATES

By means of the **Cartesian** (or **rectangular**) **coordinate system**, a highly useful correspondence can be established between ordered pairs of real numbers and the locations of points in a geometric plane.

The Cartesian coordinate system, as shown in Fig. 9.1, subdivides the plane into four regions by means of two perpendicular number lines. The

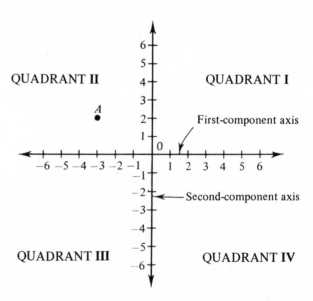

Figure 9.1

regions are called **quadrants**. The number lines are called **axes**. The axes intersect at their zero points, called the **origin** of the system. The number line which is the horizontal axis has its positive numbers associated with points to the right of the origin, and the vertical axis has its positive numbers associated with points above the origin. The four quadrants created by the axes are usually numbered in counterclockwise order, as shown in Fig. 9.1.

In this system an ordered pair of real numbers is represented by a single point. The horizontal axis serves as a scale to locate the point according to the first component of the ordered pair, and the vertical axis to locate the point according to the second component. For example, in Fig. 9.1, point A corresponds to the ordered pair $(-3, 2)$. Its location is opposite -3 on the horizontal axis and opposite 2 on the vertical axis.

The point corresponding to an ordered pair is called the **graph** of the ordered pair. In this context, the first component is usually called the **abscissa** of the point, and the second component is the **ordinate** of the point. Together, the abscissa and ordinate constitute the **coordinates** of the point.

EXAMPLES

1. Locate the point whose
   (a) abscissa is 3, ordinate is 4
   (b) abscissa is -2, ordinate is 6
   (c) abscissa -3, ordinate -1
   (d) abscissa 4, ordinate -2
   (e) abscissa 0, ordinate -6
   (f) abscissa 3, ordinate 0
   (g) abscissa -2, ordinate 0
   See Fig. 9.2.

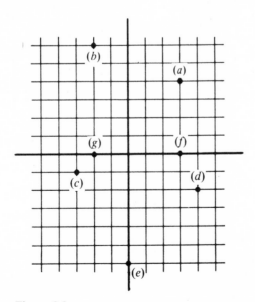

Figure 9.2

2. On the Cartesian coordinate grid of Fig. 9.3, locate the following points given by their coordinates, or ordered pairs, in which the first component is the abscissa of the point and the second is the ordinate of the point.

(a) (3, 2)

(b) (−2, 3)

(c) (−3, −3)

(d) (2, −4)

(e) (0, 0)

(f) (6, 0)

(g) (0, −5)

(h) $(2\frac{1}{2}, 3)$

(i) $(-1\frac{1}{2}, 0)$

(j) $(-\frac{1}{2}, -\frac{1}{2})$

3. Approximate the coordinates of the points plotted on the grid in Fig. 9.4.

Answers: (a) (−2, 3)

(b) (0, 3)

(c) (1, 1)

(d) $(-2\frac{1}{2}, 0)$

(e) $(-2\frac{1}{2}, -1\frac{1}{2})$

(f) (−0.8, −3.7)

(g) (2, −2)

(h) $(4\frac{1}{4}, 0)$

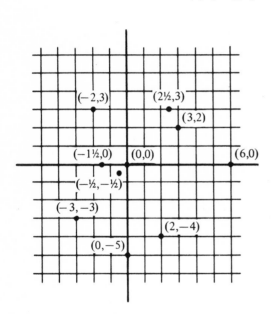

Figure 9.3

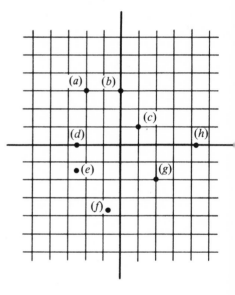

Figure 9.4

EXERCISE 9-2

Plot the following points:

1. Abscissa 3, ordinate 2.

2. Abscissa 4, ordinate −3.

3. Abscissa −5, ordinate −3.

4. Abscissa −2, ordinate −1.

5. Abscissa -2, ordinate 0.

6. $(-3, 2)$

7. $(2, 1\frac{1}{2})$

8. $(-3, -2\frac{1}{2})$

9. $(4, -3\frac{1}{2})$

10. $(-\frac{1}{2}, \frac{1}{2})$

11. In Fig. 9.5, estimate the coordinates for the points labeled A, B, C, D, and E.

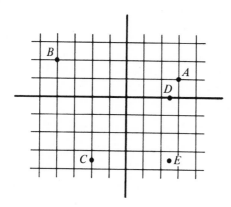

Figure 9.5

12. Draw the triangle ABC whose vertices are: A: $(2, -2)$; B: $(0, 4)$; C: $(-4, -1)$.

13. Connect these points in order: A: $(2, -3)$; B: $(4, 1)$; C: $(0, 3)$; D: $(-2, 1)$. What kind of figure is $ABCD$?

14. Draw the two diagonals in the figure of Exercise 13, and estimate the coordinates of the point of intersection.

15. The lower left corner of a square 5 units on edge is at $(-3, -1)$. If the sides are parallel to the axes, what are the coordinates of the other three corners?

16. Draw the diagonals of the square of Exercise 15, and estimate the coordinates of their point of intersection.

3. GRAPHING AN EQUATION IN TWO VARIABLES

When the replacement set for an equation in two variables is the set of real numbers, the number of ordered pairs which satisfy the equation is infinite. One useful way to display such a solution set is to plot the solutions on a Cartesian coordinate grid. Collectively, this set of points is known as the **graph of the equation**. The abscissa and ordinate of each point of the equation's graph correspond to a pair of variable replacements which satisfy the equation. Because the solution set of the equation contains infinitely many members, it is impossible to tabulate or to plot all of them. However, by plotting occasional solution pairs, and then connecting those points with a smooth continuous line, we can develop a reasonably accurate picture of the solution set of the equation.

9.1

To graph an equation in two variables:

Step 1. Solve the equation for one of its variables.

Step 2. Replace the variable in the "solution" member of the equation of Step 1 with numbers selected from the replacement set, and determine the corresponding replacement for the other variable necessary to make the equation true.

Step 3. Plot the solution pairs of Step 2 on a Cartesian coordinate grid.

Step 4. Connect the plotted points of Step 3 with a smooth continuous line.

EXAMPLES

1. Graph: $3x - y = 2$.

Step 1. Solve the equation for one of its variables:

$$3x - y = 2$$
$$y = 3x - 2$$

Step 2. Select numbers to replace the variable x: 0, 1, 2, say. Determine

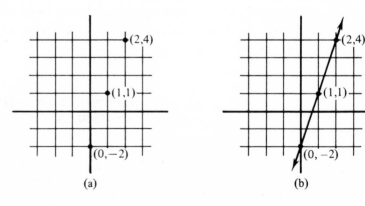

(a) (b)

Figure 9.6

the corresponding replacements for the other variable, y:

$$y = 3x - 2$$
$$x = 0: \quad y = 3(0) - 2 = -2$$
$$x = 1: \quad y = 3(1) - 2 = 1$$
$$x = 2: \quad y = 3(2) - 2 = 4$$

Step 3. The ordered-pair solutions resulting from Step 2 are $(0, -2)$, $(1, 1)$, and $(2, 4)$. They are plotted in Fig. 9.6(a).

Step 4. Connect the points of Step 3 with a smooth continuous line, as in Fig. 9.6(b).

2. Graph: $x^2 - y = -2$.

Step 1. $x^2 - y = -2$

$y = x^2 + 2$

Steps 2 and 3. These two steps may be combined by constructing a table as shown below. First the x replacements are decided upon; then the corresponding y replacements are computed. If the x replacements are recorded along the top row (even though we may sometimes substitute for y first) and the corresponding y replacements along the bottom row, then the vertical, top-bottom pairs in the table are the ordered pairs we wish to plot.

x	-2	-1	0	1	2
y	6	3	2	3	6

Step 4. See Fig. 9.7.

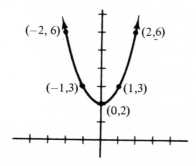

Figure 9.7

3. Graph: $3x - y = x^2 + 4$.

Step 1. $3x - y = x^2 + 4$

$y = -x^2 + 3x - 4$

Steps 2 and 3.

x	-2	-1	0	1	2	3	4
y	-14	-8	-4	-2	-2	-4	-8

Step 4.   See Fig. 9.8.

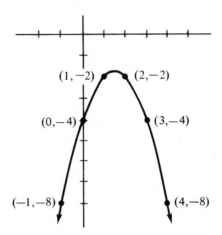

Figure 9.8

4. Graph: (a) $y = 6$; (b) $x = -2$.

(a) The solution set for $y = 6$ is a set of ordered pairs in which *every* y component is 6, no matter what the x component is. Thus, $(0, 6)$, $(-5, 6)$, and $(125, 6)$ are among the members of the solution set. Graphically, points corresponding to such a set of ordered pairs occur

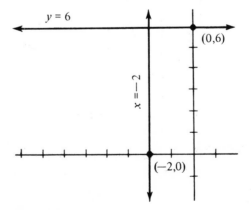

Figure 9.9

along a line that is parallel to the horizontal or x-axis, and 6 units above it. The line contains all points for which the ordinate is 6. See Fig. 9.9.

(b) The solution set for $x = -2$ is a set of ordered pairs in which every x component is -2, no matter what the y component is. Thus, $(-2, -6)$, $(-2, 6)$, $(-2, 3\frac{1}{2})$, and $(-2, 635)$ are among the members of the solution set. Graphically, points corresponding to such a set of ordered pairs occur along a line parallel to the vertical or y-axis, and 2 units to the left of it. The line contains all points for which the abscissa is -2. See Fig. 9.9.

EXERCISE 9-3

The graph of each of the following is a straight line; so only two points are necessary for its determination. However:
 (a) *Graph each equation with a minimum of five points.*
 (b) *Estimate the coordinates of two other points on the graph, and see if they satisfy the equation.*

1. $y = x$ **2.** $y = x + 1$ **3.** $x + y = 1$

4. $2x + y = 3$ **5.** $4x - 2y = 5$

6. Plot (a) $y = -x$, (b) $x + y = 3$, and (c) $2x + 2y = 4$ on the same grid. What geometric property do these lines seem to exhibit?

7. Write the equation of the straight line that is parallel to the x-axis and 4 units above it.

8. Write the equation of the straight line that is parallel to the y-axis and 8 units to its left.

Graph each of the following with no less than eight points; keep the variable replacements between $+10$ and -10 (i.e., $-10 < x < +10$, $-10 < y < 10$).

9. $y = 2x^2$ **10.** $y = x^2 - x$ **11.** $y = 2x^2 - 3$

12. $y = x^3$ **13.** $x^2 + 2y = 3$ **14.** $x^2 + y = 2x - 3$

15. $x^2 + y^2 = 16$

4. GRAPHING AN INEQUALITY IN TWO VARIABLES

The straight-line graph of a first-degree equation in two variables divides the Cartesian coordinate plane into two regions called **half-planes**. For example, the graph of the equation $y = x$ in Fig. 9.10 divides the Cartesian plane into one half-plane in which the ordinates are greater than the abscissas ($y > x$), and a second half-plane in which the ordinates are less than the abscissas ($y < x$).

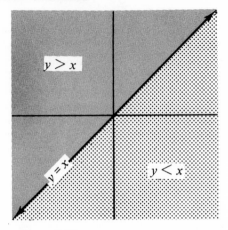

Figure 9.10

Similarly, the graph of every linear equation in two variables separates the points of the Cartesian plane into three distinct sets:

1. The set of points that corresponds to the solution set of the equation and that makes up the graph of the equation.
2. The set of points in the half-plane on one side of the graph of the equation.
3. The set of points in the half-plane on the other side of the graph of the equation.

Sets 2 and 3 correspond to the solution sets of two inequalities. The two inequalities are term-by-term duplicates of the equation whose graph is given by set 1, except that the = sign is replaced by > or <. See Fig. 9.11.

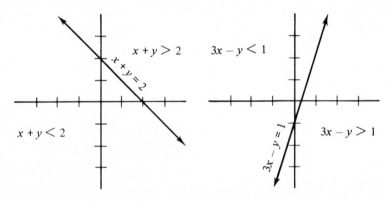

Figure 9.11

9.2

To graph the solution set of an inequality:

Step 1. Transform the given inequality into an equation by replacing the inequality sign with an = sign.

Step 2. Graph the equation of Step 1 on a Cartesian coordinate grid.

Step 3. Select any point in one of the half-planes determined by the graph of Step 2, and substitute its abscissa and ordinate for the appropriate variables of the given inequality. If the given inequality is satisfied, then the points of that half-plane correspond to the solution set of the inequality. If the inequality is not satisfied, then the points of the other half-plane correspond to the solution set of the given inequality.

EXAMPLES

1. Graph: $x + 2y > 1$.

Step 1.   Transform the inequality into an equation:

$$x + 2y = 1$$

Step 2.   Graph the equation of Step 1, as in Fig. 9.12a.

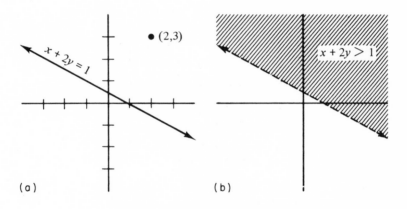

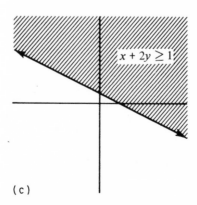

Figure 9.12

Step 3. Select (2, 3) as the check point in the half-plane above the graph of $x + 2y = 1$, and substitute for the appropriate variables in the given inequality.

$$x + 2y > 1$$

$$(2, 3): \quad 2 + 2(3) > 1 \quad \text{(true)}$$

Because the check point satisfies the inequality, the points of the half-plane above the graph of $x + 2y = 1$ correspond to the solution set of $x + 2y > 1$.

[*Note:* The graph of the solution set of an inequality in two variables is usually illustrated as a shaded half-plane, with the bounding line dashed, as shown in Fig. 9.12(b). When the inequality is compound (i.e., $\geq$ or $\leq$), then the set of points of the bounding line is included and indicated by a solid line, as shown in Fig. 9.12(c).]

2. Graph: $2x - y < 4$.

Step 1. $2x - y = 4$

Step 2. Graph $2x - y = 4$. See Fig. 9.13.

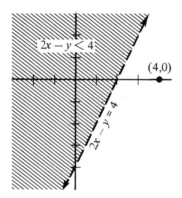

Figure 9.13

Step 3. Select (4, 0) as a convenient check point.

$$2x - y < 4$$

$$(4, 0): \quad 2(4) - 0 < 4 \quad \text{(false)}$$

The fact that (4, 0) does not satisfy the inequality implies that the graph of the solution set is the half-plane *other* than the one that includes (4, 0). See Fig. 9.13.

3. Graph: $2x \leq 9 + 3y$.

Step 1. $2x = 9 + 3y$

Step 2. Graph $2x = 9 + 3y$. See Fig. 9.14.

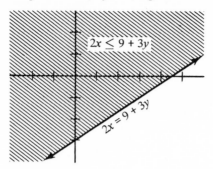

Figure 9.14

Step 3. Select $(0, 0)$ as a check point.

$$2x \leq 9 + 3y$$
$$(0, 0): \quad 2(0) \leq 9 + 3(0) \quad \text{(true)}$$

The fact that $(0, 0)$ satisfies the inequality implies that the graph of the solution set is the half-plane that includes $(0, 0)$. See Fig. 9.14.

4. Graph: $x^2 - y > -2$.

Any graph that separates the coordinate plane into two distinct plane regions may be plotted according to Program **9.2**.

Step 1. $x^2 - y = -2$

Step 2. (See Fig. 9.7, p. 207.)

Step 3. $(0, 0)$ is selected as the check point:

$$x^2 - y > -2$$
$$(0, 0): \quad 0^2 - 0 > -2 \quad \text{(true)}$$

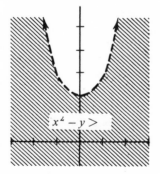

Figure 9.15

Graph the solution set of each inequality:

1. $y > 2x$

2. $y < 2x$

3. $y < x + 1$

4. $y > x + 1$

5. $y \geq 3x - 1$

6. $y \leq 3x - 1$

7. $x + y < 1$

8. $x + y > 1$

9. $2x + y \leq 3$

10. $2x + y \geq 3$

11. $y \geq 3$

12. $x \leq -2$

13. $4x < 5 + 2y$

14. $4x > 5 + 2y$

15. $y < 2x^2$

16. $y \geq 2x^2$

17. $y \leq x^2 - x$

18. $y > x^2 - x$

19. $y > x^3$

20. $y < x^3$

5. SYSTEMS OF EQUATIONS

A **system of equations** is a set of equations. The simplest system consists of a pair of linear or first-degree equations as, for example:

$$\begin{cases} 2x - y = 7 \\ x + y = 5 \end{cases}$$

The solution of such a system is an ordered pair of numbers that satisfies *both* equations. One direct way to identify this ordered pair is to plot the equations separately and then note the abscissa and ordinate of their point of intersection.

Systems of two linear equations in two variables may have one solution, no solution, or an infinite number of solutions. This becomes apparent if we study the graphs of the equations. As a pair of straight lines in a plane, the graphs may intersect *once*—as when they are not parallel or coincident (i.e., have the same path); *not at all*—as when they are parallel, in which case the equations of the system are said to be **inconsistent**; or *an infinite number of times*—as when they are coincident, in which case the equations of the system are said to be **dependent**.

It is important to remember that solutions of systems of equations arrived at graphically are, at best, approximate; they should be verified for each equation of the system.

9.3

To solve a system of linear equations graphically:

Step 1. Graph each equation of the system separately on the same grid.

Step 2. Estimate the ordered-pair coordinates of the point at which the graphs of Step 1 intersect.

Step 3. Substitute the ordered-pair of Step 2 for the appropriate variables in each equation of the system to verify that it is a common solution of the equations of the system.

EXAMPLES

1. Solve the system of equations graphically: $\begin{cases} 2x - y = 7 \\ x + y = 5 \end{cases}$.

Step 1.

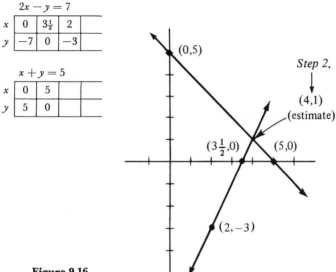

$2x - y = 7$

x	0	$3\frac{1}{2}$	2		
y	-7	0	-3		

$x + y = 5$

x	0	5		
y	5	0		

(0,5)

Step 2.

(4,1)
(estimate)

$(3\frac{1}{2},0)$ (5,0)

(2,−3)

Figure 9.16

Step 3.   Verify:

$$2x - y = 7 \qquad\qquad\qquad x + y = 5$$

$\left.\begin{array}{l} x = 4 \\ y = 1 \end{array}\right\}$ $2(4) - 1 = 7$ $\left.\begin{array}{l} x = 4 \\ y = 1 \end{array}\right\}$ $(4) + (1) = 5$

Replacing x and y with 4 and 1, respectively, satisfies both equations in the system; thus, $(4, 1)$ is the solution of the system.

2. Solve the system graphically: $\begin{cases} x = 7 \\ x + 3y - 4 = 0 \end{cases}$.

Step 1.   The graph of $x = 7$ is a line perpendicular to the x-axis along which all points have an abscissa of 7. For the other equation of the system:

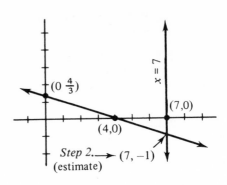

$$x + 3y - 4 = 0$$

x	0	4
y	$\frac{4}{3}$	0

Figure 9.17

$(0\,\frac{4}{3})$

$(7,0)$

$(4,0)$

$x = 7$

Step 2. ⟶ $(7, -1)$
(estimate)

Step 3. $x = 7$ $x + 3y - 4 = 0$

$\left.\begin{array}{l} x = 7 \\ y = -1 \end{array}\right\}$ $(7) = 7$ $\left.\begin{array}{l} x = 7 \\ y = -1 \end{array}\right\}$ $(7) + 3(-1) - 4 = 0$

The solution of the system is the ordered pair $(7, -1)$.

3. Solve graphically: $\begin{cases} x + 2y - 4 = 0 \\ 2x + 4y - 12 = 0 \end{cases}$

The graphs of the two equations in the system appear to be parallel lines. (See Fig. 9.18.) If so, the equations are inconsistent, and the system has no solution.

$$x + 2y - 4 = 0$$

x	0	4
y	2	0

$$2x + 4y - 12 = 0$$

x	0	6
y	3	0

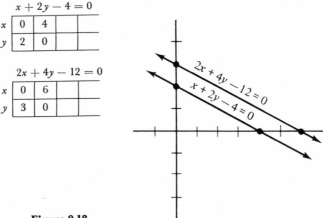

$2x + 4y - 12 = 0$

$x + 2y - 4 = 0$

Figure 9.18

4. Solve graphically: $\begin{cases} 8x - 2y = 10 \\ 12x - 3y = 15 \end{cases}$

The graphs of the two equations appear to be coincident. (See Fig. 9.19.) If so, their infinite number of points in common mean that the equations of the system are dependent, having an infinite number of common solutions.

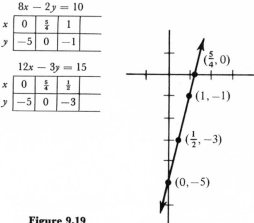

$8x - 2y = 10$

x	0	$\frac{5}{4}$	1	
y	-5	0	-1	

$12x - 3y = 15$

x	0	$\frac{5}{4}$	$\frac{1}{2}$	
y	-5	0	-3	

$(\frac{5}{4}, 0)$

$(1, -1)$

$(\frac{1}{2}, -3)$

$(0, -5)$

Figure 9.19

EXERCISE 9-5

Solve the systems of equations graphically:

1. $\begin{cases} 2x + y = 7 \\ x + y = 5 \end{cases}$
 2. $\begin{cases} 3x + y = 7 \\ 2x + y = 6 \end{cases}$

3. $\begin{cases} x + 2y - 4 = 0 \\ x + y - 1 = 0 \end{cases}$
 4. $\begin{cases} x - 2y - 1 = 0 \\ 2x + y + 8 = 0 \end{cases}$

5. $\begin{cases} x - 2y - 5 = 0 \\ x + 2y - 3 = 0 \end{cases}$
 6. $\begin{cases} x - 3y = 2 \\ 2x - 6y = 4 \end{cases}$

7. $\begin{cases} 4x + 3y + 7 = 0 \\ 2x + 5y = 0 \end{cases}$
 8. $\begin{cases} 3x - 2y = 8 \\ 6x - 4y = 10 \end{cases}$

9. $\begin{cases} x + y + 8 = 0 \\ x - y - 1 = 0 \end{cases}$
 10. $\begin{cases} 3x - y - 6 = 0 \\ 3x + 2y + 12 = 0 \end{cases}$

11. $\begin{cases} 4x - y = -8 \\ x + 2 = 0 \end{cases}$
 12. $\begin{cases} 4x + 3y - 3 = 0 \\ 2x + 1.5y - 1 = 0 \end{cases}$

13. Given: $f(x, y) = 2x - y - 4$, $g(x, y) = x - y - 1$, and $p(x, y) = f(x, y) + g(x, y)$.

(a) Graph $f(x, y) = 0$, $g(x, y) = 0$, and $p(x, y) = 0$ on the same grid. Do all three graphs intersect at the same point?

(b) Graph $m(x, y) = 0$, where $m(x, y) = f(x, y) - g(x, y)$ on the same grid. Does it intersect the three lines of part (a) at a single point?

(c) Graph $n(x, y) = 0$, where $n(x, y) = f(x, y) - 2g(x, y)$. Does it intersect the four lines of parts (a) and (b) at the same point? [*Note:* $2g(x, y)$ means "twice $g(x, y)$."]

(d) Find the solution for the system of five equations given by
$$\begin{cases} f(x, y) = 0 \\ g(x, y) = 0 \\ p(x, y) = 0 \\ m(x, y) = 0 \\ n(x, y) = 0 \end{cases}$$

6. ALGEBRAIC SOLUTION OF SYSTEMS OF TWO LINEAR EQUATIONS IN TWO VARIABLES

Two algebraic methods for solving systems of two linear equations in two variables are substitution and addition-subtraction.

I. Solution by Substitution

If we solve each equation of the system for the same variable, we obtain two equivalent expressions for that variable. For example, if the system is

$$\begin{cases} x - y = 3 \\ x + 2y = 6 \end{cases}$$

we may express the equations equivalently as

$$\begin{cases} x = 3 + y \\ x = 6 - 2y \end{cases}$$

Equating the equivalent expressions for x, we obtain:

$$(A) \qquad 3 + y = 6 - 2y$$

which is a linear equation in a single variable, y. In effect, this amounts to substituting for one variable in one equation of the system its equivalent obtained from the other equation of the system. Thus:

$$x + 2y = 6$$

$$x = 3 + y: \quad (3 + y) + 2y = 6$$

$$(B) \qquad 3 + y + 2y = 6$$

Equations (A) and (B) are clearly equivalent and may be solved. Once the y component of the ordered-pair solution is known, the x component is readily obtained from either equation of the system.

9.4

To solve a system of two linear equations in two variables by substitution:

Step 1. Solve one of the equations for one of the variables in terms of the other.

Step 2. Substitute the solution of Step 1 for the appropriate variable in the other equation.

Step 3. Solve the equation of Step 2 for one component of the ordered-pair solution of the system.

Step 4. Substitute the solution of Step 3 for the appropriate variable in one of the equations of the system, and solve for the remaining variable.

Step 5. Verify by replacing the variables of the equation of the system not used in Step 4 with the ordered pair determined in Steps 3 and 4.

EXAMPLES

1. Solve by substitution: $\begin{cases} 2x - y = 7 \\ x + y = 5 \end{cases}$.

Step 1. Solve $2x - y = 7$ for y: $y = 2x - 7$.

Step 2. Substitute $2x - 7$ for y in

$$x + y = 5$$
$$y = 2x - 7: \quad x + (2x - 7) = 5$$

Step 3.
$$x + 2x - 7 = 5$$
$$3x = 12$$
$$x = 4$$

Step 4.
$$2x - y = 7$$
$$x = 4: \quad 2(4) - y = 7$$
$$y = 1$$

Step 5. Verification:

$$x + y = 5$$
$$\left. \begin{array}{l} x = 4 \\ y = 1 \end{array} \right\} \quad (4) + (1) \overset{?}{=} 5$$
$$5 = 5$$

The ordered pair $(4, 1)$ satisfies both equations and, therefore, is the solution of the system.

2. Solve by substitution: $\begin{cases} 6x - 3y = 15 \\ x - 4y = 6 \end{cases}$.

Step 1.
$$x - 4y = 6$$
$$x = 4y + 6$$

Step 2.
$$6x - 3y = 15$$
$$x = 4y + 6: \quad 6(4y + 6) - 3y = 15$$

Step 3.
$$24y + 36 - 3y = 15$$
$$24y - 3y = 15 - 36$$
$$21y = -21$$
$$y = -1$$

Step 4.
$$x - 4y = 6$$
$$y = -1: \quad x - 4(-1) = 6$$
$$x = 6 - 4$$
$$x = 2$$

Step 5. Verification:

$$6x - 3y = 15$$

$$\left.\begin{array}{l} x = 2 \\ y = -1 \end{array}\right\} \quad 6(2) - 3(-1) \overset{?}{=} 15$$

$$12 + 3 \overset{?}{=} 15$$
$$15 = 15$$

The solution of the system is $(2, -1)$.

3. Solve by substitution: $\begin{cases} 4x + 3y = 7 \\ 8x - 3y = 41 \end{cases}$

Step 1. $4x + 3y = 7$
$$4x = 7 - 3y$$
$$x = \frac{7 - 3y}{4}$$
$$= (\tfrac{1}{4})(7 - 3y)$$

Step 2.
$$8x - 3y = 41$$
$$x = \tfrac{1}{4}(7 - 3y): \quad 8(\tfrac{1}{4})(7 - 3y) - 3y = 41$$

Step 3.
$$2(7 - 3y) - 3y = 41$$
$$14 - 6y - 3y = 41$$
$$-6y - 3y = 41 - 14$$
$$-9y = 27$$
$$y = -3$$

Step 4.
$$4x + 3y = 7$$
$$y = -3: \quad 4x + 3(-3) = 7$$
$$4x - 9 = 7$$
$$4x = 16$$
$$x = 4$$

Step 5. Verification:

$$8x - 3y = 41$$

$$\left.\begin{array}{l} x = 4 \\ y = -3 \end{array}\right\} \quad 8(4) - 3(-3) \overset{?}{=} 41$$

$$32 + 9 \overset{?}{=} 41$$
$$41 = 41$$

II. Solution by Addition-Subtraction

By an appropriate term-by-term addition or subtraction of the two linear equations of a system, an equivalent equation will result in which only one of the variables is present, that is, the other variable takes on zero as its numerical coefficient. This is referred to as *eliminating a variable*. In cases where eliminating a variable is not immediately possible, an appropriate multiplication of all terms of either or both equations will permit a subsequent addition or subtraction of the equations to eliminate one of the variables. Thus, in Example 3 (p. 220), by *adding* the two equations together, term by term, we derive an equivalent equation in which the y term is eliminated. The derived equation is readily solved:

$$
\begin{array}{r}
4x + 3y = 7 \\
8x - 3y = 41 \\
\hline
12x + 0y = 48 \\
12x = 48 \\
x = 4
\end{array}
$$

$(+)$

This solution, when substituted for the variable x in one of the two equations of the system, results in another equation in a single variable, which can be solved. In this way, the ordered-pair solution of the system is determined; we can verify it by using it to replace the variables in the other of the two equations of the system.

9.5

To solve a system of two linear equations in two variables by addition-subtraction:

Step 1. Multiply, if necessary, the terms of one or both equations by a number that will make the numerical coefficients of one of the variables in both equations either the same as or negatives of one another.

Step 2. Eliminate the variables whose coefficients are negatives of one another by adding the two equations together, term by term, or by subtracting one equation from the other, term by term, when the coefficients are the same.

Step 3. Solve the resulting equation of Step 2 for one component of the ordered-pair solution of the system.

Step 4. Substitute the solution of Step 3 for the appropriate variable in one of the equations of the system, and solve for the other component of the ordered-pair solution of the system.

Step 5. Verify the solution by replacing the variables of the equation of the system not used in Step 4 with the ordered pair determined in Steps 3 and 4.

EXAMPLES

1. Solve by the addition-subtraction method: $\begin{cases} 4x + 3y = 7 \\ 8x - 3y = 41 \end{cases}$.

Step 1.   No multiplication is necessary, since the numerical coefficients of the y terms are negatives of each other.

Step 2.   Add the two equations, term by term:

$$4x + 3y = 7$$
$$8x - 3y = 41$$

$$(+) \overline{}$$

Step 3.   Solve for x: $12x = 48$

$$x = 4$$

Step 4.   Replace the variable x in the first equation with 4; solve for y:

$$4x + 3y = 7$$
$$x = 4: \quad 4(4) + 3y = 7$$
$$3y = 7 - 16$$
$$y = -3$$

Step 5.   Verify by replacing x and y of the second equation with 4 and -3, respectively:

$$8x - 3y = 41$$

$$\left. \begin{array}{l} x = 4 \\ y = -3 \end{array} \right\} \quad 8(4) - 3(-3) \overset{?}{=} 41$$

$$32 + 9 \overset{?}{=} 41$$
$$41 = 41$$

The solution of the given system is the ordered pair $(4, -3)$.

2. Solve by the addition-subtraction method: $\begin{cases} 6x + 2y = 11 \\ 4x + 3y = 14 \end{cases}$.

Step 1.   Multiply all terms of the top equation by 2 and of the bottom equation by 3:

$$2 \cdot (6x + 2y = 11): \quad 12x + 4y = 22$$
$$3 \cdot (4x + 3y = 14): \quad 12x + 9y = 42$$

Step 2.   Subtract: $-5y = -20$

Step 3.   Solve for y: $y = 4$

Step 4.   Substitute 4 for y in one of the equations of the system, and solve for x:

$$6x + 2y = 11$$
$$y = 4: \quad 6x + 2(4) = 11$$
$$6x + 8 = 11$$

$$6x = 3$$
$$x = \tfrac{1}{2}$$

Step 5.   Verify, using the equation of the system not used in Step 4:

$$4x + 3y = 14$$

$$\left.\begin{array}{r} x = \tfrac{1}{2} \\ y = 4 \end{array}\right\} \quad 4(\tfrac{1}{2}) + 3(4) \overset{?}{=} 14$$

$$2 + 12 \overset{?}{=} 14$$

$$14 = 14$$

Solution of the system: $(\tfrac{1}{2}, 4)$.

3. Solve: $\begin{cases} 5x - 2y = 5 \\ 2x + 3y = 21 \end{cases}$

Step 1.   Multiply the top equation by 3 and the bottom equation by 2:

$$3 \cdot (5x - 2y = 5): \quad 15x - 6y = 15$$
$$2 \cdot (2x + 3y = 21): \quad 4x + 6y = 42$$

Step 2.   Add: $19x \qquad = 57$

Step 3.   Solve for x: $x \qquad = 3$

Step 4.   Substitute 3 for x in one of the equations of the system, and solve for y:

$$5x - 2y = 5$$
$$x = 3: \quad 5(3) - 2y = 5$$
$$15 - 2y = 5$$
$$-2y = -10$$
$$y = 5$$

Step 5.   Verify, using the other of the two equations of the system:

$$2x + 3y = 21$$

$$\left.\begin{array}{r} x = 3 \\ y = 5 \end{array}\right\} \quad 2(3) + 3(5) \overset{?}{=} 21$$

$$6 + 15 \overset{?}{=} 21$$

$$21 = 21$$

Solution of the system: $(3, 5)$.

7. DEPENDENT AND INCONSISTENT EQUATIONS

In Section 5 we noted that the equations of certain systems had coincident graphs and parallel graphs. These were called dependent and inconsistent equations, respectively. Greater certainty can be had about the nature of these equations if we apply Program **9.5**.

A pair of dependent equations differ one from the other, term by term, only by some numerical factor. Their solution sets are identical. For example, an equation having the same solution set as $x + 2y = 5$ can be derived if we multiply each member of the equation by some number, say, 4:

$$(A) \qquad\qquad x + 2y = 5$$

$$(B) \quad (= 4 \times (A)) \qquad 4x + 8y = 20$$

Obviously, for equations (A) and (B), any attempt at making the numerical coefficients of either variable the same by some appropriate multiplication (Step 1, Program **9.5**) will automatically make the numerical coefficients of the other variable and the constant terms the same also; hence, the result of Step 2 is bound to be $0 = 0$. Such a result is certain evidence that the two equations of the system are dependent.

EXAMPLE

Solve: $\begin{cases} 8x - 2y = 10 \\ 12x - 3y = 15 \end{cases}$.

Step 1. Multiply the top equation by 3 and the bottom equation by 2:

$$3 \cdot (8x - 2y = 10): \quad 24x - 6y = 30$$
$$2 \cdot (12x - 3y = 15): \quad 24x - 6y = 30$$

Step 2.
$$0x + 0y = 0$$
$$0 = 0$$

The original equations are dependent, and any ordered pair that satisfies one equation of the system satisfies the other.

Inconsistent equations differ from dependent equations in that the numerical coefficients of the variables differ by a certain factor, but not the constant terms. This assures that the solution sets of the two equations will be completely different. When Step 2 of Program **9.5** is performed, both the variable terms are eliminated—but not the constant terms. Such a result is certain evidence that the equations of the system are inconsistent.

EXAMPLE

Solve: $\begin{cases} 4x - 5y = 8 \\ 8x - 10y = 2 \end{cases}$.

Step 1. Multiply the top equation by 2:

$$2 \cdot (4x - 5y = 8): \quad 8x - 10y = 16$$
$$8x - 10y = 2$$

Step 2. Subtract:
$$0x + 0y = 14$$
$$0 = 14$$

The result of Step 2, $0 = 14$, is a false sentence. The equations are therefore said to be inconsistent; there is *no* ordered pair that satisfies *both* equations of the system.

EXERCISE 9-6

Solve the systems by substitution:

1. $\begin{cases} 2x + 3y = 13 \\ x + 2y = 8 \end{cases}$
2. $\begin{cases} 2x - y = 3 \\ 3x - 2y = 1 \end{cases}$

3. $\begin{cases} 4x + 2y - 5 = 0 \\ 3x + y - 2 = 0 \end{cases}$
4. $\begin{cases} x + 2y = -1 \\ 2x + 5y = 0 \end{cases}$

Solve the systems by addition-subtraction:

5. $\begin{cases} 3x - 2y = 4 \\ 3x + 2y = 8 \end{cases}$
6. $\begin{cases} 3x - 2y = 9 \\ 3x - 5y = 18 \end{cases}$

7. $\begin{cases} 3x - 4y = 6 \\ 2x - y = -1 \end{cases}$
8. $\begin{cases} 2x - 3y - 20 = 0 \\ 3x + 5y - 11 = 0 \end{cases}$

Solve the systems by either algebraic method:

9. $\begin{cases} 7x + 3y = 4 \\ 14x + 9y = 7 \end{cases}$
10. $\begin{cases} x - 2y = 4 \\ x + 4y = 1 \end{cases}$
11. $\begin{cases} 3x + 6y = 9 \\ 4x + 8y = 12 \end{cases}$

12. $\begin{cases} x - 4y = 1 \\ 3x - 8y = 7 \end{cases}$
13. $\begin{cases} 2x - 3y = 5 \\ -6y + 4x = 8 \end{cases}$
14. $\begin{cases} 6x + y - 4 = 0 \\ y + 8 = 0 \end{cases}$

15. $\begin{cases} x - 5y = -6\frac{1}{2} \\ 2x + 2y = 17 \end{cases}$
16. $\begin{cases} x + y = -2 \\ 0.3x + 0.7y = -5 \end{cases}$
17. $\begin{cases} 4x + 2y = 7 \\ y = -2x + 3 \end{cases}$

18. $\begin{cases} 6x - 5y + 4 = 0 \\ x - 10y = 1 \end{cases}$
19. $\begin{cases} \frac{2}{3}x + \frac{3}{4}y = 26 \\ \frac{1}{6}x - \frac{3}{8}y = -7 \end{cases}$
20. $\begin{cases} \dfrac{x+3}{4} = \dfrac{y}{2} + 1 \\ \dfrac{x}{2} = \dfrac{y+3}{4} + \dfrac{1}{2} \end{cases}$

21. The length of a rectangle exceeds its width by 8 ft, and the perimeter of the rectangle is 76 ft. Find the dimensions of the rectangle.

22. During one month, the charge for 25 daily papers and 5 Sunday papers was $2.00. During the next month, the paper bill was $1.95 for 27 daily and 4 Sunday papers. What is the unit cost for daily and Sunday papers?

23. Between two cities, an 8-minute phone call costs $3.15 and a 10-minute call costs $3.85. Find the basic rate for the first three minutes and the overtime rate for each additional minute.

24. A chemistry student wishes to make 100 cc of 27% acid solution by combining quantities of 20% and 30% solutions. How many cubic centimeters of each will be necessary?

25. Use one of the algebraic methods to show that for the system

$$\begin{cases} a_1 x + b_1 y = c_1 \\ a_2 x + b_2 y = c_2 \end{cases}$$

we obtain:

$$x = \frac{b_2 c_1 - b_1 c_2}{a_1 b_2 - a_2 b_1} \quad \text{and} \quad y = \frac{a_1 c_2 - a_2 c_1}{a_1 b_2 - a_2 b_1}$$

(Note that the denominators for x and y are equal.)

26. Use the expressions for x and y derived in Exercise 25 as formulas to solve the systems of Exercises 1, 7, 11, and 13.

8. GRAPHING SYSTEMS OF INEQUALITIES

Solution of systems of inequalities has taken on new importance in recent years among the applications of mathematics. As with systems of equations, the solution of a system of inequalities may be represented by the intersection of the graphs of inequalities that make up the system. When the system consists of equations, the solution is represented by a point; when the system consists of inequalities, the solutions are represented by a region.

EXAMPLES

1. Solve: $\begin{cases} x + y < 4 \\ x - y > 3 \end{cases}$.

The graph of $x + y < 4$ is shown in Fig. 9.20 as a tinted region; the graph of $x - y > 3$ is shown as a dotted region. The intersection of

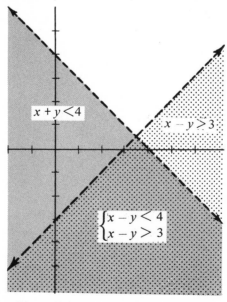

Figure 9.20

these two regions (a dotted-tinted region) represents the solution set of the system.

2. Solve: $\begin{cases} x \geq 3 \\ x - 2y > 2 \end{cases}$.

The graph of $x \geq 3$ is shown in Fig. 9.21 as a tinted region, including the bounding set of points, because $x \geq 3$ means $x > 3$ or $x = 3$. The graph of $x - 2y > 2$ is shown as a dotted region, excluding the bounding set of points, because the graph of $x - 2y = 2$ is not included in that of $x - 2y > 2$. The graph of the solution set of the system is that set of points in the dotted-tinted region, *including* the bounding set of points at the left, and *excluding* the bounding set of points above.

3. Solve: $\begin{cases} x - y \leq 5 \\ x - y \geq 1 \end{cases}$.

The graphs of the inequalities of the system are shown in Fig. 9.22. The respective bounding sets of points are included. The graph of the solution set of the system is the set of points in the dotted-tinted band, including the bounding points.

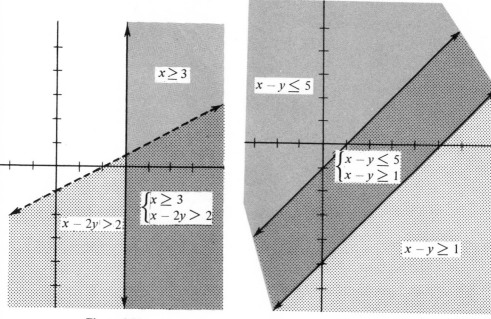

Figure 9.21   **Figure 9.22**

EXERCISE 9-7

Solve the systems graphically:

1. $\begin{cases} x + y < 3 \\ x - y < 1 \end{cases}$ **2.** $\begin{cases} x + y < 6 \\ x - y < 2 \end{cases}$

3. $\begin{cases} 2x - y > 4 \\ x - y > 1 \end{cases}$

4. $\begin{cases} 2x + 3y \geq 10 \\ x - 3y > 5 \end{cases}$

5. $\begin{cases} 3x - 2y < 8 \\ 2x + y > 3 \end{cases}$

6. $\begin{cases} 3x - 2y > 8 \\ 2x + y < 3 \end{cases}$

7. $\begin{cases} 4x + y < 0 \\ 2x - y \leq -6 \end{cases}$

8. $\begin{cases} 2x + 3y \leq 5 \\ y > 3 \end{cases}$

9. $\begin{cases} x - y > 4 \\ x - y < 1 \end{cases}$

10. $\begin{cases} 2x - y \leq 3 \\ 2x - y \geq 5 \end{cases}$

The solution set of each of the following systems is represented by the region common to all inequalities in the system. Graph the solution set for:

11. $\begin{cases} x \geq 0 \\ y \geq 0 \\ x - y \geq 0 \end{cases}$

12. $\begin{cases} x > 1 \\ y < 2 \\ x - y < 0 \end{cases}$

13. $\begin{cases} x \leq 1 \\ y \leq 3 \\ 2x + 3y \leq 0 \end{cases}$

14. $\begin{cases} x > -2 \\ y < 1 \\ 2x - y < 2 \end{cases}$

Determine the coordinates of the points of the triangular region defined by:

15. The system of inequalities of Exercise 13.

16. The system of inequalities of Exercise 14.

Write a system of inequalities to define a triangular region whose vertices are:

17. $(3, 3)$, $(6, 6)$, $(3, 6)$

18. $(5, -2)$, $(-2, -2)$, $(5, 5)$

REVIEW—CHAPTER IX

Select from among the ordered pairs given for each equation those which are solutions of the equation:

1. $3x - y = 5$; $(0, 5)$, $(3, 4)$, $(1, -2)$, $(2, -1)$

2. $2x - 4y + 3 = 0$; $(0, \frac{3}{4})$, $(-1, \frac{1}{4})$, $(3, 2\frac{1}{2})$, $(-3, \frac{3}{4})$

3. $3x - 5y = 6$; $(2, 0)$, $(\frac{4}{3}, 2)$, $(3, 4)$, $(7, 3)$

4. $2x^2 - 3y^2 = 8$; $(3, 2)$, $(-4, 0)$, $(5\frac{1}{2}, -1)$, $(\sqrt{10}, -2)$

Replace the "?" so as to make the ordered pair a solution of the given equation:

5. $4x - 5y = 11$; $(?, 3)$, $(-1, ?)$, $(?, -2)$

6. $x^2 - 4y = 4$; $(?, 3)$, $(-6, ?)$, $(?, 0)$

7. Connect the four points in order with line segments: A: $(0, 4)$; B: $(-2, 2)$; C: $(2, -1)$; D: $(4, 1)$.
(a) What kind of figure is $ABCD$?
(b) Draw the diagonals and estimate their point of intersection.

(c) Using each of the sides of the figure as a hypotenuse, visualize a right triangle in the squares of the grid, and determine the length of each side.

8. The ends of the hypotenuse of two right triangles whose legs parallel the axes are at $(-2, -3)$ and $(4, 0)$. What are the coordinates of the right-angle vertices?

9. The mirror image of an object appears to be behind the surface of the mirror at a distance equal to that between the surface of the mirror and the actual object. Consider the horizontal and vertical axes to be mirror surfaces, and give the coordinates of the two points which are "mirror images" of each of the following:

(a) $(3, 4)$ (b) $(-2, 3)$ (c) $(-4, -4)$ (d) $(3, -1)$

10. What geometric property do the graphs of the lines $3y = 2x + 3$, $6y - 4x = 5$, and $y = \frac{2}{3}x + 2$ exhibit?

11. Write the equations of the two lines that are parallel to each of the axes and intersect at $(-3, 2)$.

Graph the following:

12. $y = 2x^2 - 3x + 1$

13. $y = x^3 - x$

14. $y = \dfrac{12}{x}$

15. $x + 3y = x^2 - 7$

Graph the solution set for each inequality:

16. $y > x$

17. $3x - y \leq 4$

18. $x + y < 3$

19. $2x + 3y \leq 0$

20. $2x + y < 2$

21. $y > x^2 - 2x$

Solve the systems graphically:

22. $\begin{cases} 3x + 2y = 11 \\ 2x + 3y = 4 \end{cases}$

23. $\begin{cases} 2x + 4y + 1 = 0 \\ 4x + 5y - 4 = 0 \end{cases}$

24. $\begin{cases} x - y + 2 = 0 \\ 3x - y - 1 = 0 \end{cases}$

25. $\begin{cases} 2x - 3y = 5 \\ 6y - 4x + 10 = 0 \end{cases}$

26. $\begin{cases} 11x + 2y + 6 = 0 \\ y + 3 = 0 \end{cases}$

27. $\begin{cases} 3x - 2y + 4 = 0 \\ -6x + 4y + 7 = 0 \end{cases}$

Solve the systems algebraically:

28. $\begin{cases} 2x + 3y = 2 \\ 3x - y = 14 \end{cases}$

29. $\begin{cases} 4x - 3y + 5 = 0 \\ 3x - 5y + 1 = 0 \end{cases}$

30. $\begin{cases} 4x + 3y = 4 \\ 6x - 9y = -3 \end{cases}$

31. $\begin{cases} 4x - 7y + 32 = 0 \\ 7x + 3y - 66 = 0 \end{cases}$

32. $\begin{cases} 4x - 7y + 2 = 0 \\ 8x - 6y = 8y \end{cases}$

33. $\begin{cases} 3x + 2y + 9 = 0 \\ x + 3 = 0 \end{cases}$

34. $\begin{cases} \frac{2}{3}x + \frac{3}{4}y = 8 \\ \frac{1}{3}x + \frac{3}{8}y = 4 \end{cases}$

35. $\begin{cases} \dfrac{3x - 5}{2} + 2 = \dfrac{y}{8} \\ \dfrac{7 - 5x}{3} + \dfrac{y}{2} = \dfrac{1}{3} \end{cases}$

36. Admission charges for a church benefit were 50¢ for children and 80¢ for adults. Total receipts from the 160 paid admissions were $117.50. How many of each kind of ticket must have been sold?

37. A chemist wishes to produce 400 cc of 45% alcohol solution by using quantities of 70% solution and 30% solution. How much of each should he use?

Graph the solution set for each system:

38. $\begin{cases} x + y < 3 \\ x - y > 2 \end{cases}$ **39.** $\begin{cases} 2x - y \leq 1 \\ x - y \geq 2 \end{cases}$

40. $\begin{cases} 2x - y \geq 1 \\ 2x - y \leq 3 \end{cases}$ **41.** $\begin{cases} x \geq 1 \\ y < 2 \\ x - y < 3 \end{cases}$

X

LOGARITHMS

Before the invention of computing machines, mathematicians developed a set of "auxiliary numbers" to help in the tedious work of computing products, quotients, powers, and roots. These numbers are called **logarithms**. Logarithms continue to be of interest today because of their many applications to science, business, and industry.

Logarithms are essentially exponents and have both the properties and operational characteristics of exponents. For each base number, there is an ordered relationship between the value of the exponent and the numerical value of the power of which it is a part. An example of that relationship is illustrated in the following display in which the base number is 2. Note that as the exponent decreases in value, left to right, the numerical value of the power also decreases.

2^5	2^4	2^3	2^2	2^1	2^0	2^{-1}	2^{-2}	2^{-3}	2^{-4}	2^{-5}
32	16	8	4	2	1	$\frac{1}{2}$	$\frac{1}{4}$	$\frac{1}{8}$	$\frac{1}{16}$	$\frac{1}{32}$

The relationship between a power and its numerical equivalent may be stated in the following way:

1. EXPONENTS AND LOGARITHMS

$2^5 = 32$: the logarithm of 32 to the base of 2 is 5; or $\log_2 32 = 5$
$2^2 = 4$: the logarithm of 4 to the base of 2 is 2; or $\log_2 4 = 2$
$2^{-3} = \frac{1}{8}$: the logarithm of $\frac{1}{8}$ to the base of 2 is -3; or $\log_2 \frac{1}{8} = -3$

In general:

If $a^L = n$, where a denotes any positive number other than 1, the exponent L is called the **logarithm of n to the base a**; *symbolically,* $\log_a \mathbf{n} = \mathbf{L}$.

Thus $\log_a n = L$ and $a^L = n$ are equivalent sentences.

EXAMPLES

1. Write each of the following in logarithmic form:

(a) $2^7 = 128$.   Answer: $\log_2 128 = 7$.

(b) $3^4 = 81$.   Answer: $\log_3 81 = 4$.

(c) $5^{-2} = 0.04$.   Answer: $\log_5 0.04 = -2$.

2. Write each of the following in exponential form:

(a) $\log_3 243 = 5$.   Answer: $3^5 = 243$.

(b) $\log_{10} 100 = 2$.   Answer: $10^2 = 100$.

(c) $\log_a 25 = 2$.   Answer: $a^2 = 25$.

3. What replacement for the variable in each of the following will make the sentence true: (a) $\log_4 16 = L$; (b) $\log_a 125 = 3$; (c) $\log_4 n = 0.5$?

(a) $\log_4 16 = L$: $4^L = 16$
$4^2 = 16$
$L = 2$

(b) $\log_a 125 = 3$: $a^3 = 125$
$5^3 = 125$
$a = 5$

(c) $\log_4 n = 0.5$: $4^{0.5} = n$
$4^{1/2} = n$
$2 = n$

EXERCISE 10-1

Express each in logarithmic form:

1. $2^3 = 8$ **2.** $10^3 = 1000$ **3.** $3^5 = 243$

4. $4^3 = 64$ **5.** $16^{1/4} = 2$ **6.** $36^{0.5} = 6$

7. $7^{-2} = \frac{1}{49}$ **8.** $2^{-2} = 0.25$ **9.** $3^{-1} = 0.33$

10. $27^{-0.33} = 3$

Express each in exponential form:

11. $\log_4 64 = 3$ **12.** $\log_3 81 = 4$ **13.** $\log_{0.1} 0.001 = 3$

14. $\log_{0.5} 0.125 = 3$ **15.** $\log_8 2 = 0.33$ **16.** $\log_{10} 0.01 = -2$

17. $\log_{10} 2 = 0.301$ **18.** $\log_{10} 5 = 0.699$ **19.** $\log_{10} 17 = 0.230$

20. $\log_a 8 = 3$

What replacement for the variable in each of the following will make the sentence true?

21. $\log_3 81 = L$ **22.** $\log_2 32 = L$ **23.** $\log_a 8 = 3$

24. $\log_a 16 = 2$ **25.** $\log_5 n = 2$ **26.** $\log_4 n = -2$

27. $\log_a 6 = 0.5$ **28.** $\log_4 \frac{1}{4} = L$ **29.** $\log_8 4 = L$

30. $\log_a 216 = 3$ **31.** $\log_5 n = 3$ **32.** $\log_3 \frac{1}{27} = L$

33. $\log_a 0.64 = 2$ **34.** $\log_5 625 = L$ **35.** $\log_{11} n = -2$

36. $\log_{12} n = 0.5$

2. LOGARITHM TABLE

Our Hindu-Arabic system of number notation is said to be a decimal system because it has 10 as its base number (*decem* is the Latin word for ten). Logarithms that have 10 as a base are called **common logarithms**. Systems of logarithms having other numbers as base are not only feasible, but sometimes preferable. Since we limit our discussion in the remainder of this chapter to common logarithms, we adopt the usual practice of dropping the base subscript (as used in the notation of the previous section) when the base is 10. Thus, "log n" means "$\log_{10} n$."

Exponential form	*Logarithmic form*
$0.01 = 10^{-2}$	$\log 0.01 = -2$
$0.1 = 10^{-1}$	$\log 0.1 = -1$
$1 = 10^0$	$\log 1 = 0$
$10 = 10^1$	$\log 10 = 1$
$100 = 10^2$	$\log 100 = 2$

The equivalent exponential and logarithmic forms of several numbers in ordered sequence are displayed above. From this we can infer that common logarithms for numbers between 1 and 10 are numbers between 0 and 1. Tables such as Table II on pages 350–351, partially reproduced in Fig. 10.1, provide approximate common logarithms for numbers between 1 and 10.

N	0	1	2	3	4	5	6	7	8	9
10	0 000	043	086	128	170	212	253	294	334	374
11	414	453	492	531	569	607	645	682	719	755
12	0 792	828	864	899	934	969	*004	*038	*072	*106
13	1 139	173	206	239	271	303	335	367	399	430
14	461	492	523	553	584	614	644	673	703	732
15	1 761	790	818	847	875	903	931	959	987	*014
16	2 041	068	095	122	148	175	201	227	253	279
17	304	330	355	380	405	430	455	480	504	529
18	553	577	601	625	648	672	695	718	742	765
19	2 788	810	833	856	878	900	923	945	967	989
20	3 010	032	054	075	096	118	139	160	181	201
21	222	243	263	284	304	324	345	365	385	404
22	424	444	464	483	502	522	541	560	579	598
23	617	636	655	674	692	711	729	747	766	784
24	802	820	838	856	874	892	909	927	945	962
25	3 979	997	*014	*031	*048	*065	*082	*099	*116	*133
26	4 150	166	183	200	216	232	249	265	281	298

Proportional Parts

	43	42	41	40
1	4.3	4.2	4.1	4
2	8.6	8.4	8.2	8
3	12.9	12.6	12.3	12
4	17.2	16.8	16.4	16
5	21.5	21.0	20.5	20
6	25.8	25.2	24.6	24
7	30.1	29.4	28.7	28
8	34.4	33.6	32.8	32
9	38.7	37.8	36.9	36

	39	38	37	36
1	3.9	3.8	3.7	3.6
2	7.8	7.6	7.4	7.2
3	11.7	11.4	11.1	10.8
4	15.6	15.2	14.8	14.4
5	19.5	19.0	18.5	18.0
6	23.4	22.8	22.2	21.6
7	27.3	26.6	25.9	25.2
8	31.2	30.4	29.6	28.8
9	35.1	34.2	33.3	32.4

Figure 10.1

EXAMPLES

1. Find the value of log 1.92 using Table II.

The first two digits of 1.92, "19," identify the row in Table II (Fig. 10.1); the third digit, "2," identifies the column in which the logarithm is to be found. Thus:

$$\log 1.92 \approx .2833 \quad (\approx \text{ means "is approximately equal to"})$$

[*Note*: Decimal points are usually omitted from the table. The user is expected to know that when $1 < N < 10$, as is true when $N = 1.92$, then $.0000 < \log 1.92 < 1.0000$, as inferred from the display on p. 233.]

2. Find the value of log 1.72.

The value of log 1.72 is found at the intersection of row 17 and column 2. Thus:

$$\log 1.72 \approx .2355$$

3. From Fig. 10.1:

$\log 2.01 \approx .3032$

$\log 1.89 \approx .2765$

$\log 2.3 = \log 2.30 \approx .3617$

$\log 2 = \log 2.00 \approx .3010$

$\log 2.55 = .4065$ (The * means "use the next first digit.")

Within our decimal system of number notation, it is possible to "name" a number in a wide variety of ways. One is to express a number as a product in which one factor is *some number between one and ten,* and the other factor is *some integral power of ten.* For example:

$$128 = 1.28 \times 10^2 \qquad 3 = 3 \times 10^0$$
$$3542 = 3.542 \times 10^3 \qquad .6 = 6 \times 10^{-1}$$
$$71,000 = 7.1 \times 10^4 \qquad .481 = 4.81 \times 10^{-1}$$
$$15 = 1.5 \times 10^1 \qquad .0075 = 7.5 \times 10^{-3}$$

Numbers expressed in this special factor form are said to be in **standard form of scientific notation.** With it we may extend the use of Table II to the writing of the approximate logarithms for many other numbers. For example, the common logarithm for 12.8 may be determined by reasoning as follows:

$$\log 1.28 \approx .1072$$

or

$$1.28 \approx 10^{.1072}$$

and

$$12.8 = 1.28 \times 10^1$$
$$\approx 10^{.1072} \times 10^1$$
$$\approx 10^{1+.1072}$$
$$\approx 10^{1.1072}$$

Thus, $\log 12.8 \approx 1.1072$.

Similarly, we can develop the common logarithm for 128, using Table II and the laws of exponents:

$$128 = 1.28 \times 100 = 1.28 \times 10^2$$
$$\approx 10^{.1072} \times 10^2$$
$$\approx 10^{2+.1072}$$
$$\approx 10^{2.1072}$$

Thus, $\log 128 \approx 2.1072$.

EXAMPLES

1. $2.34 \approx 10^{.3692}$; so, $\log 2.34 \approx .3692$.

$23.4 = 2.34 \times 10^1 \approx 10^{.3692} \times 10^1 = 10^{1.3692}$; so, $\log 23.4 \approx 1.3692$.

$234 = 2.34 \times 10^2 \approx 10^{.3692} \times 10^2 = 10^{2.3692}$; so, $\log 234 \approx 2.3692$.

2. $1.61 \approx 10^{.2068}$; so, $\log 1.61 \approx .2068$.

$16.1 \approx 10^{1.2068}$; so, $\log 16.1 \approx 1.2068$.

$161 \approx 10^{2.2068}$; so, log $161 \approx 2.2068$.

$16{,}100 \approx 10^{4.2068}$; so, log $16{,}100 \approx 4.2068$.

EXERCISE 10-2

Express the following in standard form of scientific notation:

1. 4000	**2.** 350	**3.** 1836
4. 4,832,000	**5.** 3,000,000	**6.** 4275
7. 0.83	**8.** 0.624	**9.** 0.0004
10. 0.00000074	**11.** 0.0006325	**12.** 8.42

13. If $10^{0.4} = 2.51$, what would be the proper exponent, p, for each of the following?
  (a) $10^p = 25.1$ (b) $10^p = 251$ (c) $10^p = 0.251$
  (d) $10^p = 0.0251$ (e) $10^p = 2510$

Use Table II to locate the proper exponent (logarithm) having 10 as the base:

14. 3.86	**15.** 7.34	**16.** 5.03
17. 8.66	**18.** 4.1	**19.** 8
20. 6.03	**21.** 60.3	**22.** 300
23. 3000	**24.** 682	**25.** 0.631
26. 0.0631	**27.** 0.00631	

3. CHARACTERISTIC AND MANTISSA

The numeral for a common logarithm may be thought to consist of two parts, separated by a decimal point. The part to the left of the decimal point is called the **characteristic**; the part to the right of the decimal point is called the **mantissa**. For example:

$$\log 213 = \underbrace{2}_{}.\underbrace{3284}_{}$$

where 2 is the characteristic and 3284 is the mantissa.

$$\log 1.8 = \underbrace{0}_{}.\underbrace{2553}_{}$$

where 0 is the characteristic and 2553 is the mantissa.

In general, for any positive number N:

1. The mantissa of its common logarithm is related to the sequence of digits in the decimal numeral for N.

2. The characteristic of its common logarithm is related to the location of the decimal point in the decimal numeral for N.

EXAMPLES

1. When the sequence of digits for N is the same—regardless of the location of the decimal point—the mantissa is the same:

$$2.13 \approx 10^{.3284}; \quad \text{so,} \quad \log 2.13 \approx 0.3284$$

$$21.3 = 2.13 \times 10 \approx 10^{.3284} \times 10^1 = 10^{1.3284};$$
$$\text{so,} \quad \log 21.3 \approx 1.3284$$

$$213 = 2.13 \times 10^2 \approx 10^{.3284} \times 10^2 = 10^{2.3284};$$
$$\text{so,} \quad \log 213 \approx 2.3284$$

$$2130 = 2.13 \times 10^3 \approx 10^{.3284} \times 10^3 = 10^{3.3284};$$
$$\text{so,} \quad \log 2130 \approx 3.3284$$

Notice that the characteristic is 1 less than the number of digits to the left of the decimal point in the decimal numeral for N.

2. $\log 1.28 \approx .1072$

$\log 12.8 \approx 1.1072$

$\log 128 \approx 2.1072$

$\log 12{,}800 \approx 4.1072$

Common logarithms for numbers between 0 and 1 follow the same pattern as that for numbers greater than 1. For example:

$$2.13 \approx 10^{.3284}; \quad \text{so,} \quad \log 2.13 \approx .3284$$

$$.213 = 2.13 \times 10^{-1} \approx 10^{-1+.3284}; \quad \text{so,} \quad \log .213 \approx -1 + .3284$$

$$.0213 = 2.13 \times 10^{-2} \approx 10^{-2+.3284}; \quad \text{so,} \quad \log .0213 \approx -2 + .3284$$

$$.00213 = 2.13 \times 10^{-3} \approx 10^{-3+.3284}; \quad \text{so,} \quad \log .00213 \approx -3 + .3284$$

We could write $\log .213$ as $-.6716$ $(= -1 + .3284)$, $\log .0213$ as -1.6716 $(= -2 + .3284)$, $\log .00213$ as -2.6716 $(= -3 + .3284)$, and so on, but this makes computation with logarithms cumbersome. There are advantages to expressing these negative characteristics as a binomial, the difference between some number and a multiple of ten. For instance, negative characteristic -1 can be expressed as an equivalent $9 - 10$; -2 as $8 - 10$; -3 as $7 - 10$; an infrequent characteristic of -13 can be written as $7 - 20$. Thus, the negative logarithms above can also be expressed as follows:

$-1 + .3284$ as $9.3284 - 10$ (which is the equivalent of $-.6716$)

$-2 + .3284$ as $8.3284 - 10$ (which is the equivalent of -1.6716)

$-3 + .3284$ as $7.3284 - 10$ (which is the equivalent of -2.6716)

Therefore:

$$\log .213 \approx 9.3284 - 10$$

$$\log .0213 \approx 8.3284 - 10$$
$$\log .00213 \approx 7.3284 - 10$$
$$\log .000213 \approx 6.3284 - 10$$

10.1

To determine the common logarithm for a number whose decimal numeral contains three significant digits:†

Step 1. Count the number of places in the numeral that the decimal point would have to be moved in order for the numeral to represent a number between 1 and 10.

(a) If the move is to the left, the characteristic of the desired logarithm is the same as the number of places moved.

(b) If the move is to the right, the characteristic of the desired logarithm is the negative of the number of places moved, expressed as a binomial: $9 - 10$ for one place, $8 - 10$ for two places, $7 - 10$ for three places, and so on.

Step 2. Locate the mantissa of the desired logarithm in Table II at the intersection of the *row* corresponding to the first two significant digits of the decimal numeral and the *column* corresponding to its third digit.

Step 3. Combine the characteristic and mantissa found in Steps 1 and 2 for the desired logarithm.

EXAMPLES

1. Determine the common logarithm for 17.9.

Step 1.   1̲7̲,9: The decimal point would have to be moved *one* place to the *left* in order for the numeral to represent a number between 1 and 10. Thus, the characteristic is 1.

Step 2.   In 17.9, the first two significant digits are 17, the third, 9. The mantissa will be found in Table II at the intersection of the row labeled 17 and the column headed by 9. Thus, the desired mantissa is: 2529.

Step 3.   The common logarithm for 17.9 is 1.2529.

2. Determine the common logarithm for .00234.

---

†All digits in a decimal numeral are considered significant except those zeros which are used principally to locate the decimal point. For example, 206 has three significant digits; .026 has two significant digits; .004 has one significant digit. Numerals such as 200 may have three, two, or one significant digits. If 200 means 200 ones to the nearest one—three significant digits; if 20 tens to the nearest ten—two significant digits; if 2 hundreds to the nearest hundred—one significant digit.

Step 1.   .0 0 2 3 4: The decimal point would have to be moved *three* places to the *right* in order for the numeral to represent a number between 1 and 10. Thus, the characteristic is −3 or 7 − 10.

Step 2.   In .00234, the first two significant digits are 23, the third, 4. The mantissa will be found in Table II at the intersection of row 23 and the column 4: 3692.

Step 3.   The common logarithm for .00234 is 7.3692 − 10.

3. Determine the common log for .021.

Step 1.   .0 2 1 —two places, right; characteristic: 8 − 10.

Step 2.   In .021, the first two significant digits are 21, the third is assumed to be 0. The mantissa is found at the intersection of row 21 and column 0: 3222.

Step 3.   log .021 ≈ 8.3222 − 10.

4. Determine the common log for 102,000.

Step 1.   1.0 2 0 0 0,—five places, left; characteristic 5.

Step 2.   Mantissa at row 10, column 2: 0086.

Step 3.   log 102,000 ≈ 5.0086.

EXERCISE 10-3

Use Table II to determine the common logarithm for each of the following:

1. 321	**2.** 847	**3.** 63
4. 807	**5.** 42	**6.** 5.6
7. 0.37	**8.** 8.42	**9.** 5000
10. 0.031	**11.** 0.0052	**12.** 0.0257
13. 100	**14.** 797	**15.** 0.999
16. 0.001	**17.** 3040	**18.** 0.0502
19. 162,000	**20.** 5,000,000	**21.** 0.00032
22. 0.00427	**23.** 5.0600	**24.** 0.00001

4. ANTILOGARITHMS

We have seen that the logarithm of 2, for instance, is .3010. To state the inverse, we say *the* **antilogarithm** *of .3010 is 2.* To determine the antilogarithm of a given logarithm, we reverse the procedure for determining the logarithm for a given number (Program **10.1**).

10.2

To determine the antilogarithm for a logarithm, using a four-place table of mantissas:†

Step 1. Locate the mantissa of the logarithm in the table. The first two significant digits of the desired antilogarithm will be those of the row containing the mantissa, and the third significant digit will be that of the column containing the mantissa.

Step 2. Place the decimal point temporarily after the first significant digit in the antilogarithm numeral; then

(a) If the characteristic of the logarithm is positive, move the decimal point in the antilogarithm numeral the same number of places to the right as given by the characteristic. (It may be necessary to annex zeros.)

(b) If the characteristic of the logarithm is negative, move the decimal point in the antilogarithm the same number of places to the left as given by the absolute value of the characteristic. (It may be necessary to annex zeros.)

EXAMPLES

1. Determine the antilogarithm of 2.1818.

Step 1.   The mantissa is 1818. In the table (see Fig. 10.1) 1818 is in row 15 and column 2. The sequence of significant digits in the antilog of 2.1818 is 152.

Step 2.   Temporarily place the decimal point ($\wedge$) after the first digit: $1_\wedge 52$. The characteristic of the logarithm in question is positive: 2. Move the decimal point two places to the right: $1_\wedge 52$. The antilogarithm of 2.1818 is 152.

2. Determine the antilogarithm of 8.2405 — 10.

Step 1.   The mantissa is 2405. In the table (Fig. 10.1), 2405 is in row 17, column 4. The sequence of significant digits in the desired antilogarithm is 174.

Step 2.   Temporarily place the decimal point ($\wedge$) after the first significant digit: $1_\wedge 74$. The characteristic is negative: $(8 - 10)$; its absolute value is $|(8 - 10)| = |-2| = 2$. Move the decimal point two places to the left; it will be necessary to annex one zero: $01_\wedge 74$. The antilogarithm of 8.2405 — 10 is .0174.

3. Determine the number whose logarithm is 4.0792.

Step 1.   Mantissa 0792 is in row 12, column 0. The sequence of significant digits will be: 120.

---

†Table II is a four-place table, that is, mantissas are expressed with four digits. Tables of more places are available for more precise expressions of logarithms.

Step 2. The characteristic is 4. The decimal point is located four places to the right after the first digit: 12000. The number whose logarithm is 4.0792 is 12,000.

4. Determine the number whose logarithm is 8.2833 − 20.

Step 1. Mantissa 2833 is in row 19, column 2. The sequence of significant digits will be: 192.

Step 2. The characteristic is negative, 8 − 20; its absolute value is 12. The decimal point is located eleven places before first significant digit; .00000000000192 is the number whose logarithm is 8.2833 − 20. [*Note:* In cases such as this, it is better to use the standard form of scientific notation: .00000000000192 = 1.92 × 10^{-12}.]

EXERCISE 10-4

Use Table II to determine the antilogarithm for each of the following:

1. 1.4150	**2.** 0.3096	**3.** 2.5315
4. 9.5658 − 10	**5.** 8.6010 − 10	**6.** 9.6474 − 10
7. 4.8710	**8.** 3.8825	**9.** 9.8287 − 10
10. 7.9009 − 10	**11.** 2.2253	**12.** 0.0043
13. 8.7966 − 10	**14.** 6.9335 − 10	**15.** 5.3160
16. 0.5705	**17.** 9.6758 − 10	**18.** 6.9542
19. 3.9015	**20.** 0.8041	**21.** 1.7396
22. 8.0934 − 20	**23.** 2.3560 − 10	**24.** 9.4099 − 10

5. INTERPOLATION

To **interpolate** means to compute an intermediate value between two stated or tabulated values. It is possible, by means of interpolation, to extend the usefulness of the four-place table of mantissas that we have been using. For example, we note in Table II that the mantissa for 211 (and 2110) is 3243, and the mantissa for 212 (and 2120) is 3263. As far as mantissas are concerned, 211 and 2110 are the same sequence of digits; so are 212 and 2120. Because 2115 is midway between 2110 and 2120, we might assume the corresponding mantissa to be 3253, which is midway between 3243 and 3263. See Fig. 10.2 (next page).

Actually, this assumption is not completely valid, as can be shown for greater differences, e.g., the mantissa for 200 is 3010, and mantissa for 600

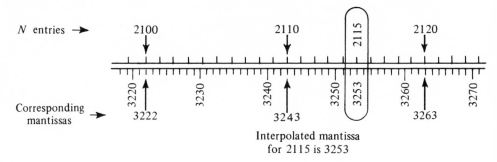

Figure 10.2

is 7782. According to this assumption, the mantissa for 400, which is midway between 200 and 600, should be $(3010 + 7782)/2$ or 5396; the actual mantissa for 400 is 6021. The inaccuracy produced by our assumption, however, over small differences, is less than the grosser inaccuracies inherent in the table itself. Consequently we can assume for practical purposes that, for intervals between tabulated entries in tables of mantissas, the variation in the antilogarithm is directly proportionate to the variation in the logarithm, and conversely.

10.3

To determine the mantissa for a four-digit number, using a four-place table of mantissas:

Step 1. Find the difference between the mantissa entry that corresponds to the number's first three digits and the next tabulated mantissa.

Step 2. Take $\frac{1}{10}$ of the difference of Step 1 and multiply that by the number's last (fourth) digit; round off to the nearest integer.

Step 3. Add the result of Step 2—the "correction"—to the smaller mantissa entry of Step 1.

EXAMPLES

1. Determine the mantissa for 1664.

Step 1.   Mantissa for 166 is 2201; the next tabulated mantissa is 2227; the difference is $2227 - 2201 = 26$.

Step 2. $\frac{1}{10} \times 26 = 2.6$; the last digit in 1664 is 4; $4 \times 2.6 = 10.4$; 10.4 rounded off to nearest integer is 10.

Step 3.   Add the "correction" of Step 2 (10) to the smaller mantissa of Step 1 (2201) for sum of 2211. The mantissa for 1664 is 2211.

2. Determine the mantissa for 2397.

Step 1.   The mantissa for 239 is 3784; the next mantissa entry is 3802; the difference is 18.

Step 2. $\frac{1}{10} \times 18 = 1.8$; 7 (last digit) $\times 1.8 = 12.6$; rounded off: 12.6 ≈ 13.

Step 3. $3784 + 13 = 3797$. The mantissa for 2397 is 3797.

3. Determine the mantissa for 1252.

Step 1.   The mantissa for 125 is 0969; the next mantissa entry is 1004; the difference is 35.

Step 2. $\frac{1}{10} \times 35 \times 2 = 7$

Step 3. $0969 + 7 = 0976$; the mantissa for 1252 is 0976.

[*Note:* In Table II, the tables of proportional parts listed at the side, which give consecutive tenth-parts of mantissa differences, may be used to simplify the computations of Step 2 in Program **10.3**.]

Often in logarithmic computation we wish to determine the antilogarithm of a logarithm whose mantissa is not listed in the table. We may do so by using a procedure that is the reverse of Program **10.3**.

10.4

To determine the sequence of significant digits corresponding to a mantissa unlisted in a four-place table:

Step 1.   Locate in the table the mantissa that is nearest, but less than, the unlisted mantissa. The first two significant digits of the desired sequence will be those of the row, and the third will be that of the column, containing this listed mantissa.

Step 2.   Subtract the mantissa of Step 1 from the unlisted mantissa.

Step 3.   Subtract the mantissa of Step 1 from the next larger listed mantissa.

Step 4.   Divide the difference of Step 2 by the difference of Step 3, and round off the quotient to the nearest tenth. This quotient digit will be the fourth significant digit of the desired sequence.

EXAMPLES

1. Determine the sequence of significant digits that corresponds to the mantissa 1537.

Step 1.   The nearest listed mantissa less than 1537 is 1523. It is in row 14 and column 2. The first three digits of the desired sequence are 142.

Step 2.   The difference between the mantissa of Step 1 (1523) and the unlisted mantissa (1537) is 14.

Step 3.   The difference between the mantissa of Step 1 (1523) and the next larger listed mantissa (1553) is 30.

Step 4.   Divide the difference of Step 2 (14) by the difference of Step 3 (30), and round off to the nearest tenth:

$$14 \div 30 = \tfrac{14}{30} = .46^+ \approx .5$$

The fourth digit in the desired sequence is 5. The sequence of significant digits that corresponds to the mantissa 1537 is 1425.

[*Note:* The table of proportional parts may be used instead of the computation of Step 4. Among the tenth-parts listed under 30, the nearest entry in value to 14 is 15. The tenth-part of 30 that is 15 is listed in the column at the far left: 5.]

2. Determine the sequence of digits for the mantissa 2920.

Step 1.   The nearest mantissa entry less than 2920 is 2900; row 19, column 5. The first three significant digits of the desired sequence are 195.

Step 2.   The difference between the listed mantissa (2900) and the unlisted mantissa (2920) is 20.

Step 3.   The difference between mantissa entries before and after the unlisted mantissa (2900 and 2923) is 23.

Step 4. $\tfrac{20}{23} = .86^+ \approx .9$. The fourth digit in the desired sequence is 9; the complete sequence: 1959.

[*Note:* A condensed version:

digit sequence	mantissa

$$x : 10 = 20 : 23$$

$$\frac{x}{10} = \frac{20}{23}$$

$$x = \frac{20}{23} \times 10$$

$$= 8.6^+ \approx 9]$$

3. Determine the digit sequence for the mantissa 3283.

Step 1.   The nearest smaller mantissa entry is 3263; row 21, column 2. The first three significant digits are 212.

Step 2. $3283 - 3263 = 20$

Step 3. $3284 - 3263 = 21$

Step 4. $\tfrac{20}{21} = .95^+$. When rounded off to the nearest tenth, $.95^+$ is 1.0. The fourth "digit" here is 10; the desired sequence of digits is 2130.

[*Note:* In the table of proportional parts under 21, the difference of 20 is nearer to the whole, 21, than to the tabulated $\tfrac{9}{10}$ part, 18.9.]

EXERCISE 10-5

Determine the common logarithm for each of the following:

1. 4.625   **2.** 0.2074   **3.** 74.37

4. 0.05603   **5.** 523.4   **6.** 67.63

7. 0.006403   **8.** 7855   **9.** 424,700

10. 0.07523   **11.** 0.9771   **12.** 8.794×10^{-6}

Determine the antilogarithm for each of the following:

13. 0.6174   **14.** $9.6558 - 10$ **15.** 2.9104

16. $8.4096 - 10$ **17.** 4.6844   **18.** 2.5012

19. 1.8604   **20.** $9.8304 - 10$ **21.** $7.2307 - 10$

22. $3.5605 - 10$ **23.** 5.9348   **24.** 9.7751

6. COMPUTING WITH LOGARITHMS: PRODUCTS

There are four fundamental laws of logarithms that are specific outgrowths of the laws of exponents (Section 6, Chapter VII). In this section we shall consider the first of these, the one that relates to products of numbers. It parallels law of exponents I: $a^m \times a^n = a^{m+n}$.

Logarithm Law I:

The logarithm of a product of positive numbers is equal to the sum of the logarithms of those numbers.

Consider the product $A \times B$. Let $\log A = x$ and $\log B = y$. Then $A = 10^x$ and $B = 10^y$, and $A \times B = 10^{x+y}$. Thus

$$\log (A \times B) = x + y \quad \text{or} \quad \log (A \times B) = \log A + \log B$$

By extension:

$$\log (A \times B \times C \times \cdots \times N) = \log A + \log B + \log C + \cdots + \log N$$

10.5

To compute the product of several positive numbers by logarithms:

Step 1. Determine the logarithm for each of the numbers.

Step 2. Add the logarithms of Step 1.

Step 3. Determine the antilogarithm of the result of Step 2 for the desired product.

EXAMPLES

1. Compute the product 3.62×47.9 by logarithms.

Step 1. $\log 3.62 \approx .5587$; $\log 47.9 \approx 1.6803$

Step 2.
$$
\begin{array}{rr}
\log 3.62 \approx & .5587 \\
\log 47.9 \approx & 1.6803 \\
\hline
(+) \\
\log (36.2 \times 47.9) \approx & 2.2390
\end{array}
$$

Step 3. The antilogarithm of 2.2390 is 173.4. Therefore, 3.62 × 47.9 is 173.4 (approximately).

2. Find the product of .00732 × .0167.

$$\log (.00732 \times .0167) = \log .00732 + \log .0167$$

$$
\begin{array}{rl}
\log .00732 \approx & 7.8645 - 10 \\
\log .0167 \approx & 8.2227 - 10 \\
(+) \overline{} \\
\log (.00732 \times .0167) \approx & 16.0872 - 20 = 6.0872 - 10
\end{array}
$$

antilog of $6.0872 - 10 \approx .0001222$

Thus,

$$.00732 \times .0167 \approx .0001222$$

3. Find the product of $(3.624) \times (-3.71) \times (.0046)$ by logarithms.

Our definition of logarithm precludes logarithms of negative numbers. However, we may use logarithms to compute a product involving one or more negative numbers by computing the product of the factors as though they were all positive numbers, then making the product negative if the number of negative factors involved is odd, positive if the number of negative factors is even:

$$(3.624) \times (-3.71) \times (.0046) = -[(3.624) \times (3.71) \times (.0046)]$$

$$\log [3.624 \times 3.71 \times .0046] = \log 3.624 + \log 3.71 + \log .0046$$

$$
\begin{array}{rl}
\log 3.624 \approx & .5592 \\
\log 3.71 \approx & .5694 \\
\log .0046 \approx & 7.6628 - 10 \\
(+) \overline{} \\
\log \text{ product} \approx & 8.7914 - 10
\end{array}
$$

antilog $8.7914 - 10 \approx .06186$

Thus,

$$(3.624)(-3.71)(.0046) \approx -.06186$$

EXERCISE 10-6

Use logarithms to compute:

1. 14 × 36

2. 17 × 360

3. 325 × 1.6

4. 428 × 0.632

5. 0.637 × 0.142

6. 0.0032 × 6.78

7. 0.0243 × 0.725

8. (−3.62) × (−0.043)

9. 526 × 0.000327

10. 426.5 × 817

11. 3.624 × 67.85

12. 0.6387 × 0.4321

13. (−6.328) × (4.26)

14. 0.836 × 0.09799

15. $327 \times 491 \times 632$

16. $524 \times 0.623 \times 0.042$

17. $(-4.32) \times (8.43) \times (-0.6745)$

18. $0.0326 \times 0.0478 \times 0.832$

19. $3.246 \times 49.37 \times 0.03265 \times 48.21$

20. $0.004263 \times 0.07327 \times 0.0004263 \times 0.0003271$

7. COMPUTING WITH LOGARITHMS: QUOTIENTS

Logarithm law II relates to quotients of numbers, and it parallels law of exponents II: $a^m \div a^n = a^{m-n}$.

Logarithm Law II:

The logarithm of a quotient of two positive numbers is equal to the logarithm of the dividend less the logarithm of the divisor.

Consider the quotient $A \div B$. Let $\log A = x$ and $\log B = y$. Then $A = 10^x$ and $B = 10^y$, and $A \div B = 10^x \div 10^y = 10^{x-y}$. Thus,

$$\log (A \div B) = x - y \quad \text{or} \quad \log (A \div B) = \log A - \log B$$

10.6

To compute the quotient of two positive numbers by logarithms:

Step 1. Determine the logarithms for the dividend and divisor.

Step 2. Subtract the logarithm of the divisor from the logarithm of the dividend.

Step 3. Determine the antilogarithm of the result of Step 2 for the desired quotient.

EXAMPLES

1. Determine the quotient, $86.3 \div 0.427$, by logarithms.

Step 1. $\log 86.3 \approx 1.9360$ (dividend); $\log .427 \approx 9.6304 - 10$ (divisor).

Step 2. Subtract (here it is necessary to adjust the minuend 1.9360 to an equivalent $11.9360 - 10$):

$$
\begin{array}{r}
\log 86.3 \approx \quad 11.9360 - 10 \\
\log .427 \approx \quad\ \ 9.6304 - 10 \\
(-) \overline{\hphantom{\quad 11.9360 - 10}} \\
\log (86.3 \div .427) \approx \quad 2.3056
\end{array}
$$

Step 3. Antilog of $2.3056 \approx 202.1$. Therefore, $86.3 \div .427 \approx 202.1$.

2. Determine the quotient $0.00426 \div 0.0798$.

$$\log(.00426 \div .0798) = \log .00426 - \log .0798$$

$\log .00426 \approx$	$7.6294 - 10 =$	$17.6294 - 20$
$\log .0798 \approx$	$8.9020 - 10 =$	$8.9020 - 10$
	$(-)$ ———————	$(-)$ ———————
$\log$ quotient $\approx$		$8.7274 - 10$

antilog $8.7274 - 10 \approx .05339$

Thus,

$$.00426 \div .0798 \approx .05339.$$

3. Compute by logarithms: $\dfrac{463 \times 2.18}{0.674}$.

$$\log \frac{463 \times 2.18}{.674} = \log(463 \times 2.18) - \log .674$$

$$= \log 463 + \log 2.18 - \log .674$$

$\log 463 \approx$	2.6656		
$\log 2.18 \approx$	$.3385$		
	$(+)$ ———————		
$\log$ numerator $\approx$	3.0041	$=$	$13.0041 - 10$
$\log .674 \approx$	$9.8287 - 10 =$		$9.8287 - 10$
	$(-)$ ———————	$(-)$	———————
$\log \dfrac{463 \times 2.18}{.674} \approx$			3.1754

antilog $3.1754 \approx 1498$

Therefore $\dfrac{463 \times 2.18}{0.674} \approx 1498.$

EXERCISE 10-7

Use logarithms to compute:

1. $448 \div 56$

2. $36.2 \div 1.78$

3. $4.29 \div 2.63$

4. $8360 \div 51.7$

5. $67.3 \div 205$

6. $(-8.37) \div (54.3)$

7. $0.427 \div 0.372$

8. $0.62 \div 0.078$

9. $0.426 \div 1.63$

10. $(-527) \div (-0.0263)$

11. $6.87 \div 6.93$

12. $0.0032 \div 0.0627$

13. $(-8.037) \div (0.426)$

14. $84.2 \div 5936$

15. $1 \div 0.8625$

16. $1 \div 4.284$

17. $42{,}880 \div 67.3$

18. $\dfrac{32.6 \times 42.9}{5.26}$

19. $\dfrac{(0.0632) \times (-0.0427)}{(-63.2)}$

20. $\dfrac{32.7 \times 4.635}{37.93 \times 5.26}$

8. COMPUTING WITH LOGARITHMS: POWERS

Logarithm law III relates to powers, and it parallels law of exponents III: $(a^n)^m = a^{m \times n}$.

Logarithm Law III:

The logarithm of a power of a positive number is equal to the product of the exponent and the logarithm of the number.

Consider A^p. Let $\log A = x$. Then $A = 10^x$, and $A^p = (10^x)^p = 10^{px}$. Thus,

$$\log (A^p) = px \quad \text{or} \quad \log (A^p) = p \log A$$

10.7

To compute the expansion of a given power of a positive number by logarithms:

Step 1. Determine the logarithm for the number to be raised to the given power.

Step 2. Multiply the logarithm of Step 1 by the exponent of the given power.

Step 3. Determine the antilogarithm of the result of Step 2 for the desired expansion.

EXAMPLES

1. Compute the expansion of $(8.62)^3$.

Step 1. $\log 8.62 \approx .9355$.

Step 2.   Multiply $\log 8.62$ by the exponent 3:

$$\log 8.62 \approx \quad .9355$$
$$(\times) \frac{3}{}$$
$$\log (8.62)^3 \approx \quad 2.8065$$

Step 3.   Antilog $2.8065 \approx 640.4$. Therefore, $(8.62)^3 \approx 640.4$.

2. Expand $(.073)^6$.

$$\log (.073)^6 = \quad 6 \log (.073)$$
$$\log .073 \approx \quad 8.8633 - 10$$
$$(\times) \frac{6}{}$$
$$\log (.073)^6 \approx \quad 53.1798 - 60 = 3.1798 - 10$$

antilog $3.1798 - 10 \approx .0000001513$

Therefore,

$$(.073)^6 \approx .0000001513, \text{ or } 1.513 \times 10^{-7}$$

3. Expand $(-3.62)^5$

$$(-3.62)^5 = -(3.62)^5$$

(The number of factors, 5, is odd; the expansion is a negative number.)

$$\log (3.62)^5 = 5 \log 3.62$$

$$\log 3.62 \approx .5587$$

$$(\times) \frac{5}{}$$

$$\log (3.62)^5 \approx 2.7935$$

$$\text{antilog } 2.7935 \approx 621.6$$

Thus,

$$(-3.62)^5 \approx -621.6$$

EXERCISE 10-8

Use logarithms to compute:

1. $(67)^3$ **2.** $(1.08)^7$ **3.** $(42.7)^4$

4. $(0.67)^5$ **5.** $(0.0426)^2$ **6.** $(4.12)^{10}$

7. $(0.00632)^3$ **8.** $(-8.42)^3$ **9.** $(4.37)^{4.1}$

10. $(6.275)^2$ **11.** $(-0.09874)^4$ **12.** $(3.67 \times 49.2)^3$

13. $(0.0637)^3 \times (0.0548)^4$ **14.** $(18.37 \times 0.0625)^6$

15. $(18.37)^6 \times (0.0625)^6$ **16.** $(3.72)^2 \times (46.3)^3 \times (0.0379)^4$

17. $(-3.62)^4 \times (0.4875)^5$ **18.** $\left(\dfrac{3.26}{48.2}\right)^3$

19. $\left(\dfrac{0.000627}{684}\right)^4$ **20.** $\left[\dfrac{32.65 \times 87.21}{(-62.4)}\right]^3$

9. COMPUTING WITH LOGARITHMS: ROOTS

In an exponential sense, roots and powers are essentially the same. That is, $\sqrt[n]{a}$ (the nth root of a) is the same as $a^{1/n}$ (a to $(1/n)$th power). Thus, logarithm law III applies equally for raising to a power and extracting a root. For convenience, however, we state a separate logarithm law and program for roots of numbers.

Logarithm Law IV:

The logarithm of a root of a positive number is equal to the logarithm of the number divided by the index of the root.

Consider $\sqrt[n]{A}$. Let $\log A = x$. Then $A = 10^x$, and $\sqrt[n]{A} = \sqrt[n]{10^x} = (10^x)^{1/n} = 10^{(1/n)(x)}$. Thus,

$$\log (\sqrt[n]{A}) = \left(\frac{1}{n}\right)(x) = \frac{1}{n} \log A = \frac{\log A}{n} = \log A \div n$$

10.8

To compute a given root of a positive number by logarithms:

Step 1. Determine the logarithm of the positive number.

Step 2. Divide the logarithm of Step 1 by the index of the given root.

Step 3. Determine the antilogarithm of the result of Step 2 as the desired root.

EXAMPLES

1. Compute $\sqrt[5]{27.4}$.

Step 1. $\log 27.4 \approx 1.4378$

Step 2. $\dfrac{\log 27.4}{5} \approx \dfrac{1.4378}{5} \approx .2876$

Step 3.   Antilog $.2876 \approx 1.939$. Therefore, $\sqrt[5]{27.4} \approx 1.939$.

2. Compute: $\sqrt[3]{.0261}$.

$$\log \sqrt[3]{.0261} = \frac{\log .0261}{3} \approx \frac{8.4166 - 10}{3}$$

[*Note:* To divide $8.4166 - 10$ by 3 would yield a quotient of $2.8055 - 3\frac{1}{3}$, which expresses the correct negative logarithm for $\sqrt[3]{.0261}$. But in this form we cannot readily find its antilogarithm from Table II. The difficulty can be avoided if we replace the logarithm binomial, *before dividing*, with an equivalent binomial whose second term is ten times the divisor. Thus, we replace $8.4166 - 10$ with the equivalent $28.4166 - 30$, then divide by 3.]

$$\log \sqrt[3]{0.261} = \frac{\log .0261}{3} \approx \frac{8.4166 - 10}{3} = \frac{28.4166 - 30}{3} = 9.4722 - 10$$

Antilog $9.4722 - 10$ is $.2966$ (approximately). Thus, $\sqrt[3]{.026} \approx .2966$.

[*Note:* The two logarithm expressions, $9.4722 - 10$ and $2.8055 - 3\frac{1}{3}$, are in fact equivalent, as can be shown:

$$
\begin{array}{ll}
9.4722 - 10: &
\begin{array}{r}
-10.0000 \\
+\ 9.4722 \\
\hline
-\ \ .5278
\end{array}
\qquad
2.8055 - 3\frac{1}{3}:
\begin{array}{r}
-3.3333 \\
+2.8055 \\
\hline
-\ .5278
\end{array}
\end{array}
$$

The antilogarithm of the former $(9.4722 - 10)$ can be found directly from Table II, while the latter $(2.8055 - 3\frac{1}{3})$ cannot.]

3. Compute: $\sqrt[3]{\dfrac{(3.27)(-62)}{.0183}}$.

Since one of the factors in the radicand is negative, and the index is odd, the root will be a negative number:

$$\sqrt[3]{\frac{(3.27)(-62)}{.0183}} = -\sqrt[3]{\frac{(3.27)(62)}{.0183}}$$

$$\log \sqrt[3]{\frac{(3.27)(62)}{.0183}} = \frac{\log 3.27 + \log 62 - \log .0183}{3}$$

$$
\begin{array}{rl}
\log 3.27 \approx & .5145 \\
\log 62 \approx & 1.7924 \\
(+) & \overline{} \\
\log \text{ numerator} \approx & 12.3069 - \mathbf{10} \\
\log .0183 \approx & 8.2625 - 10 \\
(-) & \overline{} \\
\log \text{ radicand} \approx & 4.0444
\end{array}
$$

$$4.0444 \div 3 \approx 1.3481$$

$$\text{antilog } 1.3481 \approx 22.29$$

Thus,

$$\sqrt[3]{\frac{(3.27)(-62)}{.0183}} \approx -22.29$$

EXERCISE 10-9

Use logarithms to compute:

1. $\sqrt{729}$ **2.** $\sqrt{2.25}$ **3.** $\sqrt{13.69}$

4. $\sqrt{0.0484}$ **5.** $\sqrt[3]{512}$ **6.** $\sqrt[3]{9261}$

7. $\sqrt{63.74}$ **8.** $\sqrt[3]{7.986}$ **9.** $\sqrt[3]{0.062}$

10. $\sqrt[3]{0.00274}$ **11.** $\sqrt[5]{4.24}$ **12.** $\sqrt[3]{-3.682}$

13. $\sqrt[7]{0.0263}$ **14.** $\sqrt{3.26 \times 5.71}$ **15.** $\sqrt[3]{0.026 \times 0.384}$

16. $\sqrt[5]{0.00627 \times 0.03814}$ **17.** $\sqrt{\frac{32.7}{48.5}}$ **18.** $\left(\frac{0.673}{0.284}\right)^{-1/2}$

19. $\sqrt[3]{\frac{-1.03}{0.427}}$ **20.** $\sqrt[3]{\frac{0.042 \times (3.26)^2}{1.932}}$

REVIEW—CHAPTER X

Express in logarithmic form:

1. $0.5^2 = 0.25$ **2.** $0.5^{-2} = 4$

3. $2^6 = 64$ **4.** $81^{0.5} = 9$

5. $64^{0.67} = 16$ **6.** $8^{-0.33} = 0.5$

Express in exponential form:

7. $\log_9 81 = 2$

8. $\log_8 2 = \frac{1}{3}$

9. $\log_{16} 2 = .25$

10. $\log_{10} 6.3 = .799$

11. $\log_{10} 2.6 = .415$

12. $\log_a 64 = 6$

What replacement for the variable in each of the following will make the sentence true?

13. $\log_6 36 = L$

14. $\log_a 2 = .25$

15. $\log_8 n = 4$

16. $\log_a .001 = -3$

17. $\log_{16} n = -2.25$

18. $\log_{100} 1000 = L$

Express in standard form:

19. 3864

20. 30,006

21. 40,000,000

22. 0.042

23. 0.000789

24. 0.00000000000042

From Table II, what replacement for p will make the following true?

25. $10^p = 3.75$

26. $10^p = 4$

27. $10^p = 5.09$

28. $10^p = 460$

29. $10^p = 0.46$

30. $10^p = 0.0046$

Use Table II to write the appropriate common logarithm for each of the following:

31. 863

32. 0.0082

33. 0.000001

34. 426,000

35. 31,800

36. 0.5757

37. 283.4

38. 0.006764

39. 3.397×10^{-6}

40. 8,995,000

Use Table II to write the appropriate antilogarithm for each of the following:

41. .8675

42. $8.8035 - 10$

43. 2.1139

44. $3.5024 - 10$

45. $7.0253 - 20$

46. $7.8865 - 10$

47. 0.9144

48. 4.6250

49. 1.8750

50. 3.5791

51. $6.8003 - 20$

52. 7.3504

Compute, using logarithms:

53. 427×0.697

54. $(0.0035) \times (-4.26)$

55. 0.00823×0.0647

56. $824 \times 3.62 \times 87.41$

57. $(-2.38) \times (-0.063) \times (0.4278)$

58. $83.62 \times 42.51 \times 3.879$

59. $0.0001046 \times 0.003275$

60. 0.00362 × 0.005714 × 0.0000632 × (−0.004621)

61. 4320 ÷ 61.8

62. (−3.94) ÷ (54.8)

63. (0.073) ÷ (0.0846)

64. (−428) ÷ (−0.0367)

65. 0.00476 ÷ 0.0832

66. 1 ÷ 3.762

67. $\dfrac{(-0.0432) \times (6.814)}{(-82.6)}$

68. $\dfrac{32.75 \times 106.2}{0.0361 \times 0.04287}$

69. $(56.8)^4$

70. $(0.0682)^3$

71. $(0.0003714)^5$

72. $(-3.62)^3$

73. $(4.26 \times 3.81)^2$

74. $(-6.37 \times 0.0428)^3$

75. $(-8.37)^3 \times (4.26)^2$

76. $\left(\dfrac{0.00624 \times 0.8365}{0.0724}\right)^4$

77. $\sqrt{8.365}$

78. $\sqrt[3]{0.0487}$

79. $(0.0632)^{1/5}$

80. $\sqrt[5]{0.0624 \times 0.0381}$

81. $(6.742)^{-1/3}$

82. $\sqrt[3]{\dfrac{4.62}{37.5}}$

83. $\sqrt[4]{\dfrac{(4.06)^2 \times (3.62)^3}{(-1.8)^2}}$

84. $\left[\dfrac{(3.07)^3 \times (2.634)^{1/2}}{(8.62)^4}\right]^{-1/2}$

RATIO, PROPORTION, AND PROBABILITY

A **ratio** expresses a relationship between two numbers. For the numbers a and b, their ratio may be symbolized as $a:b$, or as a fraction, a/b. Both express the ratio of a to b. The equality of two ratios is called a **proportion**. Thus, if a compares to b in the same ratio as c compares to d, we have a proportion which may be expressed as

1. RATIO AND PROPORTION

$$a:b = c:d \quad \text{or} \quad \frac{a}{b} = \frac{c}{d}$$

Arbitrarily, the numbers a and d above are called the **extremes**, and the numbers b and c are called the **means** of the proportion. When any one of the four numbers in a proportion is unknown and the other three are known, the unknown number can be determined by a procedure known as *solving a proportion*.

A.1

To solve a proportion:

Step 1. Assign a variable to represent the unknown number, and express the proportion as the equality of two ratio fractions.

Step 2. Solve the equation of Step 1 for the variable.

EXAMPLES

1. What number is in the same ratio to 11 as 3 is to 4?

Step 1. Let x reptesent the unknown number.

$$\frac{x}{11} = \frac{3}{4}$$

Step 2. $x = 11 \times \frac{3}{4}$

$x = \frac{33}{4}$

$x = 8\frac{1}{4}$

Thus, $8\frac{1}{4}$ is in the same ratio to 11 as 3 is to 4.

2. Six pounds compares to what weight in the ratio of 7 to 3?

Step 1. Let x represent the unknown weight.

$$\frac{6}{x} = \frac{7}{3}$$

Step 2. $6 = \frac{7}{3}x$

$\frac{18}{7} = x$

$2\frac{4}{7} = x$

Answer: $2\frac{4}{7}$ lb.

3. If on a map $\frac{1}{2}$ in. represents 60 mi, what length will represent 390 mi?

Step 1. Let x represent the unknown length (inches).

$$\frac{\frac{1}{2}}{60} = \frac{x}{390}$$

Step 2.   (In equations where the members are in fraction form, it is often more convenient to cross-multiply, which amounts to using a common denominator that may or may not be the least common denominator.)

$$\frac{\frac{1}{2}}{60} = \frac{x}{390}$$

$$\frac{1}{2} \times 390 = 60 \times x$$

$$195 = 60x$$

$$3\frac{1}{4} = x$$

Answer: $3\frac{1}{4}$ in.

4. Find two numbers in the ratio of $4 : 5$ whose sum is 63.

Step 1. Let x represent one unknown number; then $63 - x$ represents the other unknown number.

$$\frac{x}{63 - x} = \frac{4}{5}$$

Step 2. $5x = 4(63 - x)$

$5x = 252 - 4x$

$$9x = 252$$
$$x = 28$$
$$63 - x = 35$$

Answer: 28 and 35 are the numbers.

The cross-multiplying step in Example 3 parallels the well-known rule:

In a proportion, the product of the means is equal to the product of the extremes.

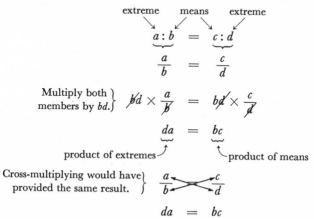

EXAMPLES

1. If a car goes 62 mi on 3 gal of gas at a certain speed, how far could it be expected to go on 7 gal at that speed?

Let x represent the unknown distance. Several proportions suggest themselves, all of which lead to the same solution, namely, $144\frac{2}{3}$ mi.

(a) 62 mi : 3 gal $= x$ mi : 7 gal.

$$62 : 3 = x : 7$$
$$3x = 62 \times 7$$
$$3x = 434$$
$$x = 144\tfrac{2}{3}$$

(b) 3 gal : 62 mi $=$ 7 gal : x mi.

$$3 : 62 = 7 : x$$
$$3x = 62 \times 7$$
$$x = 144\tfrac{2}{3}$$

(c) 3 gal : 7 gal $=$ 62 mi : x mi.

$$3 : 7 = 62 : x$$
$$3x = 7 \times 62$$
$$x = 144\tfrac{2}{3}$$

(d)   7 gal : 3 gal $= x$ mi : 62 mi.

$$7 : 3 = x : 62$$

$$3x = 7 \times 62$$

$$x = 144\tfrac{2}{3}$$

2. If tennis balls are 3 for \$2.40, how much will 17 balls cost?

Let x represent the cost of 17 tennis balls. The proportion:

$$3 \text{ balls} : \$2.40 = 17 \text{ balls} : \$x$$

$$3 : 2.40 = 17 : x$$

$$3x = 2.40 \times 17$$

$$3x = 40.80$$

$$x = 13.60$$

Answer: \$13.60.

3. If a load of 16 tons represents 125% of capacity, what is capacity?

Let x represent capacity in tons (capacity in percent $= 100\%$). The proportion:

$$16 \text{ tons} : 125\% \text{ capacity} = x \text{ tons} : 100\% \text{ capacity}$$

$$16 : 125\% = x : 100\%$$

$$(x)(125\%) = (16)(100\%)$$

$$1.25x = 16 \times 1.00$$

$$1.25x = 16$$

$$x = \frac{16}{1.25}$$

$$x = 12.8$$

Answer: 12.8 tons.

Example 3 may also be classified as a percentage problem. All percentage problems can be made into proportion problems if we express R, the rate, as a fraction $R/1$, and then replace the denominator of 1 with its equivalent, 100%. Thus, in the percentage formula in which $P =$ percentage, $B =$ base, and $R =$ rate:

$$\left.\begin{array}{l} P = B \times R \\[2mm] \dfrac{P}{B} = R \end{array}\right\} \quad \text{(percentage-formula forms)}$$

$$\frac{P}{B} = \frac{R}{1} \qquad \left(\text{expressing } R \text{ as } \frac{R}{1}\right)$$

$$\frac{P}{B} = \frac{R}{100\%} \qquad \text{(replacing 1 with } 100\% \text{—proportion form)}$$

EXAMPLE

1. In a shipment of merchandise, 22% or 715 of the items were damaged. How many items were in the shipment?

Here we are seeking the base, B. The proportion:

$$\frac{P}{B} = \frac{R}{100\%} \quad \text{or} \quad P:B = R:100\%$$

$$\frac{715}{B} = \frac{22\%}{100\%} \quad \text{or} \quad 715:B = 22\%:100\%$$

$$(100\%)(715) = (22\%)(B)$$

$$715 = 0.22B$$

$$3250 = B$$

Answer: 3250 items in the shipment.

EXERCISE A-1

Solve each of the following proportions for x:

1. $3:4 = 9:x$
2. $5:8 = x:40$
3. $3:x = 7:10\frac{1}{2}$
4. $18:5 = x:2$
5. $16:x = x:4$
6. $3.6:2.7 = x:1.5$
7. $x:6 = \frac{3}{5}:\frac{7}{10}$
8. $4:6 = x:(x+2)$
9. $x:3 = (2x-5):5$
10. $3:x = a:2$
11. $4:7b = 2b:x$
12. $a:(3x-2) = 2a:x$

Solve each of the following by proportion:

13. What number is in the same ratio to 15 as 9 is to 5?

14. $\frac{3}{5}$ compares to what number in the same ratio that $\frac{2}{15}$ compares to $\frac{1}{3}$?

15. If $1\frac{1}{4}$ in. on a map represents 60 mi, how many miles will $2\frac{7}{8}$ in. represent?

16. If $3\frac{1}{8}$ in. on a map represents 125 mi, how many inches would there be between the map representations of two points that are actually 310 mi apart?

17. If 80 ft of wire weighs 18 lb, how much will 360 ft weigh?

18. If an automobile travels 116 mi in $3\frac{1}{2}$ hrs, how far will it travel in $5\frac{1}{4}$ hrs at the same speed?

19. If 16.8 lbs of iron, when oxidized, yields 24.0 lbs of iron oxide, how many pounds of iron must be oxidized to yield 35.0 lbs of iron oxide?

20. When pictures are enlarged in photography, their dimensions remain proportional. How wide would a 4- by 5-in. snapshot be if enlarged to a picture 8 in. across its longer dimension?

21. Find two numbers whose ratio is $3:5$ and whose difference is 14.

22. If a $44.80 price represents 140% of cost, what is cost?

23. A $32,000 warehouse yields an investor $4800 a year. If he buys a second one at

$22,200 and hopes to obtain a return in rent that is $\frac{1}{5}$ larger in ratio than what he receives from his first investment, how much return does he expect?

24. Find a number which bears the same ratio to 10 as the ratio between the number increased by 3 and the number decreased by 3.

2. PROBABILITY RATIOS

Of increasing importance to the natural and social sciences, industry, and business is the **probability ratio**. This ratio expresses the comparison of the number of different ways an event might happen *in a certain way* to the *total number* of different ways it could happen. For instance, if we toss a balanced coin in the air, it will fall to the ground in either a heads-up or tails-up position. Assuming either landing position as equally likely, the chance of it falling, say, heads-up, is 1 in 2. Expressed as a ratio, that would be 1 : 2, or as a ratio fraction, $\frac{1}{2}$. The chance of guessing a number picked at random from ten equally likely numbers is 1 in 10, or 1 : 10, or $\frac{1}{10}$. The probability of guessing the outcome of a roll of a regular die, since there are six different equally likely outcomes, is 1 in 6, or $\frac{1}{6}$.

When the numbers involved are simple, probability ratios are often more easily expressed as fractions. But in more complicated situations, it is more convenient to express these ratio fractions by their decimal equivalents. Thus, a probability ratio of $\frac{1}{2}$ may be referred to as "a probability of 0.5"; a probability ratio of $\frac{1}{4}$ as "a probability of 0.25"; a probability ratio of $\frac{1}{6}$ as "a probability of 0.17 or 0.167," depending upon the desired degree of precision. The probability, then, of tossing a "head" with the flip of a coin is 0.5; the probability of rolling, say, a "two" with a regular die is 0.167.

In general, probability ratios may be expressed by the fraction

$$\frac{s}{s+f}$$

where s represents the number of different ways an event can happen in a certain way (succeed), and f is the number of different ways it can happen *not* in that certain way (fail). Consequently, $s + f$ represents the *total* number of different outcomes possible for an event.

When an event is certain to occur, as for example guessing "four" to be the outcome of a roll of a die having four dots on each of its six faces, we assign to that event a probability of 1. This is consistent with the formula, $s/(s+f)$, for in the example the "four" can result (succeed) in six different ways and fail in none, or zero ways. Thus:

$$\frac{6}{6+0} = \frac{6}{6} = 1$$

On the other hand, 0 is the probability of certain failure:

$$\frac{0 \leftarrow \text{(no chance of success)}}{0 + f} = \frac{0}{f} = 0$$

When an event is certain either to occur or not occur (e.g., either one wins a raffle or he doesn't), the sum of the success probability and the failure probability is 1. For instance, if there are 40 different tickets sold on a raffle and we hold one of them, our success probability is 1 in 40, or 0.025; our failure probability is 39 in 40, or 0.975. The sum of these two probabilities is $0.025 + 0.975 = 1.000$. Thus:

$$\frac{s}{s+f} + \frac{f}{s+f} = \frac{s+f}{s+f} = 1$$

From this we draw an important relationship between the probabilities of success (p_s) and failure (p_f) for a given event:

$$p_s + p_f = 1$$
$$p_s = 1 - p_f$$
$$p_f = 1 - p_s$$

EXAMPLES

1. What is the probability of guessing the correct value and suit of a card drawn at random from a regular deck of playing cards?

   There are 52 different playing cards in a regular deck. The chance of guessing correctly a card drawn at random from the deck is 1 in 52, a probability of $\frac{1}{52}$ or $p_s = 0.019$.

2. What is the probability of guessing (a) the correct suit (i.e., clubs, spades, hearts, diamonds) of a card drawn at random from a complete deck of playing cards and (b) the correct value?

   (a) Since there are 13 cards of each suit in a deck of 52 playing cards, the chance of guessing the correct suit of a card drawn at random is 13 in 52, or $\frac{13}{52}$, or $\frac{1}{4}$, a p_s of 0.25.

   (b) Since the cards in each suit bear the same sequence of values, it follows that there are four aces, four kings, . . . , four 3's, and four 2's in the complete deck. Consequently, the chance of guessing the correct value (without regard to suit) of any card drawn at random from the deck is 4 in 52, or 1 in 13; a probability of 0.077.

The probability found in Example 2(b), for guessing the value apart from the suit of a playing card drawn at random from a regular deck of playing cards, could have been arrived at in a different way. Suppose the guess is that a randomly drawn card will be an ace. There is 1 chance in 52 that it will be the ace of spades, 1 chance in 52 that it will be the ace of clubs, 1 chance in 52 that it will be the ace of hearts, and 1 chance in 52 that it will be the ace of diamonds. The probabilities:

$\frac{1}{52} + \frac{1}{52} + \frac{1}{52} + \frac{1}{52} = \frac{4}{52}$ or $0.017 + 0.017 + 0.017 + 0.017 = 0.078$

(The discrepancy between 0.078 and the 0.077 of Example 2(b) is due to rounding errors.) Thus, we may generalize:

If an event A has a probability of p_a and event B has a probability of p_b, the probability that either A or B will occur is $p_a + p_b$.

EXAMPLES

1. What is the probability of rolling either a "three" or a "four" with a balanced die?

The probability of rolling a "three" is $\frac{1}{6}$.
The probability of rolling a "four" is $\frac{1}{6}$.
The probability of rolling *either* a "three" or "four" is $\frac{1}{6} + \frac{1}{6} = \frac{2}{6} = \frac{1}{3}$ $= 0.33$.

2. In a bag there are 20 balls: 4 red, 2 black, and 14 white. What is the probability of selecting either a red or a black ball from the bag?

There are 4 chances in 20 of selecting a red ball: $p_r = \frac{4}{20} = 0.20$.
There are 2 chances in 20 of selecting a black ball: $p_b = \frac{2}{20} = 0.10$.
The probability of drawing either a red or a black ball is the sum of the two probabilities: $p_r + p_b = 0.20 + 0.10 = 0.30$.

[*Note: Successfully* drawing either a red or black ball is the same as *failure* to draw a white ball. The probability of drawing a white ball is 14 in 20, or 0.70. If we denote this as p_s, then p_f (drawing a nonwhite ball, i.e., a red or black one) is

$$p_f = 1 - p_s$$
$$p_f = 1 - 0.70$$
$$p_f = 0.30]$$

EXERCISE A-2

1. An "Extra Sensory Perception" set of cards contains five different geometric symbols. What is the probability of guessing the symbol on a card drawn at random from the pack if each symbol is equally represented in the pack?

2. What is the probability that the guesser will be wrong when guessing the symbol on a card drawn at random from the pack referred to in Exercise 1?

3. The social chairman of a college class is to be selected by lot from among six male and three female students willing to serve. What is the probability that the chairman will be male?

4. If the social chairman in Exercise 3 turned out to be a male, and the vice-chairman is also to be drawn by lot from the unsuccessful students in the first draw, what is the probability that the vice-chairman will also be a male?

5. Had the social chairman referred to in Exercise 3 been a female, what is the probability that her vice-chairman would also be female, assuming the selection would have been made as indicated in Exercise 4?

6. In a college conference wrestling tournament, there are 14 entrants in a given weight classification, including last year's champion. If the champion is the first to draw his opening-round opponent by lot, what is the probability that he will not draw any given wrestler in that weight class?

7. A teacher wishes to place a supply of red and white tags of equal size in a bag in such quantities as to make the probability that a student drawing a red one at random from the bag will be 0.40. What combinations might the teacher include if he wants to have at least 60 tags, but not more than 75, in the bag?

8. What is the probability of drawing either a 3, 4, or 5 of hearts from a regular deck of 52 playing cards?

9. What is the probability of drawing a 3, 4, or 5 of any suit from a regular deck of 52 playing cards?

10. In a bag there are 18 balls: 6 red, 2 white, and 10 black. What is the probability of drawing a black ball?

11. In the bag of balls described in Exercise 10, what is the probability of selecting either a red or white ball?

12. Cards are numbered in order from 1 to 49, then mixed. What is the probability of drawing a card whose number is exactly divisible by 5?

13. If a first event has a probability of p_1 and a second event has a probability of p_2, then the probability of these two events happening in that sequence is the product of their probabilities: $p_1 p_2$. In the situation described in Exercise 3, what is the probability that the chairman–vice-chairman team will be male–female, respectively?

14. What is the probability of drawing an ace of any suit, followed by a king of the same suit, from a regular deck of playing cards?

15. In the bag of balls described in Exercise 10, what is the probability of drawing first a red ball and then a black ball?

3. MATHEMATICAL EXPECTATION

If one's chances of winning a prize of say $100 is 3 in 20, or a probability of 0.15, a fair price to pay to become eligible to win the prize would be $15. The use of the word "fair" here means that, if the activity were repeated many times under the same circumstances, in the long run both the player and the banker would break even. The $15 eligibility price in this situation is called **mathematical expectation**. Thus, if the prize for success (winning) is represented by A, and the probability of success is p, then the mathematical expectation, E, is defined to be

$$E = pA$$

In other words, in a game that is equally "fair" for player and banker alike, the ratio of the cost of eligibility to the value of the potential prize should be equal to the ratio of the successful outcomes of the game to the total number of possible outcomes, i.e., the probability of success. Thus the proportion

$$\frac{E}{A} = \frac{s}{s+f} = p_s$$

When the cost of becoming eligible to win a prize is greater than the mathematical expectation, the player has what amounts to a long-run losing proposition; when the cost of becoming eligible is less than the mathematical expectation, the player has what amounts to a long-run winning proposition.

EXAMPLES

1. There are 20 different numbers on a "wheel of chance," and the prize is a 60¢ article. (a) What is the mathematical expectation? (b) If the cost to play is 5¢, will the player be involved in a long-run winning or losing proposition?

   (a) The probability of guessing the winning number is 1 in 20, or 0.05. The mathematical expectation for an award worth 60¢ is

   $$E = pA$$
   $$E = (0.05)(60¢)$$
   $$E = 3¢$$

   (b) If we compare the mathematical expectation (3¢) with the cost of eligibility (5¢), we can see that the player is involved in a long-run losing proposition. His average loss per game is 2¢ (5¢ − 3¢).

2. A prize worth $3.50 is offered on a raffle for which the tickets are priced at 25¢ each. After the sale of how many tickets does the raffle become "unfair" to the buyer?

   Here the prize is $3.50 and the mathematical expectation is 25¢. Using $E/A = p$, we get $\$0.25/\$3.50 = \frac{1}{14}$. Hence, the raffle is "fair" to both buyer and seller when 14 tickets have been sold, because at that point the buyer's probability of winning is the same as the ratio of the value of his expectation to the value of the prize. For each ticket sold after the 14th, the raffle becomes progressively less "fair" for the buyer.

EXERCISE A-3

1. On a raffle, 500 tickets have been sold and the prize is worth $30. What is the mathematical expectation per ticket?

2. A person holds 8 tickets among 4000 in a drawing for a $1000 award. What is his mathematical expectation?

3. Ten different playing cards, ace through ten, are placed in random order, face down. The prize for guessing the value of any given card is $1. What would be a "fair" price to pay for each call?

4. If tickets sold for 10¢ each on the raffle described in Exercise 1, would the raffle be "fair" for the ticket buyer?

5. On a raffle, 150 tickets were sold at 10¢ each, or 3 for 25¢. If the prize is worth $13.50, is the raffle favorable or unfavorable to the ticket buyer?

6. At a carnival, the prize for guessing which of 3 shells the pea is under is worth 30¢. If the price to play is 15¢, does the game favor the guesser or the operator?

7. A game at a carnival advertises, "Play until you win." What is the value of the mathematical expectation here?

8. Prizes on a "wheel of chance" are merchandise certificates worth $5, which cost the operator $4 each. If there are 45 different numerals on the wheel and chances cost 10¢ each, is the game advantageous to the player or operator?

9. In a football pool, where ties don't count, the allocation of points makes the probability of picking a winning team in advance of the game, for all practical purposes, 1 in 2. The operators of the pool offer 7 to 1 odds (a prize of 7 times the price of eligibility or the amount bet) for picking four winners. To make the pool "fair," what should the odds be?

10. According to insurance tables, the probability of a person of age 20 dying within a year is 0.00243. What is the mathematical expectation (basic premium) if an insurance company agrees to a contract (policy) that calls for a payment of $3000 if that person should die within the year?

REVIEW—UNIT A

Solve each of the following proportions for x :

1. $6 : 8 = 18 : x$

2. $3 : x = 9 : 13\frac{1}{2}$

3. $9 : x = x : 4$

4. $x : 5 = \frac{3}{4} : \frac{7}{12}$

5. $x : 4 = (3x - 2) : 6$

6. $3 : 2a = 5a : x$

Solve each of the following by proportion :

7. What number bears the same ratio to 25 as 11 does to 5?

8. If $1\frac{7}{8}$ in. on a map represents 60 mi, what distance will $2\frac{1}{4}$ in. represent?

9. If 60 ft of wire weighs 27 lbs, how much will 380 ft weigh?

10. If 11.9 lbs of iron, when oxidized, yields 17.0 lbs of iron oxide, how many pounds of iron must be oxidized to yield 46.0 lbs of iron oxide?

11. Find two numbers whose ratio is 3 : 7 and whose difference is 36.

12. An $18,600 building yields an investor $2790 a year. If he buys a second one at $22,200 and hopes to obtain a return that is $\frac{1}{3}$ larger in ratio than what he receives from his first investment, how much return does he expect?

13. A children's game is played with a "spinner" on a card. If there are eight different numerals at which the pointer can stop, what is the probability of the pointer stopping at the highest value?

14. The social chairman of a college class is to be selected by lot from among eight male and four female students willing to serve. What is the probability that the chairman will be female?

15. Had the social chairman in Exercise 14 been a female, what is the probability that her vice-chairman would also be female, assuming the vice-chairman is to be drawn by lot from the unsuccessful candidates for the chairmanship?

16. A teacher wishes to place a supply of red and white tags of equal size in a bag in such quantities as to make the probability of a student drawing a red one at random from the bag 0.60. What combinations might the teacher include if he wants to have at least 70 tags, but not more than 85 in the bag?

17. What is the probability of drawing a 6, 7, 8, or 9 of any suit from a regular deck of 52 playing cards?

18. In a bag there are 24 balls: 8 red, 6 white, and 10 black. What is the probability of selecting either a red or white ball?

19. If a first event has a probability of p_1 and a second event has a probability of p_2, then the probability of these two events happening in that sequence is the product of their probabilities: $p_1 p_2$. In the situation described in Exercise 14, what is the probability that the chairman–vice-chairman team will be male–female, respectively?

20. In the bag of balls described in Exercise 18, what is the probability of drawing first a red ball and then a black ball? (cf. Exercise 19.)

21. On a raffle, 1000 tickets have been sold and the prize is worth $45. What is the mathematical expectation?

22. Ten different playing cards, ace through ten, are placed in random order, face down. The prize for guessing the value of any given card is $5. What would be a fair price to pay for each call?

23. On a raffle, 300 tickets were sold at 5¢ each, or 6 for 25¢. If the prize is worth $13.50, is the raffle favorable or unfavorable to the ticket buyer?

24. A carnival stand allows you to toss rings until you win. If the value of the prize is 25¢, what is the mathematical expectation?

25. In a football pool, where ties don't count, the allocation of points makes the probability of picking a winning team in advance of the game, for all practical purposes, 1 in 2. The operators of the pool offer 10 to 1 odds (a prize of 10 times the price of eligibility or the amount bet) for picking five winners. To make the pool "fair," what should the odds be?

SUPPLEMENTARY UNIT B

PROGRESSIONS

1. SEQUENCES

A set of numbers arranged so that there is a first number, second number, third number, and so on, is called a **sequence**. Sequences may or may not have a last number, that is, they may possess either a finite or infinite number of terms. Also, sequences may or may not possess some form of regularity or pattern among the terms. Although patterned sequences are endless in their variety, in this unit we shall consider but two: the basic and useful arithmetic and geometric progressions.

2. ARITHMETIC PROGRESSIONS

An **arithmetic progression** (A.P.) is a sequence of numbers in which each number after the first is obtained by adding a specified number to the preceding number in the sequence. This specified number is referred to as the **common difference**.

EXAMPLES

1. 2, 5, 8, 11, 14 is an arithmetic progression having a common difference of 3.

2. $-4, -5, -6, -7$ is an A.P. with a common difference of -1.

3. 2.7., 3.8., 4.9., 6.0, 7.1 is an A.P. with a common difference of 1.1.

4. $\sqrt{2}, 2\sqrt{2}, 3\sqrt{2}, 4\sqrt{2}$ is an A.P. with a common difference of $\sqrt{2}$.

5. $\frac{1}{2}, \frac{1}{4}, 0, -\frac{1}{4}, -\frac{1}{2}$ is an A.P. with a common difference of $-\frac{1}{4}$.

3. GENERAL TERM OF AN ARITHMETIC PROGRESSION

The general term of an arithmetic progression is represented by an expression which, upon proper substitution, yields any specified term in the progression. Thus, if we let a represent the first number of an arithmetic progression and d the common difference, then

The first term is a;

the second term is $(a) + d$, or $a + d$;

the third term is $(a + d) + d$, or $a + 2d$;

the fourth term is $(a + 2d) + d$, or $a + 3d$;

the fifth term is $(a + 3d) + d$, or $a + 4d$;

the sixth term is $a + 5d$;

.

.

.

the twentieth term is $a + 19d$;

.

.

.

the nth term is $a + (n - 1)d$.

If we designate the nth term of an arithmetic progression by the letter l (which suggests "last term"), we have an expression, or general term, which represents any term in an arithmetic progression:

$$l = a + (n - 1)d$$

EXAMPLES

1. Find the seventh term of an arithmetic progression in which the first term is 11 and the common difference is 4.

Here $n = 7$, $a = 11$, $d = 4$; l as the seventh term in the A.P. is:

$$l = a + (n - 1)d$$
$$l = 11 + (7 - 1)(4)$$
$$l = 11 + (6)(4)$$
$$l = 35$$

2. Find the 135th term of an A.P. where $a = \frac{1}{2}$, $d = -\frac{1}{4}$.

In this progression $a = \frac{1}{2}$, $d = -\frac{1}{4}$, $n = 135$, and

$$l = a + (n - 1)d$$
$$l = \frac{1}{2} + (135 - 1)(-\frac{1}{4})$$
$$l = \frac{1}{2} + (134)(-\frac{1}{4})$$
$$l = \frac{1}{2} - 33\frac{1}{2}$$
$$l = -33$$

3. Find the 13th term of the A.P.:   7, 17, 27, . . .

Here $a = 7$, $n = 13$, $d = 17 - 7 = 10$, and

$$l = a + (n - 1)d$$
$$l = 7 + (13 - 1)(10)$$
$$l = 7 + 120$$
$$l = 127$$

The general term can often complete the data for an arithmetic progression. Note that, if three of the variables are appropriately replaced in $l = a + (n - 1)d$, the value of the fourth variable is determined.

EXAMPLES

1. What is the first term of an eight-term A.P. in which the common difference is 6 and the 8th term is 17?

Here $a = ?$, $l = 17$, $n = 8$, and $d = 6$.

$$l = a + (n - 1)d$$
$$17 = a + (8 - 1)(6)$$
$$17 = a + 42$$
$$-25 = a$$

2. What is the common difference in a ten-term A.P. in which the first term is 0.32 and the last term is 0.59?

Here $a = 0.32$, $l = 0.59$, $d = ?$, and $n = 10$.

$$l = a + (n - 1)d$$
$$0.59 = 0.32 + (10 - 1)d$$
$$0.59 - 0.32 = 9d$$
$$0.27 = 9d$$
$$0.03 = d$$

3. How many terms are there in an A.P. in which the first and last terms are $8\frac{1}{4}$ and $12\frac{1}{2}$, respectively, and in which the common difference is $\frac{1}{8}$?

Here $a = 8\frac{1}{4}$, $l = 12\frac{1}{2}$, $d = \frac{1}{8}$, and $n = ?$

$$l = a + (n - 1)d$$
$$12\tfrac{1}{2} = 8\tfrac{1}{4} + (n - 1)(\tfrac{1}{8})$$
$$12\tfrac{1}{2} - 8\tfrac{1}{4} = \frac{n - 1}{8}$$
$$4\tfrac{1}{4} = \frac{n - 1}{8}$$
$$(8)(4\tfrac{1}{4}) = n - 1$$
$$34 = n - 1$$
$$35 = n$$

4. ARITHMETIC MEANS

The terms of a progression that occur between any two given terms (called the **extremes**) in a sequence are called the **means.** When the problem is to determine a certain number of arithmetic means between two given numbers (i.e., the complete sequence of numbers, including the inserted means and extremes, forms an A.P.), we may again resort to the general term, find the appropriate d value, and then compute the desired terms.

Note in Example 2 below that the arithmetic mean for two numbers is the same as that which is generally referred to as the "average of two numbers."

EXAMPLES

1. Insert three arithmetic means between 6 and 41.

The resulting A.P. will be a five-term sequence: 6, __, __, __, 41. So $n = 5$, $a = 6$, $l = 41$, and $d = ?$

$$l = a + (n - 1)d$$
$$41 = 6 + (5 - 1)d$$
$$41 = 6 + 4d$$
$$35 = 4d$$
$$8\tfrac{3}{4} = d$$

The five-term A.P. is: 6, $14\tfrac{3}{4}$, $23\tfrac{1}{2}$, $32\tfrac{1}{4}$, 41; the desired means are: $14\tfrac{3}{4}$, $23\tfrac{1}{2}$, $32\tfrac{1}{4}$.

2. Insert one arithmetic mean between 86 and 70.

The resulting A.P. will be: 86, __, 70. Thus, $a = 86$, $n = 3$, $l = 70$, and $d = ?$

$$l = a + (n - 1)d$$

$$70 = 86 + (3 - 1)d$$

$$70 = 86 + 2d$$

$$70 - 86 = 2d$$

$$-\tfrac{16}{2} = d$$

$$-8 = d$$

The three-term A.P. is: 86, 78, 70; the desired mean is 78.

[*Note:* The "average" of 86 and 70 is

$$\frac{86 + 70}{2} = \frac{156}{2} = 78$$

Thus the mean or middle term of a three-term arithmetic progression, a, $a + d$, $a + 2d$, is the term $a + d$, which we can also determine by "averaging" the first and third terms, a and $a + 2d$:

$$\text{"average"} = \frac{(a) + (a + 2d)}{2} = \frac{a + a + 2d}{2} = \frac{2a + 2d}{2} = a + d]$$

EXERCISE B-1

Add three more terms to each of the sequences so that the whole sequence forms an arithmetic progression:

1. $19, 25, 31, \ldots$

2. $6, -2, -10, \ldots$

3. $106, 54, 2, \ldots$

4. $\tfrac{1}{2}, \tfrac{5}{6}, 1\tfrac{1}{6}, \ldots$

5. $\tfrac{2}{3}, \tfrac{5}{12}, \tfrac{1}{6}, \ldots$

6. $0.04, 0.017, \ldots$

For each of the following progressions, find the term indicated in parentheses:

7. $5, 11, 17, \ldots$, (20th)

8. $17, 13, 9, \ldots$, (16th)

9. $4, 6\tfrac{1}{2}, 9, \ldots$, (21st)

10. $3\tfrac{1}{2}, 1, -1\tfrac{1}{2}, \ldots$, (17th)

11. $\tfrac{1}{4}, \tfrac{1}{6}, \tfrac{1}{12}, \ldots$, (25th)

12. $0.08, 0.22, 0.36, \ldots$, (11th)

13. $\sqrt{2}, 0, -\sqrt{2}, \ldots$, (18th)

14. $1.314, 1.248, 1.182, \ldots$, (26th)

In the following, a represents the first term of an A.P., n the number of terms, l the last term, and d the common difference.

15. $a = 2, d = 3, l = 38, n = ?$

16. $a = -93, l = 97, n = 20, d = ?$

17. $l = 1\tfrac{1}{2}, n = 11, d = -\tfrac{3}{4}, a = ?$

18. $a = -\tfrac{1}{3}, d = -\tfrac{1}{3}, l = -3, n = ?$

19. $d = \frac{3}{8}, n = 17, l = 6\frac{3}{4}, a = ?$

20. $l = 36.6, a = 1.4, n = 12, d = ?$

21. $d = -0.2, n = 17, l = 1.1, a = ?$

22. $a = -3.6, d = -0.4, l = -8.8, n = ?$

23. $n = 14, l = -8\frac{13}{21}, a = \frac{2}{3}, d = ?$

24. $a = x, d = x - 2a, l = 8x - 14a, n = ?$

25. Insert three arithmetic means between 3 and 27.

26. Insert four arithmetic means between $\frac{2}{3}$ and $3\frac{1}{6}$.

27. Insert three arithmetic means between 7 and -37.

28. Insert four arithmetic means between 0.04 and -0.09.

For what replacements for x will the following three-term sequences be arithmetic progressions?

29. $3x, 4x + 2, 8x - 2$

30. $\frac{1}{x}, x - 2, 1 + \frac{2}{x}$

5. SUM OF TERMS IN ARITHMETIC PROGRESSION

To derive a general formula for the sum of the terms of an arithmetic progression, we may proceed as follows. Let s represent the sum of n terms in an A.P. in which a is the first term, d is the common difference, and l is the nth term. Then

$$s = a + (a + d) + (a + 2d) + (a + 3d) + \cdots + (l - 2d) + (l - d) + l$$

By virtue of the commutative property, the sum s may be written with the addends in reverse order:

$$s = l + (l - d) + (l - 2d) + (l - 3d) + \cdots + (a + 2d) + (a + d) + a$$

Now, if we add the corresponding terms of each of these two equations, we get:

$$s + s = [(a) + (l)] + [(a + d) + (l - d)] + [(a + 2d) + (l - 2d)]$$
$$+ \cdots + [(l - d) + (a + d)] + [(l) + (a)]$$

or

$$2s = (a + l) + (a + l) + (a + l) + \cdots + (a + l) + (a + l)$$

Since there are n terms in the sequence, there will be n addends of $(a + l)$ in the foregoing sum, that is:

$$2s = n(a + l) \quad \text{or} \quad s = \frac{n}{2}(a + l)$$

In words: The sum of n terms of an arithmetic progression is equal to half the product of the number of terms and the sum of the first and last (or nth) terms.

EXAMPLES

1. Find the sum of the six terms of the A.P.: 8, 12, 16, 20, 24, 28.

In this A.P., $n = 6$, $a = 8$, and $l = 28$. Using the formula $s = \frac{n}{2}(a + l)$, we find

$$s = \tfrac{6}{2}(8 + 28)$$
$$s = 3(36)$$
$$s = 108$$

Therefore, $8 + 12 + 16 + 20 + 24 + 28 = 108$, which can be verified by the usual method of adding.

2. Find the sum of the first one-hundred counting numbers.

Here $a = 1$, $n = 100$, and $l = 100$.

$$s = \frac{n}{2}(a + l)$$
$$s = \tfrac{100}{2}(1 + 100)$$
$$s = 50(101)$$
$$s = 5050$$

6. ELEMENTS OF AN ARITHMETIC PROGRESSION

An alternative formula for the sum of n terms of an arithmetic progression can be obtained by substituting the equivalent for l, in terms of a, n, and d, into $s = \frac{n}{2}(a + l)$:

$$\begin{cases} s = \frac{n}{2}(a + l) \\ l = a + (n - 1)d \end{cases}$$

$$s = \frac{n}{2}[a + (a + (n - 1)d)]$$

$$s = \frac{n}{2}[a + a + (n - 1)d]$$

$$s = \frac{n}{2}[2a + (n - 1)d]$$

The five parts of an arithmetic progression that we have been representing by s, a, n, d, and l are called the **elements** of an arithmetic progression. If *any three* of these five elements are known, by appropriate use of the formulas

$$l = a + (n - 1)d, \quad s = \frac{n}{2}(a + l), \quad \text{and} \quad s = \frac{n}{2}[2a + (n - 1)d]$$

the remaining two elements can be determined.

EXAMPLES

1. The first term of an A.P. is 12, the last is 18, and the sum is 75. Find the common difference and number of terms in the progression.

Use $s = \frac{n}{2}(a + l)$ to find n:

$$s = \frac{n}{2}(a + l)$$

$$75 = \frac{n}{2}(12 + 18)$$

$$75 = 15n$$

$$5 = n$$

Use $l = a + (n - 1)d$ to find d:

$$l = a + (n - 1)d$$

$$18 = 12 + (5 - 1)d$$

$$18 - 12 = 4d$$

$$\tfrac{3}{2} = d$$

2. In an A.P. the sum is 585, the first term is -3, and the common difference is 6. Find the number of terms and the last term.

Here $s = 585$, $a = -3$, and $d = 6$. To find n, we use $s = \frac{n}{2}[2a + (n - 1)d]$:

$$s = \frac{n}{2}[2a + (n - 1)d]$$

$$585 = \frac{n}{2}[2(-3) + (n - 1)6]$$

$$585 = \frac{n}{2}[-6 + 6n - 6]$$

$$585 = \frac{n}{2}[6n - 12]$$

$$585 = 3n^2 - 6n$$

$$0 = 3n^2 - 6n - 585$$

$$0 = n^2 - 2n - 195$$

$$0 = (n - 15)(n + 13)$$

$$n - 15 = 0 \qquad n + 13 = 0$$

$$n = 15 \qquad\qquad n = -13$$

An A.P. of -13 terms makes no sense; so we reject the solution, -13, and accept 15 as the number of terms in the A.P. To find the value of the last term, we substitute 15 in:

$$l = a + (n - 1)d$$
$$l = -3 + (15 - 1)6$$
$$l = -3 + (14)(6)$$
$$l = 81$$

3. The third term of an eight-term arithmetic progression is 8; the seventh term is -4. Find the remaining elements of this progression.

Solution of a system of equations is necessary here. If the third term is 8 and the seventh is -4, then according to $l = a + (n - 1)d$, we arrive at the following pair of equations:

$$8 = a + (3 - 1)d \quad \text{and} \quad -4 = a + (7 - 1)d$$

$$\text{or} \quad \begin{cases} a + 2d = 8 \\ a + 6d = -4 \end{cases}$$

Subtracting the bottom equation from the top, and proceeding according to Program **9.5**:

$$-4d = 12$$
$$d = -3$$

Since $a + 2d = 8$, and if $d = -3$:

$$a + 2(-3) = 8$$
$$a = 14$$

The eighth term, or l, must be:

$$l = a + (n - 1)d$$
$$l = 14 + (8 - 1)(-3)$$
$$l = 14 + (7)(-3)$$
$$l = -7$$

Finally, to find s:

$$s = \frac{n}{2}(a + l)$$
$$s = \tfrac{8}{2}[14 + (-7)]$$
$$s = 4(7)$$
$$s = 28$$

EXERCISE B-2

In each of the following, some of the elements of an A.P., a, n, d, l, and s, are given. Find the remaining elements.

1. $a = 7, n = 12, l = 40$

2. $a = 7, d = 4, l = 79$

3. $n = 10, d = 2, l = 25$

4. $a = 3, n = 14, d = 5$

5. $a = 10, l = 74, s = 714$

6. $a = 21, n = 24, s = 2160$

7. $n = 22, d = -5, s = -869$

8. $n = 38, l = 46, s = 342$

9. $d = \frac{1}{2}, l = 0, s = -\frac{105}{2}$

10. $a = 6.1, d = -0.3, s = 64.6$

11. The second term of a seven-term A.P. is 8, and the sixth term is 20. Find the values of the remaining elements.

12. A charity raffle involves 60 tickets numbered consecutively, in multiples of 5, from 5 to 300, inclusive. If each purchaser pays in cents the number on the ticket he draws, how much will be realized from the raffle?

13. A man plans to save $100 this year and, in each succeeding year, $50 more than the year before. If he follows this plan, how much will he have accumulated at the end of 15 years?

14. A regular savings program of $500 deposited at the beginning of each year at 5% simple interest will accumulate to what amount at the end of the 14th year?

15. An object falls 16 ft during the first second, 48 ft during the second second, 80 ft during the third, and so on.
(a) How far does it fall during the twentieth second?
(b) How far will it have fallen after twenty seconds?
(c) How long will it take to drop 22,500 ft?

16. The ancient Greeks were interested in "triangular numbers," the first four of which were represented as •, ∴, ∴∴, ∴∴∴ (1, 3, 6, 10). How many dots would there be in the representation of:
(a) the eighth triangular number?
(b) the forty-third triangular number?

7. GEOMETRIC PROGRESSIONS

A **geometric progression** (G.P.) is a sequence of numbers in which each term, after the first, is obtained by *multiplying* the preceding term by a specified number. This specified number is referred to as the **common ratio**.

EXAMPLES

1. 1, 2, 4, 8, 16, 32 is a geometric progression having a common ratio of 2.

2. $-3, 6, -12, 24, -48$ is a G.P. with a common ratio of -2.

3. 6, 0.6, 0.06, 0.006, 0.0006 is a G.P. with a common ratio of 0.1.

4. $\sqrt{2}, 2, 2\sqrt{2}, 4, 4\sqrt{2}, 8$ is a G.P. with a common ratio of $\sqrt{2}$.

5. $-\frac{1}{5}, -\frac{1}{25}, -\frac{1}{125}, -\frac{1}{625}$ is a G.P. with a common ratio of $\frac{1}{5}$.

8. GENERAL TERM OF A GEOMETRIC PROGRESSION

The general term of a geometric progression is represented by an expression which, upon appropriate replacement of its variables, yields any specified term in the progression. Thus, if we denote the first term of a geometric progression by a, the common ratio by r, then

the first term is a;
the second term is $r \times a$, or ar;
the third term is $r \times (ar)$, or ar^2;
the fourth term is $r \times (ar^2)$, or ar^3;
the fifth term is ar^4;

.
.
.

the twentieth term is ar^{19};

.
.
.

the nth term is ar^{n-1}.

If we designate the nth term of a geometric progression by l, we have an expression, or general term, which represents any term in a geometric progression:

$$l = ar^{n-1}$$

EXAMPLES

1. Find the fifth term of a geometric progression in which the first term is 3 and the common ratio is 2.

Here $a = 3, r = 2$, and $n = 5$; l as the fifth term in the G.P. is:

$$l = ar^{n-1}$$
$$l = (3)(2)^{5-1}$$

$$l = (3)(2)^4$$
$$l = 48$$

2. Find the sixth term of the G.P.: $-2, \frac{1}{2}, -\frac{1}{8}, \ldots$.

Here $a = -2$, $n = 6$, and $r = \frac{1}{2} \div (-2) = -\frac{1}{4}$; the sixth term ($l$) is:

$$l = ar^{n-1}$$
$$l = (-2)(-\tfrac{1}{4})^{6-1}$$
$$l = (-2)(-\tfrac{1}{4})^5$$
$$l = (-2)(-\tfrac{1}{1024})$$
$$l = \tfrac{1}{512}$$

3. Find the tenth term of the G.P. in which the first term is 1.3 and the common ratio is 1.1.

Here $a = 1.3$, $n = 10$, and $r = 1.1$; because of the high power to which one factor must be raised, computing l is best done by logarithms:

$$l = ar^{n-1}$$
$$l = (1.3)(1.1)^{10-1}$$
$$l = (1.3)(1.1)^9$$

$$
\begin{array}{rcr}
\log 1.1 \approx & & .0414 \\
& & 9 \\
& (\times) & \overline{} \\
9 \log 1.1 \approx & & .3726 \\
\log 1.3 \approx & & .1139 \\
& (+) & \overline{} \\
\log l \approx & & .4865
\end{array}
$$

$$l = 3.066 \quad \text{(approximately)}$$

When there is incomplete information about a geometric progression, the general term may be used at times to complete that information.

EXAMPLES

1. What is the first term of a six-term G.P. in which the ratio is $\sqrt{3}$ and the sixth term is 27?

Here $a = ?$, $l = 27$, $r = \sqrt{3}$, and $n = 6$. We find a by:

$$l = ar^{n-1}$$
$$27 = a(\sqrt{3})^{6-1}$$
$$27 = a(\sqrt{3})^5$$
$$\frac{27}{9\sqrt{3}} = a$$
$$\sqrt{3} = a$$

2. What is the common ratio in an eight-term G.P. in which the fourth term is 1 and the eighth term is $\frac{625}{1296}$?

The G.P. involved here looks something like this:

$$\text{—, —, —, } 1, \text{ —, —, —, } \tfrac{625}{1296}$$

The common ratio r, which must be common throughout, can be found if we consider a part of the progression, namely, the last five terms:

$$1, \text{ —, —, —, } \tfrac{625}{1296}$$

in which $a = 1$, $l = \tfrac{625}{1296}$, $n = 5$; thus

$$l = ar^{n-1}$$

$$\frac{625}{1296} = (1)(r)^{5-1}$$

$$\frac{625}{1296} = r^4$$

$$\frac{5^4}{6^4} = r^4$$

$$\frac{5}{6} = r$$

3. How many terms are there in a G.P. in which the first and last terms are 16 and $\tfrac{1}{64}$, respectively, and in which the common ratio is $\tfrac{1}{2}$?

Here $a = 16$, $l = \tfrac{1}{64}$, $r = \tfrac{1}{2}$, and n can be found thus:

$$l = ar^{n-1}$$

$$\tfrac{1}{64} = (16)(\tfrac{1}{2})^{n-1}$$

$$\frac{1}{64 \times 16} = \left(\frac{1}{2}\right)^{n-1}$$

$$\frac{1}{2^6 \times 2^4} = \left(\frac{1}{2}\right)^{n-1}$$

$$(\tfrac{1}{2})^{10} = (\tfrac{1}{2})^{n-1}$$

$$10 = n - 1$$

$$11 = n$$

9. GEOMETRIC MEANS

The **geometric means** are those terms which occur between two extreme terms so that all of the terms form a geometric progression. Essentially, the task of determining the values of a set of geometric means is one of finding the appropriate common ratio which may then be used to generate the progression, including the desired means.

EXAMPLES

1. Insert two geometric means between 7 and 189.

The resulting G.P. will be a four-term sequence: 7, __, __, 189. Thus, $a = 7$, $l = 189$, and $n = 4$; and r may be found by:

$$l = ar^{n-1}$$

$$189 = 7(r)^{4-1}$$

$$27 = r^3$$

$$3 = r$$

The four-term G.P., then, is 7, 21, 63, 189, and the desired means are 21 and 63.

2. Insert three geometric means between 11 and 891.

The resulting G.P. will have five terms; so $n = 5$, $a = 11$, and $l = 891$.

$$l = ar^{n-1}$$

$$891 = 11(r)^{5-1}$$

$$81 = r^4$$

$$+3, -3 = r$$

Thus there are two G.P.'s which will fit the given data, one having a common ratio of $+3$: (11, **33, 99, 297,** 891), and the other having a common ratio of -3: (11, **−33, 99, −297,** 891). The respective geometric means are shown in boldface.

EXERCISE B-3

Add three more terms to each of the sequences so that the whole sequence forms a geometric progression:

1. 2, 6, 18, . . .

2. 12, 6, 3, . . .

3. 1, −3, 9, . . .

4. $\frac{3}{4}$, $-\frac{3}{8}$, $\frac{3}{16}$, . . .

5. −3, 0.6, −0.12, . . .

6. 3, $3\sqrt{2}$, 6, . . .

For each of the following progressions, find the term indicated in parentheses:

7. 1, 3, 9, . . . , (6th)

8. 3, 6, 12, . . . , (7th)

9. $\frac{1}{2}$, 1, 2, . . . , (10th)

10. $\frac{1}{2}$, $-\frac{1}{10}$, $\frac{1}{50}$, . . . , (8th)

11. $\sqrt{2}$, −2, $2\sqrt{2}$, . . . , (11th)

12. $\sqrt[3]{2}$, $\sqrt[3]{4}$, 2, . . . , (15th)

13. i, −1, −i, 1, i, . . . , (16th)

14. 5, −1, 0.2, . . . , (9th)

In the following, a represents the first term of a G.P., n is the number of terms, l the last term, and r the common ratio.

15. $l = 48$, $a = 3$, $r = 2$, $n = ?$

16. $l = 243, a = 3, n = 5, r = ?$

17. $l = -1250, n = 5, r = 5, a = ?$

18. $l = 81, a = \frac{1}{3}, r = 3, n = ?$

19. $l = -2916, r = 3, n = 7, a = ?$

20. $l = -\frac{1}{32}, n = 8, a = 4, r = ?$

21. $l = 24.3, n = 6, r = 3, a = ?$

22. $l = \frac{1}{972}, a = \frac{3}{4}, r = -\frac{1}{3}, n = ?$

23. $l = -162, a = 2\sqrt{3}, n = 8, r = ?$

24. $l = 64, a = \sqrt[3]{2}, r = \sqrt[3]{2}, n = ?$

25. Insert three geometric means between 4 and 324.

26. Insert four geometric means between 160 and 5.

27. Insert four geometric means between $-\frac{2}{3}$ and $5\frac{1}{16}$.

28. Insert five geometric means between 5 and 625.

For what replacements for x will each of the following three-term sequences be a geometric progression?

29. $x - 2, 2x + 1, 7x - 4.$

30. $3x - 1, 5x + 1, 10x + 2$

10. SUM OF A FINITE NUMBER OF TERMS IN GEOMETRIC PROGRESSION

Corresponding to the formula for the sum of n terms in arithmetic progression, there is a formula for the sum of n terms in geometric progression. It can be arrived at in this way. The sum (s) of the first n terms in a geometric progression, in which the first term is a and the common ratio is r, may be written

$$s = a + ar + ar^2 + ar^3 + \cdots + ar^{n-2} + ar^{n-1}$$

If we multiply both members of this equation by r, we get

$$rs = ar + ar^2 + ar^3 + ar^4 + \cdots + ar^{n-1} + ar^n$$

Subtracting the second equation from the first yields the equation:

$$s - rs = a - ar^n$$

which may be solved for s:

$$s(1 - r) = a(1 - r^n)$$

$$s = a\left(\frac{1 - r^n}{1 - r}\right), \qquad (r \neq 1)$$

Thus we have derived a formula for finding the sum of the first n terms of a geometric progression.

EXAMPLES

1. Find the sum of 1, 2, 4, 8, 16, 32, 64, 128.

The terms here form a geometric progression in which $a = 1$, $r = 2$, and $n = 8$; the sum can be found as follows:

$$s = a\left(\frac{1 - r^n}{1 - r}\right)$$

$$s = 1\left(\frac{1 - 2^8}{1 - 2}\right)$$

$$s = 1\left(\frac{-255}{-1}\right)$$

$$s = 255$$

2. Find the sum of the ten terms that form a geometric progression in which the first term is 0.35 and the common ratio is 1.6.

Logarithms will be useful here; $a = 0.35$, $r = 1.6$, $n = 10$.

$$s = a\left(\frac{1 - r^n}{1 - r}\right)$$

$$s = (0.35)\left(\frac{1 - 1.6^{10}}{1 - 1.6}\right) \qquad \begin{cases} \log 1.6 \approx .2041 \\ 10 \log 1.6 \approx 2.0410 \\ 1.6^{10} \approx 109.9 \end{cases}$$

$$s \approx (0.35)\left(\frac{1 - 109.9}{1 - 1.6}\right)$$

$$s \approx (0.35)\left(\frac{108.9}{0.6}\right) \qquad \begin{cases} \log 108.9 \approx \quad 2.0370 \\ \log 0.35 \approx \quad 9.5441 - 10 \\ \qquad (+) \dfrac{}{11.5811 - 10} \\ \log 0.6 \approx \quad 9.7782 - 10 \\ \qquad (-) \dfrac{}{} \\ \log s \approx \quad 1.8029 \\ \qquad s \approx 63.51 \end{cases}$$

$$s = 63.51 \quad \text{(approx.)}$$

11. ELEMENTS OF A GEOMETRIC PROGRESSION

The elements of a geometric progression are the five parts we have been representing by s, a, n, r, and l. Given any three of these five elements, appropriate use of the formulas:

(1) $l = ar^{n-1}$

(2) $s = a\left(\dfrac{1 - r^n}{1 - r}\right)$

(3) $s = \dfrac{a - rl}{1 - r}$

makes it possible to determine the remaining two elements.† Formula (3) is a variation of formula (2):

$$s = a\left(\dfrac{1 - r^n}{1 - r}\right)$$

$$\left.\begin{array}{l} l = ar^{n-1} \\ rl = ar^n \end{array}\right\} \quad \begin{array}{l} = \dfrac{a - ar^n}{1 - r} \\[2mm] = \dfrac{a - rl}{1 - r} \end{array}$$

EXAMPLES

1. A geometric progression has a sum of $10\frac{1}{12}$, a first term of $\frac{1}{12}$, and a last term of $6\frac{3}{4}$. Find the remaining elements.

Here $a = \frac{1}{12}$, $l = 6\frac{3}{4}$, $s = 10\frac{1}{12}$, $n = ?$, $r = ?$ Using $s = \dfrac{a - rl}{1 - r}$ to find r:

$$10\tfrac{1}{12} = \dfrac{(\frac{1}{12}) - (r)(6\frac{3}{4})}{1 - r}$$

$$\dfrac{121}{12} - \dfrac{121}{12}r = \dfrac{1}{12} - \dfrac{27}{4}r$$

$$121 - 121r = 1 - 81r$$

$$121 - 1 = 121r - 81r$$

$$120 = 40r$$

$$3 = r$$

Using $l = ar^{n-1}$ to find n:

$$6\tfrac{3}{4} = \tfrac{1}{12}(3)^{n-1}$$

$$(12)(\tfrac{27}{4}) = 3^{n-1}$$

$$81 = 3^{n-1}$$

$$3^4 = 3^{n-1}$$

$$4 = n - 1$$

$$5 = n$$

2. The last two terms of a four-term G.P. are $\frac{4}{175}$ and $\frac{8}{875}$. Find the remaining elements.

The common ratio, r, can be found if we divide the last term by the one before it:

---

†However, when r and l, and a and r, are paired as unknowns, a higher-degree equation will often result; n, of course, must be a positive integer.

$$r = \frac{8}{875} \div \frac{4}{175} = \frac{\overset{2}{\cancel{8}}}{875} \times \frac{175}{\underset{5}{\cancel{4}}} = \frac{2}{5}$$

Since $n = 4$, $r = \frac{2}{5}$, and $l = \frac{8}{875}$, a can be found by:

$$l = ar^{n-1}$$

$$\frac{8}{875} = a\left(\frac{2}{5}\right)^{4-1}$$

$$\frac{125}{8} \times \frac{8}{875} = a$$

$$\frac{1}{7} = a$$

Finally:

$$s = a\left(\frac{1 - r^n}{1 - r}\right)$$

$$s = \frac{1}{7}\left(\frac{1 - \left(\frac{2}{5}\right)^4}{1 - \frac{2}{5}}\right)$$

$$s = \frac{1}{7}\left(\frac{\frac{609}{625}}{\frac{3}{5}}\right)$$

$$s = \frac{29}{125}$$

12. SUM OF AN INFINITE NUMBER OF TERMS IN GEOMETRIC PROGRESSION

If we analyze the formula for the sum of a finite number of terms in geometric progression

$$s = a\left(\frac{1 - r^n}{1 - r}\right)$$

we can see that, when r is less than 1 but greater than -1, the r^n term will become numerically smaller and smaller as n, the number of terms and the exponent of r, increases. Thus, if $r = \frac{1}{3}$, then $r^2 = \frac{1}{9}$, $r^3 = \frac{1}{27}$, $r^4 = \frac{1}{81}$, etc. As the number of terms in the geometric progression increases indefinitely, the value of the r^n term (for r's between -1 and $+1$, i.e., $-1 < r < +1$) tends toward zero, and the formula for the sum of an *infinite* number of terms in geometric progression becomes

$$S_\infty = a\left(\frac{1 - 0}{1 - r}\right) = \frac{a}{1 - r}$$

Thus for some progressions, though the number of terms is infinite, their sum will be finite.

EXAMPLES

1. Find the sum of the infinite geometric progression $\frac{1}{2}, \frac{1}{4}, \frac{1}{8}, \frac{1}{16}, \dots$

Here the first term is $\frac{1}{2}$, the common ratio is $\frac{1}{2}(-1 < \frac{1}{2} < +1)$, and the number of terms is infinite; so:

$$S_\infty = \frac{a}{1 - r}$$

$$S_\infty = \frac{\frac{1}{2}}{1 - \frac{1}{2}} = \frac{\frac{1}{2}}{\frac{1}{2}} = 1$$

2. Express the infinite repeating decimal 0.31616161616... as a fraction.

Here 0.31616161616... may be written as a sum of terms that form a G.P., after the first digit:

$$0.31616161616\ldots = 0.3 + \underbrace{0.016 + 0.00016 + 0.0000016 + \cdots}_{\text{G.P.}}$$

Considering 0.016 as the first term of a G.P., and $r = 0.01$, then

$$S_\infty = \frac{a}{1 - r}$$

$$S_\infty = \frac{0.016}{1 - 0.01} = \frac{0.016}{0.99} = \frac{16}{990}$$

and

$$0.31616161616\ldots = 0.3 + \frac{16}{990}$$

$$= \frac{3}{10} + \frac{16}{990} = \frac{297 + 16}{990} = \frac{313}{990}$$

EXERCISE B-4

In each of the following, some of the elements of a G.P., a, n, r, l, and s, are given. Find the remaining elements.

1. $n = 6, r = 2, l = 32$

2. $a = \frac{1}{2}, n = 6, l = 16$

3. $a = 128, n = 7, r = -\frac{1}{2}$

4. $a = 1, r = 3, l = 243$

5. $n = 6, r = -\frac{1}{2}, s = \frac{21}{2}$

6. $r = \frac{1}{2}, l = \frac{1}{4}, s = \frac{63}{4}$

7. $a = 1\frac{1}{2}, l = 96, s = 190\frac{1}{2}$

8. $a = -96, r = -\frac{3}{2}, s = 399$

9. Write the seven terms that form a geometric progression if the third term is 81 and the sixth term is -3.

10. Write the six terms that form a geometric progression if the second term is $\frac{5}{12}$ and the fifth term is $\frac{2}{75}$.

11. A man plans to save $3 during January, $6 during February, $12 during March, and so on, each month's savings being double that of the month before.
(a) How much must he save during August?
(b) How much will he have saved by the end of the year?

12. A certain type of bacteria doubles its number every three hours. How many times as great will their number be after 24 hrs?

13. If there are n of the bacteria referred to in Exercise 12 in a culture at the beginning of a time interval, how many would there be $19\frac{1}{2}$ hrs later?

14. For figuring depreciation on certain items, accountants sometimes use a "constant per cent" method, in which the value of the item is considered to depreciate the same per cent each year, based on the item's book value of the previous year. According to this method, what would be the book value of an item after 5 yrs of use if it cost $1000 originally, and if depreciation is figured at a constant 20% a year of its previous year's book value? [*Note:* 20% *off* is the same as 80% *of.*]

15. A ball is dropped from a height of 27 ft, and on each rebound, it comes up to a height that is two-thirds of its previous fall. How far will it have traveled by the time it reaches the top of its fifth rebound?

16. Theoretically, how far will the ball of Exercise 15 travel before it comes to rest?

Find the sums of each of the following infinite geometric progressions:

17. 8, 4, 2, . . .

18. 27, -9, 3, . . .

19. 6, 0.6, 0.06, . . .

20. $\sqrt{2}$, 1, $\frac{1}{2}\sqrt{2}$, . . .

Express the following infinitely repeating decimals as equivalent fractions:

21. 0.7777 . . .

22. 0.373737 . . .

23. 0.22333 . . .

24. 4.603111 . . .

REVIEW—UNIT B

Add three additional terms to each of the sequences so that the whole sequence forms an an arithmetic progression:

1. 38, 44, 50, . . .

2. 98, 51, 4, . . .

3. $\frac{5}{6}$, $\frac{5}{12}$, 0, . . .

For each of the following progressions, find the term indicated in parentheses:

4. 6, 11, 16, . . . , (18th)

5. 7, $9\frac{1}{2}$, 12, . . . , (21st)

6. $\frac{11}{12}$, $\frac{2}{3}$, $\frac{5}{12}$, . . . , (25th)

7. $\sqrt{3}$, 0, $-\sqrt{3}$, . . . , (23rd)

In the following, a stands for the first term of an A.P., n for the number of terms, l the last term, and d the common difference.

8. $a = 9, d = 2, l = 57, n = ?$

9. $l = 2\frac{1}{2}, n = 13, d = -\frac{1}{2}, a = ?$

10. $d = \frac{2}{7}, n = 15, l = 4\frac{3}{7}, a = ?$

11. $d = -0.3, n = 19, l = 0.7, a = ?$

12. $n = 17, l = -8\frac{17}{20}, a = \frac{3}{4}, d = ?$

13. Insert three arithmetic means between 5 and 29.

14. Insert three arithmetic means between 3 and -33.

15. For what replacements for x will the three-term sequence $2x + 1, 4x - 2, 3x + 4$ be an arithmetic progression?

In each of the following, some of the elements of an A.P., a, n, d, l, and s, are given. Find the remaining elements.

16. $a = 7, n = 11, l = 47$

17. $n = 18, d = 3, l = 56$

18. $a = 9, l = 86, s = 570$

19. $n = 25, d = -\frac{5}{2}, s = 750$

20. $d = -\frac{3}{4}, l = \frac{3}{4}, s = \frac{117}{2}$

21. The second term of a six-term A.P. is -3, and the fifth term is -15. Find the values of the remaining elements: $a, d, l,$ and s.

22. A man plans to save \$50 this year and, in each succeeding year, \$50 more than the year before. If he follows this plan, how much will he have accumulated at the end of 15 years?

23. An object falls 16 ft during the first second, 48 ft during the second second, 80 ft during the third, and so on.
(a) How far does it fall during the sixteenth second?
(b) How far will it have fallen after sixteen seconds?
(c) How long will it take to drop 19,600 ft?

Add three additional terms to each of the sequences so that the complete sequence forms a geometric progression:

24. $7, 14, 28, \ldots$

25. $8, -16, 32, \ldots$

26. $-7, 2.1, -0.63, \ldots$

For each of the following progressions, find the term indicated in parentheses:

27. $4, 12, 36, \ldots, $ (6th)

28. $\frac{1}{3}, \frac{2}{3}, \frac{4}{3}, \ldots, $ (10th)

29. $\sqrt{3}, -3, 3\sqrt{3}, \ldots, $ (11th)

30. $-1, -i, 1, i, -1, \ldots, $ (16th)

In the following, a stands for the first term of a G.P., n for the number of terms, l the last term, and r the common ratio.

31. $l = 80, a = 5, r = 2, n = ?$

32. $l = -1875, n = 5, r = 5, a = ?$

33. $l = -1458, r = 3, n = 7, a = ?$

34. $l = 72.9, n = 6, r = 3, a = ?$

35. $l = -64, a = 2\sqrt{2}, n = 10, r = ?$

36. Insert three geometric means between 7 and 567.

37. Insert four geometric means between $-\frac{3}{5}$ and $7\frac{58}{81}$.

38. For what replacements for x will the three-term sequence $x - 1, 2x + 1, 8x - 5$ be a geometric progression?

In each of the following, some of the elements of a G.P., a, n, r, l, and s, are given. Find the remaining elements.

39. $n = 5, r = 3, l = 243$ **40.** $a = -4, n = 7, r = -2$

41. $n = 6, r = -\frac{3}{4}, s = 120\frac{1}{4}$ **42.** $a = 4096, l = 1, s = 5461$

43. Write the seven terms that form a geometric progression if the third term is 8 and the sixth term is -1.

44. A man plans to save $5 during January, $10 during February, $20 during March, and so on, each month's savings being double that of the month before.
(a) How much must he save during August?
(b) How much will he have saved by the end of the year?

45. If there are n of a certain bacteria which double their number every 4 hrs, how many will there be 22 hrs later?

46. A ball is dropped from a height of 128 ft, and on each rebound, it comes up to a height that is half of its previous fall. How far will it have traveled by the time it reaches the top of its fifth rebound?

Find the sums of each of the following infinite geometric progressions.

47. $10, 5, 2\frac{1}{2}, \ldots$ **48.** $8, 0.8, 0.08, \ldots$

Express the following infinitely repeating decimals as equivalent fractions.

49. $0.88888\ldots$ **50.** $0.326666\ldots$

SUPPLEMENTARY UNIT

ELEMENTS OF SET THEORY

The word *set* is a common term in the English language. When used as a noun, it is widely understood to indicate a collection of items—a set of dishes, a set of golf clubs, a set of circumstances, and so on. In this book, the word has been used in an occasional mathematical context, e.g., a set of numbers, a set of points, a set of solutions. Actually, the concept of set has fundamental significance in mathematics, and it has been the subject of much serious study by mathematical scholars. In this unit, we discuss some of the formal notation of sets, and a few basic relationships and operations among sets.

A set has no precise mathematical definition; it is interpreted to mean simply a collection of things, real or imagined, finite or infinite in number. The "things" that make up a set are called the **elements** or **members** of the set. For example, all persons who have read a portion of this book make up a set, and you are a member or element of the set. The fingers of your right hand form a set, and your thumb is a member of that set.

The symbol $\in$ means *is an element (or member) of*. The symbol $\notin$ means *is not an element (or member) of*. Thus, if N represents the set of all teams in the National Football League, then

Baltimore *Colts* $\in N$

University of Texas *Longhorns* $\notin N$

1. BASIC NOTATION

289

Also, if C represents the set of all counting numbers, then

$$427 \in C$$

$$\tfrac{1}{4} \notin C$$

$$-7 \notin C$$

A set is said to be **well-defined** if it is possible to make perfectly clear whether or not a given item is a member of the set. "The set of letters, a, b, c" is an example of a well-defined set because, for any specified letter of the alphabet, we can readily decide whether or not it is a member of the set. However, "The set of all tall people" is not a well-defined set because there is no clear-cut way to decide whether or not a given person is tall. On the other hand, "The set of all people who are taller than 6 ft 5 in." is a well-defined set.

The clearest way to define or *specify* a set is by the **roster method**—a tabulation of all of the elements of the set, between braces:

$$\{a,\ b,\ c\}$$

$$\{2,\ 4,\ 6,\ 8,\ 10\}$$

$$\{\text{Mon., Tues., Wed., Thurs.}\}$$

When the roster method is not practical (e.g., the set of all U.S. presidents), we may use the **rule method**—a statement that clearly defines the elements of the set:

$$\{\text{U.S. vice-presidents, 1800–1950}\}$$

$$\{\text{Citizens of Japan}\}$$

Often the rule method of specifying a set employs set-builder notation (recall pp. 65–66), that is:

$$\{x\,|\,\sim\!\sim\!\sim\!\sim\!\sim\!\sim\}$$

where the $\sim\!\sim\!\sim\!\sim\!\sim$ is replaced by a statement which clearly defines every element in the set. For example:

$$\{x\,|\,x \text{ is a counting number less than 5}\}$$

(Read: "The set of all x such that x is a counting number less than 5.") This same set expressed by the roster method would be

$$\{1,\ 2,\ 3,\ 4\}$$

EXAMPLES

1. If $A = \{x\,|\,x$ is a counting number less than 9$\}$, then $A = \{1, 2, 3, \ldots, 8\}$ and

$$3 \in A, \qquad 10 \notin A, \qquad 6 \in A, \qquad 9 \notin A$$

2. If $B = \{x|x$ is a letter of the alphabet between m and $t\}$, then $B = \{n,$ $o, p, q, r, s\}$, and

$$n \in B, \qquad r \in B, \qquad a \notin B, \qquad z \notin B$$

It is possible to conceive of a set without any members, e.g., the set of all men living today who weigh more than a ton, or the set of all counting numbers between $\frac{1}{2}$ and $\frac{3}{4}$. Such sets are examples of the **empty set**, sometimes referred to as the **null set**. The usual symbol for the empty set is either $\{\quad\}$ or $\varnothing$.

EXERCISE C-1

Use the roster method to specify the elements of each of the following sets.

1. Letters of the word *aunt*.

2. Months of the year with initial letter J.

3. Squares of first four counting numbers.

4. Odd counting numbers less than 12.

5. Different letters of the word *mathematics*.

6. Alphabet letters between m and s.

7. Even numbers greater than 98.

8. Integers less than 2.

9. Letters appearing twice in the word *uncle*.

10. Integers whose squares are -5.

11. $\{x|x$ is a counting number less than 7$\}$

12. $\{x|x$ is the sum of 7 and 12$\}$

13. $\{x|x$ is a counting number greater than 7$\}$

14. $\{n|n$ is a day of the week$\}$

15. $\{p|p$ is a month of the year with initial letter $R\}$

16. $\{m|m$ is an integer whose square is 4$\}$

Decide whether each of the following is true or false if $A = \{2, 4, 6, 8\}$, $B = \{x|x$ is an integer$\}$, and $C = \{$rational numbers$\}$.

17. $9 \notin B$	**18.** $3 \in A$	**19.** $3 \in C$
20. $24 \in A$	**21.** $2\frac{1}{4} \in A$	**22.** $2\frac{1}{4} \notin B$
23. $2\frac{1}{4} \notin C$	**24.** $-6 \in B$	**25.** $-6 \in C$
26. $-6 \in A$	**27.** $0.8 \in C$	**28.** $0.8 \notin B$
29. $-2.8 \notin C$	**30.** $1026 \in B$	**31.** $\frac{8}{4} \in B$
32. $\frac{4}{8} \in B$	**33.** $\sqrt{8} \notin C$	**34.** $2 + 3i \in C$

2. SET RELATIONS

Sets are **equal** to each other if, and only if, they have exactly the same elements (the order of their listing is not important). For example, $\{3, 8, 17\}$ and $\{8, 3, 17\}$ are equal sets, but $\{3, 8, 17\}$ and $\{2, 8, 17\}$ are not equal sets. The sets, $\{x \mid x$ is a counting number$\}$ and $\{1, 2, 3, 4, 5, \ldots\}$, are equal sets.

When the elements of one set can be matched one-to-one with the elements of another set, so that each element of each set has one and only one matched element in the other set, the sets are **equivalent**. The following three sets are equivalent, that is, each has the same number of elements:

$$\{a, \quad b, \quad c, \quad d\}$$
$$\updownarrow \quad \updownarrow \quad \cdot\updownarrow \quad \updownarrow$$
$$\{p, \quad 6, \quad \tfrac{1}{2}, \quad s\}$$
$$\updownarrow \quad \updownarrow \quad \updownarrow \quad \updownarrow$$
$$\{at, \quad go, \quad to, \quad my\}$$

Notice that equal sets are always equivalent sets, but equivalent sets are not necessarily equal sets.

If the elements of a set can be counted, and the counting eventually comes to an end, then the set is a **finite** set. On the other hand, when the membership of a set is such that counting the elements can produce no final number, the set is an **infinite** set.

$\{$hours of a person's life$\}$ finite

$\{$grains of sand in the world$\}$ finite

$\{$counting numbers$\}$ or $\{1, 2, 3, 4, \ldots\}$ infinite

$\{$integers$\}$ or $\{\ldots, -2, -1, 0, 1, 2, 3, \ldots\}$ infinite

The basic criterion for a set to be finite is that its membership *can* be counted; how long it would take or how difficult it would be to do so is not relevant. Memberships of infinite sets can never be completely counted.

When each (which implies every) element of a set is also a member of another set, the first set is a **subset** of the second set. The symbol for subset inclusion is $\subseteq$.

EXAMPLES

1. $\{2, 3, 4\} \subseteq \{1, 2, 3, 4, 5\}$

[Read: "The set whose elements are 2, 3, 4 is included in the set whose elements are 1, 2, 3, 4, 5."]

2. $\{$red, blue$\} \subseteq \{$red, white, blue$\}$

3. $\{$integers$\} \subseteq \{$rational numbers$\}$

4. $\{a, b, c, d\} \subseteq \{a, b, c, d\}$

Although every set may be thought of as a subset of itself, as suggested by Example 4, p. 292, a **proper subset** of a given set is a subset which lacks at least one element of the given set. To emphasize this condition, the inclusion symbol is modified to $\subset$. Thus, $M \subset N$ means "M is a proper subset of N." The symbol $\not\subset$ represents the negation of $\subset$. The symbol $\subset$ may be substituted for $\subseteq$ in the first three examples above, but not in the fourth example. The set $\{a, b, c, d\}$ is a subset of $\{a, b, c, d\}$, but not a proper subset. Also, the empty set, $\{\ \ \}$, is a proper subset of every set other than itself.

3. VENN DIAGRAMS

Drawings, called **Venn diagrams**, are often useful for illustrating and determining relationships among sets. The complete set under a given discussion is generally referred to as the **universal set**. For example, if we are discussing students of a certain college, all of the students attending that college may be considered to make up the universal set. All of the sophomores in that college are then a proper subset of the universal set. An appropriate Venn diagram to indicate this relationship between the subset of sophomores (A) and the universal set of all students in the college (U) is shown in Fig. C.1.

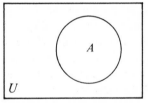

Another subset of students in the college would be the set of all seniors. A Venn diagram showing that the seniors (B) and sophomores (A) are nonoverlapping subsets of the universal set of all students in the college (U) is shown in Fig. C.2. When two sets have no element in common, as sets A and B in Fig. C.2, the sets are said to be **disjoint sets**.

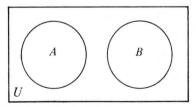

Figure C.3 illustrates a case in which:

$$U = \{\text{all students in the college}\}$$

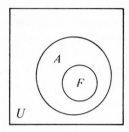

$A = \{$all sophomores in the college$\}$

$F = \{$all female sophomores in the college$\}$

Note that F is a proper subset of A, that is, $F \subset A$.

Figure C.4 illustrates a case in which:

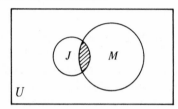

$U = \{$all students in the college$\}$

$J = \{$all juniors in the college$\}$

$M = \{$all male students in the college$\}$

The shaded part of Fig. C.4, where the J and M regions overlap, represents the set of all male juniors in the college. The rectangular region in Fig. C.4 not included in either the J or M region represents the set of all female nonjuniors who are students in the college.

In constructing Venn diagrams—which are simply visual aids—the shapes of the regions and their relative sizes are not particularly important. It is important, however, to show which sets overlap and which sets are disjoint.

EXERCISE C-2

Answer True or False.

1. $\{a, b, c, d\}$ and $\{d, c, b, a\}$ are equal sets.

2. $\{a, b, c, d\}$ and $\{d, c, b, a\}$ are equivalent sets.

3. $\{$Mon., Tues., Wed.$\}$ and $\{$Thurs., Fri., Sat.$\}$ are equal sets.

4. $\{$Mon., Tues., Wed.$\}$ and $\{$Thurs., Fri., Sat.$\}$ are equivalent sets.

5. $\{$All Orientals, living and dead$\}$ is an infinite set.

6. $\{$Molecules in a giant rock$\}$ is a finite set.

7. $\{$Integers less than 3$\}$ is a finite set.

8. {Rational numbers between $\frac{1}{2}$ and $\frac{4}{5}$} is an infinite set.

9. {Rectangles enclosing an area of 16 sq in.} is an infinite set.

10. {Squares enclosing an area of 16 sq in.} is an infinite set.

11. $\{a, b, c\} \subseteq \{a, b, c, d\}$

12. $\{a, b, c\} \subset \{a, b, c, d\}$

13. $\{a, b, c\} \subset \{c, a, b\}$

14. {integers} $\subset$ {rational numbers}

15. {negative numbers} $\not\subset$ {rational numbers}

16. $\{2, 4, 6, 8\} \not\subseteq$ {even integers}

17. $\{\ \ \} \subset \{6, 2, 4\}$

18. $\{\ \ \} \subset \{\ \ \}$

19. $\{0\}$ and $\{\ \ \}$ are equivalent sets.

20. $\{\ \ \} \subset$ {negative numbers}

21. $\{a, b, c\}$ and $\{1, 2, 3\}$ are disjoint sets.

22. $\{1, 2, 3\}$ and $\{x|x$ is an even counting number$\}$ are disjoint sets.

23. {even counting numbers} and {odd counting numbers} are disjoint sets

24. Equal sets are never disjoint sets.

Draw Venn diagrams to illustrate the following set relations. Consider the set at the right to be the universal set.

25. $\{2, 3, 4\} \subset \{1, 2, 3, 4, 5, 6\}$

26. $\{a, b, c, d\} \subset \{a, b, c, d, \ldots, m, n\}$

27. {Md., Fla., Cal.} $\subset$ {states of U.S.A.}

28. {to, at, is, if} $\subset$ {two-letter words}

29. {squares} $\subset$ {rectangles} $\subset$ {four-sided polygons}

30. {he, it, they} $\subset$ {pronouns} $\subset$ {words in English language}

31. {integers} $\subset$ {rational numbers} $\subset$ {complex numbers}

32. $\{A\} \subset \{B\} \subset \{C\} \subset \{D\}$

Draw Venn diagrams to show:

33. $U =$ {rational numbers}, $A =$ {even numbers}, $B =$ {counting numbers}. What numbers are represented by the overlapping portions of the A and B regions?

34. $U =$ {all people}, $A =$ {males}, $B =$ {people over 6 ft 2 in.}. How would you describe the people represented by the overlapping portions of the A and B regions?

35. $U =$ {letters of the alphabet}, $A = \{a, b, c, d, e\}$, $B = \{c, d, e, f, g\}$, $C = \{g, h\}$. What letters are represented in the region for U but not in any of the regions for A, B, and C?

36. For sets of n elements, there are 2^n different subsets, including the empty set and the set itself. For example, the two-element set $\{a, b\}$ has 2^2 or 4 different subsets: $\{\ \}, \{a\}, \{b\}, \{a, b\}$. Verify this generalization for sets of elements up to six in number.

4. SET OPERATIONS

When the memberships of two sets are combined to form a set, we have a binary operation referred to as the **union** of the two sets. The symbol for set union is $\cup$. For example, if $A = \{2, 3, 4\}$ and $B = \{6, 7, 8, 9\}$, then $A \cup B = \{2, 3, 4, 6, 7, 8, 9\}$. Thus $A \cup B$ (read "A union B") is another set containing all of the members of A and all of the members of B.

At times, the two sets under the operation of union are not disjoint, that is, each contains one or more common elements. For example, if $C = \{a, e, i, o, u\}$ and $D = \{a, b, c, d, e\}$, then $C \cup D = \{a, b, c, d, e, i, o, u\}$. The common elements (here a and e) appear just once in the united set.

In general,

If A and B are subsets of a universal set, the union of A and B ($A \cup B$) is the set of all elements belonging to A or B or both.

EXAMPLES

1. $M = \{1, 3, 5, 7, 9\}, N = \{2, 5, 6\}$

$M \cup N = \{1, 2, 3, 5, 6, 7, 9\}$

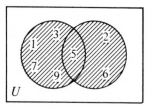

2. $P = \{\text{Mon., Tues., Wed., Thurs.}\}, Q = \{\text{Tues., Wed.}\}$

$P \cup Q = \{\text{Mon., Tues., Wed., Thurs.}\}$

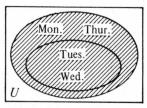

[*Note:* If $Q \subset P$, then $P \cup Q = P$.]

3. $A = \{2, 4, 6, 8\}$, $B = \{1, 2, 3\}$, $C = \{6, 8, 10\}$

$A \cup B = \{1, 2, 3, 4, 6, 8\}$

$(A \cup B) \cup C = \{1, 2, 3, 4, 6, 8, 10\} = A \cup B \cup C$

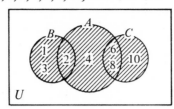

4. $X = \{a, b, c, d\}$, $Y = \varnothing$

$X \cup Y = \{a, b, c, d\}$

The **intersection** of two sets is the set of elements that are members of both sets. The symbol for set intersection is $\cap$. For example, if $C = \{a, b, c, d, e\}$ and $D = \{a, e, i, o, u\}$, then $C \cap D = \{a, e\}$. Thus $C \cap D$ (read "C interesect D") is a set containing those elements that are common to sets C and D.

When two sets are disjoint, i.e., have no common element, the intersection of the two sets is the empty set. For instance, if $R = \{2, 4, 6, 8, \ldots\}$ and $S = \{1, 3, 5, 7, \ldots\}$, then $R \cap S = \varnothing$. In general,

If A and B are subsets of a universal set, the intersection of A and B $(A \cap B)$ is the set of elements belonging to both A and B.

EXAMPLES

1. $M = \{1, 3, 5, 7, 9\}$, $N = \{2, 5, 6\}$

$M \cap N = \{5\}$

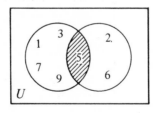

2. $P = \{\text{Mon., Tues., Wed., Thurs.}\}$, $Q = \{\text{Tues., Wed.}\}$

$P \cap Q = \{\text{Tues., Wed.}\}$

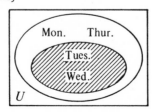

3. $J = \{1, 2, 3, 4\}$, $K = \{2, 3, 5, 7\}$, $L = \{1, 2, 5, 8\}$

$J \cap L = \{1, 2\}$

$J \cap K = \{2, 3\}$

$K \cap L = \{2, 5\}$

$(J \cap K) \cap L = \{2, 3\} \cap \{1, 2, 5, 8\} = \{2\} = J \cap K \cap L$

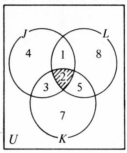

4. $X = \{a, b, c, d\}$, $Y = \varnothing$

$X \cap Y = \varnothing$

The **complement** of a given set is the set of all elements in the universal set that are *not* members of the given set. Thus, if $U = \{1, 2, 3, 4, 5, 6\}$ and subset $A = \{2, 4, 6\}$, then the complement of set A (symbolized as A') is the set $\{1, 3, 5\}$. Also if $U = \{\text{red, white, blue}\}$ and $K = \{\text{blue}\}$, then $K' = \{\text{red, white}\}$.

EXAMPLES

1. $U = \{a, b, c, d, e, f\}$, $T = \{a, b\}$

$T' = \{c, d, e, f\}$

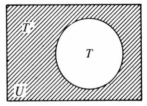

[*Note:* $T \cup T' = U$, and $T \cap T' = \varnothing$.]

2. $U = \{5, 10, 15, 20, 25, \ldots\}$

$H = \{10, 20, 30, 40, \ldots\}$

$H' = \{5, 15, 25, 35, \ldots\}$

$H \cup H' = U$, $H \cap H' = \varnothing$

EXERCISE C-3

Specify the elements in $A \cup B$ and $A \cap B$ if:

1. $A = \{a, b, c, d\}$, $B = \{b, d, f, h\}$

2. $A = \{7, 8, 16\}$, $B = \{13, 16, 19, 25\}$

3. $A = \{3, 4, 5, 6\}$, $B = \{2, 4, 6, 8\}$

4. $A = \{m, n, k\}$, $B = \{f, g, h, \ldots, m\}$

5. $A = \{\text{Mon., Tues., Wed.}\}$, $B = \{\text{Wed., Tues., Mon.}\}$

6. $A = \{m, a, d, e\}$, $B = \{d, a, m, e\}$

7. $A = \{\text{counting numbers}\}$, $B = \{\text{even counting numbers}\}$

8. $A = \{\text{months of year}\}$, $B = \{\text{months of 30 days}\}$

9. $A = \{1, 2, 3, \ldots, 6\}$, $B = \{7, 8, 9, \ldots, 12\}$

10. $A = \{a, b, c, d\}$, $B = \{7, 6, 5\}$

If A is a subset of universal set U, tabulate the elements of the complement set, A′ :

11. $U = \{2, 3, 4, 5, 6\}$, $A = \{2, 4, 6\}$

12. $U = \{a, b, c, d, e, f, g\}$, $A = \{b, c, d, e, f\}$

13. $U = \{\text{integers}\}$, $A = \{\text{even integers}\}$

14. $U = \{\text{integers}\}$, $A = \{\text{counting numbers}\}$

15. $U = \{a, b, c, d, e\}$, $A = \{\quad\}$

16. $U = \{k, t, p, r, s\}$, $A = \{s, r, t, k, p\}$

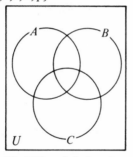

Repeat the Venn diagram in Fig C.12 and shade regions of it to show:

17. $A \cup B$ **18.** $A \cup C$

19. $B \cup C$ **20.** $A \cap B$

21. $B \cap C$ **22.** $A \cap C$

23. A' **24.** B'

25. $(A \cup B)'$ **26.** $(A \cap C)'$

Use the Venn diagram, Fig. C.12, to show:

27. $A \cup B = B \cup A$

28. $C \cap A = A \cap C$

29. $A \cup (B \cup C) = (A \cup B) \cup C$

30. $A \cap (B \cap C) = (A \cap B) \cap C$

31. $A \cap (B \cup C) = (A \cap B) \cup (A \cap C)$

32. $A \cup (B \cap C) = (A \cup B) \cap (A \cup C)$

REVIEW—UNIT C

Use the roster method to specify the elements in each of the following sets.

1. {Negative integers between -3 and 0}

2. {Cubes of -3, 2, $-\frac{1}{4}$}

3. $\{x | x$ is an integer greater than 10}

4. $\{m | m$ is a color in the U.S. flag}

5. $\{t | t$ is a counting number less than 1}

6. $\{x | x$ is the negative of $-4\}$

Decide: true or false.

7. $23 \in \{30, 31, 32, 33\}$

8. $6 \in \{$even numbers$\}$

9. Lincoln $\in \{$U.S. presidents since 1900$\}$

10. $49 \notin \{$squares of even numbers$\}$

11. $\{a, b, c, d\}$ and $\{p, q, r, s\}$ are equal sets.

12. Equivalent sets are always equal sets.

13. {triangles of area 18 sq in.} is an infinite set.

14. {circles of radius 3 in.} is an infinite set.

15. $\{2, 3, \frac{1}{2}, 0.7\} \nsubseteq \{$integers$\}$

16. $\{\ \ \} \subseteq \{1, 2, 3, 4\}$

17. $\{\ \ \} \subset \{\ \ \}$

18. $\{2, 4\} \subset \{2, 4, 6\} \subseteq \{2, 4, 6, 8\}$

19. The intersection of disjoint sets is the empty set.

20. {integers} and {rational numbers} are disjoint sets.

21. $A \cap A' = \{\ \ \}$

22. There are 16 different subsets that may be formed from a set of 4 elements.

If $A = \{h, i, j, k\}$, $B = \{a, b, h, i\}$, $C = \{c, d, e, f\}$, and $D = \{e, f, h, i\}$. tabulate the elements of the following sets:

23. $A \cup B$ **24.** $A \cap B$

25. $B \cup C$ **26.** $B \cap C$

27. $C \cap D$ **28.** $C \cup D$

29. $(A \cup C) \cup D$ **30.** $A \cap (C \cap D)$

If the universal set $U = \{3, 5, 7, 9, 11, 13\}$, and $A = \{3, 5\}$, $B = \{7, 9, 11, 13\}$, $C = \{3, 5, 7\}$, tabulate the elements of the following sets:

31. B' **32.** C'

33. $A' \cup B'$ **34.** $(A \cup B)'$

35. $A' \cap C'$ **36.** $(A \cap C)'$

If $U = \{all\ women\}$, $A = \{all\ women\ who\ have\ been\ to\ Europe\}$, *and* $B = \{all\ women$
who have been to Canada}, *describe the members of the set:*

37. $A \cup B$ **38.** $A \cap B$

39. B' **40.** $(A \cup B)'$

BINOMIAL THEOREM

1. BINOMIAL EXPANSION

Special ways to expand certain algebraic expressions, among them the square and cube of binomials, were discussed in Chapter V. A technique for writing the expansion of binomials to *any* integral power is developed in this unit.

Our approach is inductive. We list the expansions of $(a + b)^n$ for $n = 1, 2, 3, 4$ and 5, and then seek consistent properties, or patterns, in the expansions.

$(a + b)^1 = a + b$

$(a + b)^2 = a^2 + 2ab + b^2$

$(a + b)^3 = a^3 + 3a^2b + 3ab^2 + b^3$

$(a + b)^4 = a^4 + 4a^3b + 6a^2b^2 + 4ab^3 + b^4$

$(a + b)^5 = a^5 + 5a^4b + 10a^3b^2 + 10a^2b^3 + 5ab^4 + b^5$

We observe the following properties:

1. The first term in each expansion is a^n.
2. The last term in each expansion is b^n.
3. There are $n + 1$ terms in each expansion.
4. The exponents of a decrease by one with each successive term in the expansion, starting with a^n in the first term and ending with a^0 (which as a factor is unwritten, $a^0 = 1$) in the last term.

302

5. The exponents of b increase by one with each successive term in the expansion, starting with b^0 (which as a factor is unwritten, $b^0 = 1$) in the first term and ending with b^n in the last term.

6. The sum of the exponents for each term is n.

7. The numerical coefficient of the second term in each expansion is n.

8. After the first term, the numerical coefficient of any term is the product of the numerical coefficient and the exponent of the a factor of the previous term, divided by one-more-than-the-exponent of the b factor of that term. (*Example:* In the expansion of $(a + b)^5$, the second term is $5a^4b$, and the third term is $10a^3b^2$; the numerical coefficient of the third term (10) is the product of the coefficient of the previous term (5) and the exponent of the a factor of that term (4), divided by one-more-than-the-exponent of the b factor (2); or $(5)(4)/2 = 10$.)

9. The sequence of numerical coefficients from the first term to the middle of the expansion is the same as the sequence of numerical coefficients, only in reverse order, from the middle of the expansion to the last term. (Or, coefficients equidistant from the extremities of the expansion are equal.)

If we assume that what has been consistent for $n = 1, 2, 3, 4$, and 5 will also be consistent for $n = 6$, the expansion of $(a + b)^6$ should have seven terms (property 3) and should be as follows:

The first term should be $a^n : a^6$ (property 1)

The second term should be $na^{n-1}b : 6a^{6-1}b^{0+1} = 6a^5b$ (properties 4, 5, 7)

The third term should be $\dfrac{(6)(5)}{2}a^{5-1}b^{1+1} : 15a^4b^2$ (properties 4, 5, 8)

The fourth term should be $\dfrac{(15)(4)}{3}a^{4-1}b^{2+1} : 20a^3b^3$ (properties 4, 5, 8)

The fifth term should be $\dfrac{(20)(3)}{4}a^{3-1}b^{3+1} : 15a^2b^4$ (properties 4, 5, 8)

The sixth term should be $\dfrac{(15)(2)}{5}a^{2-1}b^{4+1} : 6ab^5$ (properties 4, 5, 8)

The seventh (last) term should be $b^n : b^6$ (property 2)

Thus, according to our inductive reasoning, we say

$$(a + b)^6 = a^6 + 6a^5b + 15a^4b^2 + 20a^3b^3 + 15a^2b^4 + 6ab^5 + b^6$$

That this expansion is, in fact, correct can be verified if we compute the product of the six factors, $(a + b)(a + b)(a + b)(a + b)(a + b)(a + b)$, in the usual way. What we have developed intuitively here is expressed more generally in the

BINOMIAL THEOREM: For any positive integer n:

$$(a + b)^n = a^n + na^{n-1}b + \frac{(n)(n-1)}{2!}a^{n-2}b^2 + \frac{(n)(n-1)(n-2)}{3!}a^{n-3}b^3$$

$$+ \frac{(n)(n-1)(n-2)(n-3)}{4!}a^{n-4}b^4 + \cdots$$

$$+ \frac{(n)(n-1)(n-2) \cdots (n-r+2)}{(r-1)!}a^{n-r+1}b^{r-1}†$$

$$+ \cdots + nab^{n-1} + b^n$$

[*Note*: $r!$ means the product $1 \times 2 \times 3 \times 4 \times 5 \times \cdots \times r$; $0! = 1! = 1$. For instance, $3! = 1 \times 2 \times 3 = 6$; $5! = 1 \times 2 \times 3 \times 4 \times 5 = 120$. The symbol $r!$ is read "r-factorial"; $5!$ is "five-factorial"; etc.]

EXAMPLES

1. Expand $(3x + 2y)^4$.

We use $(a + b)^4$ as our pattern:

$$(a + b)^4 = a^4 + 4a^3b + 6a^2b^2 + 4ab^3 + b^4$$

Then we replace a with $3x$ and b with $2y$:

$$(3x + 2y)^4 = (3x)^4 + 4(3x)^3(2y) + 6(3x)^2(2y)^2 + 4(3x)(2y)^3 + (2y)^4$$
$$= 81x^4 + 4(27x^3)(2y) + 6(9x^2)(4y^2) + 4(3x)(8y^3) + 16y^4$$
$$= 81x^4 + 216x^3y + 216x^2y^2 + 96xy^3 + 16y^4$$

2. Expand $(x - 2y)^5$.

The pattern:

$$(a + b)^5 = a^5 + 5a^4b + 10a^3b^2 + 10a^2b^3 + 5ab^4 + b^5$$

Replace a with x and b with $-2y$:

$$(x - 2y)^5 = (x)^5 + 5(x)^4(-2y) + 10(x)^3(-2y)^2 + 10(x)^2(-2y)^3$$
$$+ 5(x)(-2y)^4 + (-2y)^5$$
$$= x^5 + 5(x^4)(-2y) + 10(x^3)(4y^2) + 10(x^2)(-8y^3)$$
$$+ 5(x)(16y^4) + (-32y^5)$$
$$= x^5 - 10x^4y + 40x^3y^2 - 80x^2y^3 + 80xy^4 - 32y^5$$

3. Write the first four terms of the expansion $(x^2 - 2y)^{12}$.

The pattern:

$$(a + b)^{12} = a^{12} + 12a^{11}b + 66a^{10}b^2 + 220a^9b^3 + \cdots$$

The first four terms of

$$(x^2 - 2y)^{12} = (x^2)^{12} + 12(x^2)^{11}(-2y) + 66(x^2)^{10}(-2y)^2$$
$$+ 220(x^2)^9(-2y)^3 + \cdots$$
$$= x^{24} - 24x^{22}y + 264x^{20}y^2 - 1760x^{18}y^3 + \cdots$$

---

†See Section 2, following for discussion of this term.

4. Write the expansion of $(1.02)^7$ by the binomial theorem, carrying it to four terms.

Let $a = 1$ and $b = 0.02$; then $(a + b) = 1 + 0.02 = 1.02$, and $(a + b)^7 = (1.02)^7$:

$$(a + b)^7 = a^7 + 7a^6b + 21a^5b^2 + 35a^4b^3 + \cdots$$

$\left. \begin{matrix} a = 1 \\ b = 0.02 \end{matrix} \right\}$
$$= (1)^7 + 7(1)^6(0.02) + 21(1)^5(0.02)^2$$
$$+ 35(1)^4(0.02)^3 + \cdots$$
$$= 1 + (7)(1)(0.02) + (21)(1)(0.0004)$$
$$+ (35)(1)(0.000008) + \cdots$$
$$= 1 + 0.14 + 0.0084 + 0.000280 + \cdots$$

As the expansion proceeds, the increasingly higher powers of the b factor—here 0.02—makes the value of the terms progressively smaller. The sum of the first four terms is 1.14868; by five-place logarithms, $(1.02)^7$ is 1.1487.

EXERCISE D-1

Expand.

1. $(a + b)^8$ **2.** $(a - b)^9$ **3.** $(x - 2a)^5$

4. $(3a + b)^4$ **5.** $(3x - 2y)^6$ **6.** $(x^2 - y)^7$

7. $(\frac{1}{2}x - \frac{1}{3}y)^5$ **8.** $\left(\dfrac{a}{3} - b\right)^4$ **9.** $(1.03)^4$

10. $(x - \sqrt{y})^4$ **11.** $(\sqrt{a} - \sqrt{b})^3$ **12.** $\left(\dfrac{1}{\sqrt{2}} + \sqrt{2}\right)^5$

Write and simplify the first four terms in the expansions of the following:

13. $(a - 4b)^8$ **14.** $(3a + 2b)^{12}$ **15.** $(a - 4b)^{10}$

16. $(a - 2)^{17}$ **17.** $(2x + 1)^{14}$ **18.** $\left(1 + \dfrac{a}{b}\right)^{16}$

19. $(\sqrt{3} - 2)^{12}$ **20.** $(\sqrt{2} + \sqrt{3})^{18}$ **21.** $(1.03)^{10}$

22. $(2.01)^8$

2. THE rth TERM OF THE BINOMIAL EXPANSION

It is possible to write directly any particular term of a binomial expansion, say, the 24th term of $(x - 3z)^{81}$, because of a consistent property in every binomial expansion. The exponent of the second of the two binomial terms is always one less than the ordinal number of the term in the expansion. For example, in $(a + b)^6$, (see p. 303) the b factor in the expansion carries an exponent of 2 in the third term, an exponent of 4 in the fifth term, and an exponent of 6 in the seventh term.

Thus, in the general binomial expansion given in the previous section, the rth term has the factor b^{r-1} in it. And since the exponents of a and b in each term add to n, it follows that, if the exponent for the b factor in the rth term is $r - 1$, then the exponent for the a factor in that term must be the difference between n and $r - 1$, or $n - (r - 1)$, or $n - r + 1$. The numerical coefficient of the term in the binomial expansion having as its literal factors $a^{n-r+1}b^{r-1}$ is

$$\frac{(n)(n - 1)(n - 2) \cdots (n - r + 2)}{(r - 1)!}$$

Consequently, we have a formula for finding any particular (rth) term in the expansion of the binomial $(a + b)^n$:

$$\frac{(n)(n - 1)(n - 2) \cdots (n - r + 2)}{(r - 1)!} a^{n-r+1}b^{r-1}$$

EXAMPLES

1. What is the tenth term of $(a + b)^{41}$?

In the tenth term of the expansion, the b factor will have an exponent that is one less than 10, or 9. The exponent of the a factor in that term must be $41 - 9$, or 32. Hence the complete literal part in the tenth term of the expansion of $(a + b)^{41}$ is $a^{32}b^9$. The proper numerical coefficient is determined by the substitution of 41 for n and 10 for r in the formula

$$\frac{(n)(n - 1)(n - 2) \cdots (n - r + 2)}{(r - 1)!}$$

It is best to start by evaluating the last factor in the numerator, $n - r + 2$: $41 - 10 + 2 = 33$. (This factor will always be one more than the exponent of the a factor in the desired term—in this case, the exponent is 32—as should be evident from the formula: one more than $n - r + 1$ is $n - r + 2$.)

$$\frac{(n)(n - 1)(n - 2) \cdots (n - r + 2)}{(r - 1)!} = \frac{(41)\overset{13}{\cancel{(40)}}\overset{19}{\cancel{(39)}}\cancel{(38)}(37)\overset{5}{\cancel{(36)}}\overset{17}{\cancel{(35)}}\overset{11}{\cancel{(34)}}\cancel{(33)}}{\cancel{(9)}\cancel{(8)}\cancel{(7)}\cancel{(6)}\cancel{(5)}\cancel{(4)}\cancel{(3)}\cancel{(2)}(1)}$$

Therefore, the tenth term of the expansion of $(a + b)^{41}$ is

$$(41)(13)(19)(37)(5)(17)(11)a^{32}b^9 \quad \text{or} \quad 350{,}343{,}565a^{32}b^9$$

2. Write the twelfth term of the expansion $(2x - y)^{15}$.

The twelfth term in the expansion for $(a + b)^{15}$ will have an exponent of 11 attached to the b factor, and $15 - 11$, or 4, attached to the a factor. The numerical coefficient will be

$$\frac{(n)(n - 1)(n - 2) \cdots (n - r + 2)}{(r - 1)!}$$

or

$$\frac{\overset{7}{(15)\cancel{(14)}(13)\cancel{(12)}\cancel{(11)}\cancel{(10)}\cancel{(9)}\cancel{(8)}\cancel{(7)}\cancel{(6)}\cancel{(5)}}}{\cancel{(11)}\cancel{(10)}\cancel{(9)}\cancel{(8)}\cancel{(7)}\cancel{(6)}\cancel{(5)}\cancel{(4)}\cancel{(3)}\cancel{(2)}(1)}$$

or $(15)(7)(13)$; the twelfth term in the expansion of $(a + b)^{15}$ is (15) $(7)(13)a^4b^{11}$ or $1365a^4b^{11}$. If we replace a with $2x$, and b with $-y$ in $1365a^4b^{11}$, we will have the twelfth term in the expansion of $(2x - y)^{15}$:

$$1365a^4b^{11} = 1365(2x)^4(-y)^{11} = 1365(16x^4)(-y^{11})$$
$$= -21{,}840x^4y^{11}$$

3. How much does the sixth term in the binomial expansion of $(1.02)^7$ contribute to the total value of the expansion?

The sixth term of $(a + b)^7$ is

$$\frac{\overset{3}{(7)\cancel{(6)}\cancel{(5)}\cancel{(4)}\cancel{(3)}}}{\cancel{(5)}\cancel{(4)}\cancel{(3)}\cancel{(2)}(1)}a^2b^5 = 21a^2b^5$$

Substituting 1 for a and 0.02 for b, we obtain the sixth term of the expansion for $(1 + 0.02)^7$:

$$21a^2b^5 = 21(1)(0.02)^5$$
$$= (21)(1)(0.0000000032)$$
$$= 0.0000000672$$

The sixth term of the expansion $(1 + 0.02)^7$ is 0.0000000672; the seventh and eighth terms each contribute considerably less to the total value because of their higher powers of the factor 0.02.

EXERCISE D-2

Write the term indicated in expansions of the following:

1. Sixth term of $(a + 2)^{12}$

2. Fifth term of $(x - 3)^{14}$

3. Fourth term of $(2x - 3)^{10}$

4. Seventh term of $(x + \frac{1}{2})^{12}$

5. Fifth term of $(1 + 0.03)^8$

6. Fifth term of $(1 - 0.02)^7$

7. Tenth term of $(a - b)^{12}$

8. Fourteenth term of $(3x + y)^{15}$

9. What term in the expansion of $(a + 3b)^9$ contains the factor b^3?

10. In the binomial expansion of $(0.97)^9$, or $(1 - 0.03)^9$, what is the value of the fourth term?

REVIEW—UNIT D

Expand.

1. $(a + b)^9$

2. $(x - 2a)^4$

3. $(2x - 3y)^6$

4. $(\frac{1}{3}a - \frac{1}{2}b)^5$ **5.** $(1.04)^4$ **6.** $(\sqrt{x} + \sqrt{y})^3$

Write and simplify the first four terms in the expansions of the following:

7. $(a - 3b)^9$ **8.** $(a + 4b)^{10}$ **9.** $(2x - 1)^{14}$

10. $(\sqrt{3} + 2)^{12}$ **11.** $(1.02)^{10}$

Write the term indicated in the expansions of the following:

12. Sixth term of $(a - 2)^{12}$ **13.** Fourth term of $(2x + 3)^{10}$

14. Fifth term of $(1 + 0.03)^9$ **15.** Tenth term of $(a + b)^{12}$

16. What term in the expansion of $(x + 2y)^8$ contains the factor y^3?

SUPPLEMENTARY UNIT E

EXPONENTIAL EQUATIONS

Equations in which the variable occurs in the position of the exponent are called **exponential equations**. Ordinarily, the use of logarithms is necessary to solve such equations. However, when it is possible to express both members of the equation as powers of the same base, the use of logarithms can be avoided. At root is the generalization that if $A = B$, then $A^x = B^x$, and if $A^x = A^y$, then $x = y$.

1. SOLVING EXPONENTIAL EQUATIONS NOT REQUIRING THE USE OF LOGARITHMS

E.1

To solve an exponential equation in which the use of logarithms is unnecessary:

Step 1. Express both members of the equation as powers having the same base.

Step 2. Equate the exponents of the members of the equation of Step 1 and solve; the solution set of this equation will also be the solution set of the given exponential equation.

Step 3. Verify by substituting the solution(s) obtained in Step 2 for the variable in the given exponential equation.

309

EXAMPLES

1. Solve the equation: $2^{x+1} = 16$.

Step 1.   Since $16 = 2^4$, both members of the equation can be expressed as powers of 2:

$$2^{x+1} = 2^4$$

Step 2.   Equate exponents, and solve:

$$x + 1 = 4$$
$$x = 3$$

Step 3.   Verify by substituting the solution obtained (3) for the variable in the given exponential equation:

$$2^{x+1} = 16$$
$$x = 3: \quad 2^{(3)+1} \overset{?}{=} 16$$
$$2^4 \overset{?}{=} 16$$
$$16 = 16$$

Therefore, the solution set of $2^{x+1} = 16$ is $\{3\}$.

2. Solve the equation: $3^{x-7} = 27^x$.

Step 1.   Since $27 = 3^3$, then $27^x = (3^3)^x = 3^{3x}$; the given equation may be expressed as:

$$3^{x-7} = 3^{3x}$$

Step 2. $$x - 7 = 3x$$
$$-7 = 3x - x$$
$$-\tfrac{7}{2} = x$$

Step 3.   Verification: $3^{x-7} = 27^x$

$$x = -\tfrac{7}{2}: \quad 3^{(-7/2)-7} \overset{?}{=} 27^{-7/2}$$
$$3^{-21/2} \overset{?}{=} 27^{-7/2}$$
$$3^{-21/2} \overset{?}{=} (3^3)^{-7/2}$$
$$3^{-21/2} = 3^{-21/2}$$

Therefore, $3^{x-7} = 27^x$ has one solution: $-\tfrac{7}{2}$.

3. Solve: $5^{x^2-4} = (\tfrac{1}{125})^x$.

$$5^{x^2-4} = (\tfrac{1}{125})^x$$
$$5^{x^2-4} = [(125)^{-1}]^x$$
$$5^{x^2-4} = 5^{-3x}$$
$$x^2 - 4 = -3x$$
$$x^2 + 3x - 4 = 0$$
$$(x + 4)(x - 1) = 0$$
$$x = -4, \quad x = 1$$

Verification:

$$5^{x^2-4} = (\tfrac{1}{125})^x \qquad\qquad 5^{x^2-4} = (\tfrac{1}{125})^x$$

$$x = -4: \quad 5^{(-4)^2-4} \overset{?}{=} (\tfrac{1}{125})^{-4} \qquad x = 1: \quad 5^{(1)^2-4} \overset{?}{=} (\tfrac{1}{125})^1$$

$$5^{16-4} \overset{?}{=} \left(\frac{1}{5^3}\right)^{-4} \qquad\qquad 5^{1-4} \overset{?}{=} \left(\frac{1}{5^3}\right)^1$$

$$5^{12} \overset{?}{=} (5^{-3})^{-4} \qquad\qquad 5^{-3} = 5^{-3}$$

$$5^{12} = 5^{12}$$

The solution set of $5^{x^2-4} = (\tfrac{1}{125})^x$ is $\{-4, 1\}$.

EXERCISE E-1

Solve and verify:

1. $2^{x+2} = 32$ **2.** $3^{x+1} = 243$ **3.** $3^{x-1} = \tfrac{1}{81}$

4. $2^{2x-1} = \tfrac{1}{128}$ **5.** $3^{2x} = 81^2$ **6.** $5^{x-2} = 25^x$

7. $4^{x-1} = (\tfrac{1}{16})^x$ **8.** $9^{x+1} = (\tfrac{1}{27})^x$ **9.** $(\sqrt{5})^{x-1} = 25^x$

10. $(\sqrt{2})^{x-4} = (\sqrt[3]{2})^{x+1}$ **11.** $\dfrac{(\sqrt{2})^{x-1}}{2} = (\sqrt[3]{2})^x$

12. $3^{x^2-4} = \left(\dfrac{1}{27}\right)^x$ **13.** $4^{x^2} = \dfrac{8^x}{2}$

14. $(27)^{x^2-2x} = \dfrac{(81)^{2x}}{(9)^4}$ **15.** $\dfrac{(5)^{x^2}}{(25\sqrt{5})^{x+2}} = 5(\sqrt{5^5})^x$

16. $\sqrt{10^{x^2}} = 10^x$

2. SOLVING EXPONENTIAL EQUATIONS REQUIRING THE USE OF LOGARITHMS

If we apply Program **E.1** to the exponential equation

$$5^x = 3$$

we cannot readily carry out Step 1 without the use of logarithms. Since $\log 5 \approx .6990$ and $\log 3 \approx .4771$, then $3 \approx 10^{.4771}$, $5 \approx 10^{.6990}$, and $5^x \approx (10^{.6990})^x$ or $10^{.6990x}$. Now it is possible to express both members of the equation as powers having the same base:

$$5^x = 3$$

$$(10^{.6990})^x \approx 10^{.4771}$$

$$10^{.6990x} \approx 10^{.4771}$$

$$.6990x \approx .4771$$

$$x \approx \frac{.4771}{.6990}$$

$$x \approx .6825$$

This procedure can be expressed in such a way as to also include instances covered by Program **E.1**.

E.2

To solve an exponential equation:

Step 1. Express each member of the equation as an equivalent product, quotient, power, or root (not sum or difference), and then express each member logarithmically.

Step 2. Solve the logarithmic equation of Step 1; the solution set of this equation will also be the solution set of the original exponential equation.

Step 3. Verify by substituting the solution(s) obtained in Step 2 for the variable in the given equation.

EXAMPLES

1. Solve the equation: $3^x = 4$.

Step 1. 3^x expressed logarithmically is $\log 3^x$, or by logarithm law III, $x \log 3$; 4 is expressed as $\log 4$.

$$3^x = 4$$

$$\log 3^x = \log 4$$

$$x \log 3 = \log 4$$

Step 2.   Solve for x:

$$x = \frac{\log 4}{\log 3} \approx \frac{.6021}{.4771} \approx 1.262$$

[*Note:* We may compute the quotient 1.262 either by dividing .6021 by .4771 arithmetically, or by logarithms:

$$
\begin{aligned}
\log .6021 &\approx & 9.7797 - 10 \\
\log .4771 &\approx & 9.6786 - 10 \\
& (-) \overline{} \\
\log \text{ quotient} &\approx & .1011
\end{aligned}
$$

$$\text{antilog } .1011 \approx 1.262$$

Since .6021 is $\log 4$, then $\log .6021$ may be thought of as a "log of a log" or $\log (\log 4)$, or simply "log log 4."]

Step 3.   Verification: $3^x = 4$

$$x = 1.262: \qquad 3^{1.262} \overset{?}{=} 4$$
$$(10^{.4771})^{1.262} \overset{?}{=} 10^{.6021}$$
$$10^{.6021} = 10^{.6021}$$

2. Solve: $5^{x-2} = 3^2 + 4$.

Step 1. It is necessary to transform the right member so that it can be expressed logarithmically. [*Note*: $\log(3^2 + 4) \neq \log 3^2 + \log 4$.]

$$5^{x-2} = 3^2 + 4$$
$$5^{x-2} = 9 + 4$$
$$5^{x-2} = 13$$
$$\log 5^{x-2} = \log 13$$

Step 2. $\qquad (x-2)(\log 5) = \log 13$

$$x \log 5 - 2 \log 5 = \log 13$$
$$x = \frac{\log 13 + 2 \log 5}{\log 5}$$
$$x \approx \frac{1.1139 + 2(.6990)}{.6990}$$
$$x \approx \frac{2.5119}{.6990}$$
$$x \approx 3.593$$

[*Note*: If the value of x is computed by means of four-place tables, it is necessary to round off the numerator 2.5119 to 2.512 (four places) before entering Table II for the appropriate mantissa.]

Step 3. $\qquad\qquad 5^{x-2} = 3^2 + 4$

$$x = 3.593: \quad 5^{(3.593)-2} \overset{?}{=} 3^2 + 4$$
$$5^{1.593} \overset{?}{=} 13$$
$$\text{antilog}\,[1.593\,(\log 5)] \overset{?}{=} 13$$
$$\text{antilog}\,[1.1135] \overset{?}{=} 13$$
$$12.99 \approx 13$$

(The discrepancy is due to rounding errors and the approximate nature of logarithms.)

3. Solve: $.6^x = 83.2^{2x-5}$.

The logarithm of .6 is negative. It is advisable to express the logarithm first in the usual binomial form $(9.7782 - 10)$, and then as an equivalent negative numeral: $-.2218$.

$$.6^x = (83.2)^{2x-5}$$
$$\log .6^x = \log (83.2)^{2x-5}$$

$$x \log .6 = (2x - 5)(\log 83.2)$$
$$x(9.7782 - 10) \approx 2x \log 83.2 - 5 \log 83.2$$
$$x(-.2218) \approx 2x(1.9201) - 5(1.9201)$$
$$-.2218x \approx 3.8402x - 9.6005$$
$$-4.0620x \approx -9.6005$$
$$x \approx \frac{-9.6005}{-4.0620}$$
$$x \approx 2.364$$

Verification: $.6^x = 83.2^{2x-5}$

$x = 2.364:$ $.6^{2.364} \overset{?}{=} 83.2^{2(2.364)-5}$

$.6^{2.364} \overset{?}{=} 83.2^{-.272}$

$2.364 \log .6 \overset{?}{=} -.272 \log 83.2$

$(2.364)(-.2218) \overset{?}{=} -.272(1.9201)$

$-.524 \approx -.522$

(The discrepancy is due to rounding errors and the approximate nature of logarithms.)

EXERCISE E-2

Solve:

1. $8.3^x = 68.89$ **2.** $94^x = 4.547$ **3.** $44^x = 6.633$

4. $7^x = 192$ **5.** $10^x = 3.82$ **6.** $4^{x+1} = 3$

7. $3^x = 5^{x+2}$ **8.** $7^x = 3^{2x-4}$ **9.** $0.5^x = 26^{x-2}$

10. $16^{x^2} = 85$ **11.** $2^{3x} = 3^{2x+1}$ **12.** $3^x = \pi$

13. $(3.75)^x = (4.73)^{x-1}$ **14.** $3^{x+2} = 2^{2x-1}$ **15.** $3^{2x+1} = 5^{x+1}$

16. $5^{x+2} = 4^{2x-1}$ **17.** $16^{3x-1} = (2^7)(4.2)^x$ **18.** $\dfrac{3^{2x}}{2} = \dfrac{5^4}{3}$

19. $\dfrac{5^{x+2}}{5} = \dfrac{2^{2x}}{2}$ **20.** $\begin{cases} 3^{x+y} = 1000 \\ 3^{x-y} = 100 \end{cases}$

REVIEW—UNIT E

Solve and verify:

1. $2^{x+2} = 128$ **2.** $3^{x-1} = \frac{1}{243}$

3. $3^{3x} = 27^3$ **4.** $9^{x-1} = (\frac{1}{81})^x$

5. $(\sqrt{3})^{x-1} = 27^x$ **6.** $\dfrac{(\sqrt{3})^{x+1}}{3} = (\sqrt[3]{3})^x$

7. $(9)^{x^2} = \dfrac{3^x}{3}$

8. $\dfrac{(3)^{x^2}}{3(\sqrt{3})^{3x}} = (27)^3(3\sqrt{3})^x$

Solve for x:

9. $6.8^x = 46.24$

10. $86^x = 9.274$

11. $10^x = 4.61$

12. $5^x = 3^{x+2}$

13. $0.5^x = 26^{x+2}$

14. $5^{3x} = 7^{2x-1}$

15. $(53.1)^{1-x} = 129.7$

16. $5^{x+2} = 3^{2x+1}$

17. $16^{3x+1} = 2^7(4.2)^x$

18. $\dfrac{5^{x+3}}{5} = \dfrac{3^{2x+2}}{3}$

MEASURES OF CENTRAL TENDENCY AND DISPERSION

1. MEASURES OF CENTRAL TENDENCY

Many situations and relationships which occur in our daily experience can be handily characterized by formulas. But not all situations and relationships are so well-behaved. Some, particularly in the social sciences, involve masses of data which appear to obey no particular formula, but from which quantitative conclusions need to be drawn. Techniques for coping with such problems involving mass data are found in a study of statistics. In this unit we consider two of the most basic measures of statistics—those of central tendency and variability or dispersion.

One of the ways to characterize a mass of data is to pick a datum about which the others seem to cluster—some central value which typifies the total collection. In everyday language, we find the word "average" appropriate for this idea. There are three commonplace averages, or **measures of central tendency**. Depending upon the circumstances, one of these measures is often more appropriate than the others, though each will reflect a certain basic characteristic of the total distribution.

2. MEAN

The first of the three measures of central tendency is the **mean** (or more precisely, the **arithmetic mean**)

316

which most people associate with the word average. Here, however, "average" is not so restricted, but rather it is held appropriate for all three measures of central tendency.

F.1

To find the arithmetic mean of a distribution of numbers:

Step 1. Compute the sums of the numbers in the distribution.

Step 2. Divide the sum of Step 1 by the number of numbers in the distribution.

EXAMPLES

1. Find the arithmetic mean of three test grades: 83, 87, 94.

Step 1. $83 + 87 + 94 = 264$

Step 2. $264 \div 3 = 88$ (mean)

2. Find the arithmetic mean of the heights of a starting basketball team whose players are: 6 ft 5 in., 6 ft 8 in., 5 ft 10 in., 6 ft 0 in., 5 ft 11 in.

Step 1.

$$
\begin{array}{r}
6 \text{ ft} \quad 5 \text{ in.} \\
6 \quad 8 \\
5 \quad 10 \\
6 \quad 0 \\
5 \quad 11 \\
\hline
28 \text{ ft } 34 \text{ in.} = 30 \text{ ft } 10 \text{ in.} = 370 \text{ in.}
\end{array}
$$

(+)

Step 2. 370 in. $\div 5 = 74$ in. $= 6$ ft 2 in. (mean)

3. Politician Smith submitted the following list of nine contributions to his campaign: $2000, $400, $400, $275, $250, $190, $184, $170, $136. Find the arithmetic mean of these contributions.

Step 1. $2000 + 400 + 400 + 275 + 250 + 190 + 184 + 170 + 136 = $4005

Step 2. $4005 \div 9 = $445 (mean)

3. MEDIAN

Example 3 in Section 2 illustrates the chief shortcoming of the arithmetic mean as an indicator of a characteristic datum in a distribution: the mean is influenced by extreme values in the distribution. In the situation described in Example 3, the mean contribution of $445 was not typical of the nine contributions; rather it was heavily affected by the nontypical $2000 contribution which did much to pull the mean up to $445.

Under the circumstances, it would have been more appropriate to use the second of the three measures of central tendency, the **median**, which is the middle number in an ordered sequence of numbers. In Example 3, where there are nine contributions, the middle contribution of the distribution is the fifth largest (or smallest): $250. It can be seen that this middle datum of $250 is exceeded by four other contributions ($2,000, $400, $400, $275), and itself exceeds four other contributions ($190, $184, $170, $136). It should be clear that treating the data in this way frees the median from undue influence by extreme (either high or low) numbers in the distribution.

When there is an even number of numbers in the distribution, there can be no one middle number; instead, there will be a *pair* of middle numbers. In such cases, the arithmetic mean (Program **F.1**) of this pair is considered to be the median of the distribution.

F.2

To find the median of a distribution of numbers:

Step 1. Arrange the numbers of the distribution in order, either increasing or decreasing.

Step 2. Designate the middle number of the ordered distribution of Step 1 as the median. (If there is an odd number of numbers (n), the middle number will be the $\left(\dfrac{n+1}{2}\right)$th number from either end of the distribution; if there is an even number of numbers, the middle number is considered to be the arithmetic mean of the middle pair of numbers.)

EXAMPLES

1. Find the median of 18, 16, 13, 19, 21, 17, 16, 15, 15.

Step 1.   Arrange the numbers in order:

$$13, 15, 15, 16, 16, 17, 18, 19, 21$$

Step 2.   9 numbers$\Big\}$ $\left(\dfrac{9+1}{2}\right) = $ 5th number (median)
$n = 9$

2. Find the median of 17, 14, 12, 13, 12, 19, 21, 28.

Step 1.   12, 12, 13, $\underline{14, 17}$, 19, 21, 28.

Step 2.   There is an even number of numbers in the distribution; so we choose the middle pair and find its arithmetic mean. The middle pair: $\underline{14, 17}$. By Program **F.1**:

$$\frac{14 + 17}{2} = \frac{31}{2} = 15\tfrac{1}{2} \quad \text{(median)}$$

3. Find the arithmetic mean and the median salary for the Nepot Corporation whose annual payroll for its 21 employees is as follows:

J. Nepot, Pres........................	$40,000
M. Nepot, V. Pres.	$32,000
K. Nepot, Sect. Treas..............	$30,000
S. Burns, Manager	$ 4,000
M. White, Asst. Mgr...............	$ 3,500
16 employees each at	$ 2,800

Arithmetic Mean:

$$\$40,000 + 32,000 + 30,000 + 4,000 + 3,500 + (16 \times \$2,800)$$
$$= \$154,300 = \text{total annual payroll}$$

$$\$154,300 \div 21 \approx \$7348 \quad \text{(mean salary)}$$

Median: $n = 21$;

$$\frac{n+1}{2} = \frac{21+1}{2} = 11 \quad \text{(11th salary is the median)}$$

The salary distribution:

1	2	3	4	5	6	7
40,000	32,000	30,000	4000	3500	2800	2800

8	9	10	11	12	13	14
2800	2800	2800	2800	2800	2800	2800
			MEDIAN			

15	16	17	18	19	20	21
2800	2800	2800	2800	2800	2800	2800

4. MODE

The third of the three measures of central tendency is the **mode**, the most frequently repeated value in the distribution. It is by far the crudest measure of central tendency and is of limited usefulness. However, it is sometimes of value, particularly when dealing with certain discrete but numerous data. For instance, a hat manufacturer or a haberdasher is most interested in **modal** sizes; politicians and other members of our society who are particularly concerned with majorities and pluralities are greatly interested in the mode of certain distributions. In an election (provided there are no special requirements, such as a majority of all votes cast) the mode indicates the winning candidate or issue.

F.3

To find the mode of a distribution:

Step 1. Make a tally of the data in the distribution.

Step 2. Designate the most frequently occurring datum as the mode.

EXAMPLES

[*Note:* For simplicity, the following examples and exercises involve few data each and, in this respect, are not really typical of instances in which the mode is appropriate.]

1. Find the mode of the following distribution: 16, 12, 18, 12, 24, 18, 16, 18, 22, 14, 16, 12, 18.

Step 1.   12: / / /
             14: /
             16: / / /
             18: / / / /
             20:
             22: /
             24: /

Step 2.   The number 18 appears most frequently in the distribution; hence, it is the mode.

2. Find the mode of the distribution: 14, 16, 24, 18, 10, 12, 20.

Step 1.   10: /
             12: /
             14: /
             16: /
             18: /
             20: /
             22:
             24: /

Step 2.   No number occurs more than once; hence, there is *no mode* in this distribution.

3. Find the modal hat size carried by a department store at inventory time if the inventory showed the following sizes: $7\frac{1}{8}$, $6\frac{7}{8}$, $6\frac{3}{4}$, $6\frac{5}{8}$, $6\frac{1}{2}$, 7, $7\frac{1}{8}$, $6\frac{3}{8}$, $6\frac{7}{8}$, $7\frac{1}{8}$, $7\frac{1}{4}$, $6\frac{7}{8}$, $6\frac{3}{4}$, $6\frac{7}{8}$, $7\frac{1}{8}$, $6\frac{3}{8}$.

Step 1. $6\frac{3}{8}$: / /
             $6\frac{1}{2}$: /
             $6\frac{5}{8}$: /
             $6\frac{3}{4}$: / /
             $6\frac{7}{8}$: / / / /
             7 : /
             $7\frac{1}{8}$: / / / /
             $7\frac{1}{4}$: /

Step 2.   There are two sizes of greatest frequency in the distribution ($6\frac{7}{8}$ and $7\frac{1}{8}$); hence, there are *two* modes. Such a distribution is referred to as a *bimodal distribution*.

EXERCISE F-1

Find the mean, median, and mode for each of the following sets of data:

1. 8, 2, 6, 12, 10, 8, 10, 6, 14, 16, 8, 2, 8, 12, 4, 8, 10.

2. 9, 14, 12, 13, 16, 15, 17, 18, 21, 10, 6, 8, 11.

3. 37, 42, 42, 41, 37, 46, 44, 37, 42, 40, 38, 37, 37, 39, 43, 40.

4. 393, 392, 391, 389, 392, 388, 391, 394, 392, 390, 388, 386, 389, 388, 392, 387, 388. [*Hint:* Subtract 386 from each to facilitate computation.]

5. −17, −10, −8, 0, 2, 6, 10, −19, 10, −3, −6, −17, −6, −17, 4, 6, −17, 0, 10, 8, −20.

6. $\frac{2}{3}, \frac{3}{4}, \frac{3}{5}, \frac{7}{8}, \frac{1}{2}, \frac{3}{5}, \frac{3}{8}, \frac{7}{9}, \frac{5}{8}, \frac{2}{9}$.

7. (Treat as ungrouped data.)

Test score	18	19	20	21	22	23	24	25
Frequency	2	3	5	7	9	11	7	3

8. Construct a distribution of fifteen numbers using integers between 19 and 30 in which the mean is 21, the median 20, and the mode 19.

For each of the following, decide the most appropriate measure of central tendency by the nature of the circumstances; then compute this "average."

9. Mother, father, grandmother, and ten children attended a birthday party. Their ages were as follows: 36, 34, 62, 8, 9, 7, 9, 2, 6, 6, 9, 8, and 7. What was the "average" age of those attending the party?

10. In the sports department of a store, 25 pairs of basketball shoes were sold in one day. Their sizes were: 6, 7, 7, 8, 8, 9, 9, 9, 9, 10, 10, 10, 10, 11, 11, 11, 11, 11, 11, 11, 12, 12, 12, 12, 13. What was the "average" size shoe sold that day?

11. A class of fifteen students weighed a piece of metal on an analytic balance with the following results expressed in grams: 0.1236, 0.1235, 0.1237, 0.1235, 0.1239, 0.1231, 0.1234, 0.1236, 0.1240, 0.1236, 0.1233, 0.1241, 0.1229, 0.1231, 0.1236. The best estimate of the weight of the piece of metal is the "average" of these. Determine that "average."

12. Eleven students added a long column of figures quickly and got the following results: 2362, 2365, 2368, 2374, 2362, 2365, 2362, 2362, 2374, 2359, 2367. On the basis of the "average" result, what is your estimate of the correct sum?

5. MEASURES OF VARIABILITY (DISPERSION)

Although we can never fully describe a mass of data, two basic descriptions are vital: one is the description of the data in terms of "average" (measures of central tendency), and the other is a description in terms of the **variability** or spread of data. Consider the two eleven-item distributions, A and B, which both have the same mean, median, and mode at 17.

$$A: \quad 13 \quad 14 \quad 16 \quad 17 \quad 17 \quad 17 \quad 17 \quad 18 \quad 19 \quad 19 \quad 20$$
$$B: \quad 6 \quad 9 \quad 12 \quad 15 \quad 17 \quad 17 \quad 17 \quad 19 \quad 21 \quad 26 \quad 28$$

Though both distributions agree in their various measures of central tendency, they certainly are not alike. The spread or dispersion of data in B is much greater. One measure of this spread is the **range**, which is found by subtracting the smallest number in the distribution from the largest. Thus the range for A is 7 $(20 - 13)$, and the range for B is 22 $(28 - 6)$.

Another way to measure the variability or spread of a distribution is in terms of deviation from the mean. The **deviation from the mean** for a given number in a distribution is the difference obtained by subtracting the arithmetic mean of the distribution from that number. Thus, for numbers in the distribution that are greater than the mean, the deviation is positive; for numbers less than the mean, the deviation is negative. For example, if the mean of a distribution is 18, a datum of 20 has a deviation from the mean of $+2$ $(20 - 18)$; a datum of 15 has a deviation from the mean of -3 $(15 - 18)$.

Since each number in the distribution has its own deviation from the mean, the arithmetic mean of all these deviations will provide a measure of spread or dispersion for the distribution. However, when we undertake the first step of this averaging process—adding all the deviations, which are positive and negative numbers—the sum of the deviations is invariably zero. To overcome this difficulty, we may elect either of two choices: work with the positive differences or absolute values of the deviations (which is done in computing average deviation, Program **F.4**), or work with the squares of the deviations (which is done in computing standard deviation, Program **F.5**).

One simple measure of dispersion is the **average deviation** (AD)†; it is defined to be the arithmetic mean of the absolute values of the deviations from the mean of the distribution. An average deviation of 6 for a distribution implies that the numbers of the distribution differ from the mean by 6 units—on the average (mean).

F.4

To find the average deviation (AD) of a distribution:

Step 1. Compute the arithmetic mean of the distribution.

Step 2. Compute the positive difference (which is equivalent to the absolute value) between each number of the distribution and the mean found in Step 1.

Step 3. Add the differences (deviations) of Step 2.

Step 4. Divide the sum of the differences found in Step 3 by the number of numbers in the distribution for the desired AD of the distribution.

EXAMPLES

1. Find the average deviation for distribution A on p. 321.

---

†Stated more precisely: mean deviation.

Step 1	*Step 2* (positive deviations from 17)
13	4
14	3
16	1
17	0
17	0
17	0
17	0
18	1
19	2
19	2
20	3
187	16 ← *Step 3*

Mean: $187 \div 11 = 17$.

Step 4. $16 \div 11 \approx 1.45$ (AD)

2. Find the average deviation for distribution B on p. 321.

Step 1	*Step 2* (positive deviations from 17)
6	11
9	8
12	5
15	2
17	0
17	0
17	0
19	2
21	4
26	9
28	11
187	52 ← *Step 3*

Mean: $187 \div 11 = 17$.

Step 4. $52 \div 11 \approx 4.73$ (AD)

3. Find the average deviation for a distribution C: 42, 57, 56, 46, 51, 50, 53, 53.

Step 1†	*Step 2* (positive deviations from 51)
42	9
57	6
56	5
46	5
51	0
50	1
53	2
53	2
408	30 ← *Step 3*

†Unless further work is to be done with the distribution, it is not necessary to arrange the numbers in order in computing the mean or average deviation.

Mean: $408 \div 8 = 51$.

Step 4. $30 \div 8 = 3.75$ (AD)

If the dispersions of the three distributions (A, B, C) in these examples are compared, the following conclusions may be drawn:

Distribution A (AD ≈ 1.45) is the most compact, least spread of the three; each number of this distribution deviates, on the average, approximately 1.45 units from the mean of the distribution.

Distribution B (AD ≈ 4.73) is the most spread, and possesses the highest degree of variability of the three; each number of this distribution deviates, on the average, approximately 4.73 units from the mean of the distribution.

Distribution C (AD $= 3.75$) is more dispersed than A, less dispersed than B; each number of this distribution deviates, on the average, 3.75 units from the mean of the distribution.

The **standard deviation** (σ), another measure of dispersion, has wider use than average deviation, particularly in the more complex phases of statistics. The standard deviation of a distribution is defined as the square root of the arithmetic mean of the squared deviations from the mean of the distribution. Extracting the square root of the arithmetic mean compensates for the squaring of the deviations.

F.5

To find the standard deviation of a distribution:

Step 1. Compute the arithmetic mean of the distribution.

Step 2. Compute the deviation of each number of the distribution from the mean.

Step 3. Square each of the deviations of Step 2.

Step 4. Add the squared deviations of Step 3.

Step 5. Divide the sum of Step 4 by the number of numbers in the distribution.

Step 6. Extract the square root of the quotient of Step 5 for the desired standard deviation of the distribution.

EXAMPLES

1. Find the standard deviation for distribution A of Example 1, p. 323.

Step 1	*Step 2* (Deviations from 17)	*Step 3* (Deviations squared)
13	−4	16
14	−3	9
16	−1	1

17	0	0
17	0	0
17	0	0
17	0	0
18	+1	1
19	+2	4
19	+2	4
20	+3	9
187		44 ← *Step 4*

Mean: $187 \div 11 = 17$.

Step 5. $44 \div 11 = 4$

Step 6. $\sqrt{4} = 2 \quad (\sigma)$

2. Find the standard deviation for distribution B of Example 2, p. 323.

Step 1	*Step 2* (Deviations)	*Step 3* (Deviations squared)
6	−11	121
9	− 8	64
12	− 5	25
15	− 2	4
17	0	0
17	0	0
17	0	0
19	+ 2	4
21	+ 4	16
26	+ 9	81
28	+11	121
187		436 ← *Step 4*

Mean: $187 \div 11 = 17$.

Step 5. $436 \div 11 \approx 39.64$

Step 6. $\sqrt{36.64} \approx 6.3 \quad (\sigma)$

3. Find the standard deviation for distribution C of Example 3, p. 323.

Step 1	*Step 2* (Deviations)	*Step 3* (Deviations squared)
42	−9	81
57	+6	36
56	+5	25
46	−5	25
51	0	0
50	−1	1
53	+2	4
53	+2	4
408		176 ← *Step 4*

Mean: $408 \div 8 = 51$.

Step 5. $176 \div 8 = 22$

Step 6. $\sqrt{22} \approx 4.7$ (σ)

EXERCISE F-2

Determine for each of the following distributions the (a) range, (b) average deviation, and (c) standard deviation.

1. 4, 7, 10, 13, 15, 15, 15, 17, 19, 24, 26.

2. 84, 82, 86, 83, 87, 89, 91, 90, 85, 88.

3. The distribution of Exercise 7 of F-1, p. 320.

4. The distribution of Exercise 4 of F-1.

5. The distribution of Exercise 5 of F-1.

6. The distribution of Exercise 2 of F-1.

7. In a normal distribution, approximately 68% of the cases occur within one standard deviation ($\pm \sigma$) of the mean; approximately 95% of the cases occur within two standard deviations ($\pm 2\sigma$) of the mean; and approximately 99% of the cases occur within three standard deviations ($\pm 3\sigma$) of the mean. If the following scores: 20, 21, 21, 21, 21, 22, 22, 22, 22, 22, 22, 23, 23, 23, 23, 24, are assumed to be truly representative of a much larger set of scores that are normally distributed, find (a) the range for the middle 68% of all scores in the normal distribution, (b) the range for the middle 95% of all scores in the normal distribution, and (c) the range of the scores in the normal distribution that will include all but the two extreme $\frac{1}{2}$% of the scores.

8. (Refer to Exercise 7.) If we assume the heights of men in the United States to be normally distributed about a mean of 5 ft 8 in., with a standard deviation of 3 in., what are a male's chances in 100 of being *either* taller than 6 ft 2 in. *or* shorter than 5 ft 2 in.?

REVIEW—UNIT F

Find the mean, median, and mode for each of the following sets of data:

1. 15, 11, 9, 13, 17, 19, 21, 11, 5, 13, 3, 7, 17, 13, 19, 13, 21.

2. 62, 65, 67, 59, 68, 62, 53, 61, 63, 57, 54, 65, 62, 65, 58, 65.

3. $-13, -2, 9, -6, -13, -20, -8, 0, 4, -13, -18, 2, -2, 0, 12, -16, 14, -13, 4$.

4. (Treat as ungrouped data.)

Test score	14	15	16	17	18	19	20	21	22
Frequency	0	2	2	3	8	10	14	4	2

For each of the following, decide the most appropriate measure of central tendency by the nature of the problem; then compute this "average."

5. A class of fifteen students weighed a piece of metal on an analytic balance with the following results expressed in grams: 0.1749, 0.1748, 0.1750, 0.1748, 0.1752, 0.1744, 0.1747, 0.1749, 0.1753, 0.1749, 0.1746, 0.1754, 0.1742, 0.1744, 0.1749. The best estimate of the weight of the metal is the "average" of these. What is that "average"?

6. Eleven students added a long column of figures quickly and arrived at the following results: 3375, 3378, 3371, 3387, 3375, 3378, 3375, 3375, 3387, 3372, 3380. Based on the "average" of these, what is your estimate of the correct sum?

Determine for each of the following distributions the (a) range, (b) average deviation, and (c) standard deviation.

7. 8, 11, 14, 17, 19, 19, 19, 21, 23, 28, 30.

8. The distribution of Exercise 3 above.

9. The distribution of Exercise 4 above.

ANSWERS
TO ODD-NUMBERED
PROBLEMS

EXERCISE 1-1

1. Yes. **19.** Commutative—multiplication. **21.** Associative—multiplication. **23.** Distributive. **25.** Distributive. **27.** Associative—multiplication. **29.** Associative—addition.

EXERCISE 1-3

1. 2589. **3.** Impossible with nonnegative rationals. **5.** 154.26. **7.** 0.0264. **9.** 0.0001081. **11.** $\frac{1}{8}$. **13.** $\frac{1}{10}$. **15.** $\frac{13}{209}$. **17.** Impossible with nonnegative rationals. **19.** $5\frac{5}{12}$. **21.** Impossible with nonnegative rationals. **23.** -8. **25.** $+5$. **27.** $-\frac{2}{3}$. **29.** -1.8. **31.** $+2.6$. **33.** $+4\frac{2}{5}$. **35.**

$-\frac{7}{8}$	$-\frac{2}{3}$	-0.4	0	$+\frac{1}{2}$	$+\frac{5}{6}$	$+1.3$

EXERCISE 1-4

1. $+19$. **3.** $+530$. **5.** $+170.717$. **7.** $-8\frac{7}{15}$. **9.** $+61$. **11.** $+9.38$. **13.** $+912$. **15.** $+\frac{1}{4}$. **17.** $+19\frac{11}{40}$. **19.** $-4\frac{11}{20}$. **21.** -40.848. **23.** -28. **25.** -59. **27.** $+1.3$. **29.** $+111.93$. **31.** $-\frac{61}{56}$.

EXERCISE 1-5

1. $+6$. **3.** -12. **5.** $+0.4$. **7.** $+50.82$. **9.** $+1449$. **11.** $-60,508$.

13. $+\frac{4}{5}$.   **15.** $-\frac{1}{126}$.   **17.** $-\frac{49}{6}$.   **19.** $+52\frac{11}{15}$.   **21.** -10.355.
23. $+40\frac{4}{7}$.   **25.** -338.785.

EXERCISE 1-6

1. -6.   **3.** $+21$.   **5.** -4.   **7.** $+\frac{3}{8}$.   **9.** $+0.0000294$.   **11.** $+144$.
13. $-\frac{18}{55}$.   **15.** -24.   **17.** $+\frac{3}{40}$.   **19.** -864.   **21.** -18.   **23.** $+25\frac{3}{8}$.

EXERCISE 1-7

1. $+2$.   **3.** $+7$.   **5.** $+672$.   **7.** $+138$.   **9.** $-1\frac{7}{8}$.   **11.** $+1\frac{3}{4}$.   **13.** $+\frac{1}{4}$.
15. -15.   **17.** $+0.71$.   **19.** -0.00162.   **21.** $+2$.   **23.** $-1\frac{1}{3}$.   **25.** $-\frac{1}{27}$.
27. $+$.   **29.** $-$.   **31.** $-$.

REVIEW—CHAPTER 1

31. Commutative—multiplication.   **33.** Commutative—addition.   **35.** Associative—addition.   **37.** $+7$.   **39.** $+7.83$.   **41.** $+3\frac{1}{2}$.   **43.** -71.   **45.** $-\frac{3}{16}$.
47. 230.21.   **49.** $+28$.   **51.** $+156$.   **53.** $+0.5$.   **55.** $+3080.29$.
57. $-2\frac{489}{1232}$.   **59.** $+12$.   **61.** $+147.9$.   **63.** $-96,501$ **65.** $-\frac{5}{36}$.
67. $-72\frac{7}{40}$.   **69.** $-34\frac{9}{10}$.   **71.** -130.   **73.** -18.   **75.** -0.1684.
77. $+1.9752$.   **79.** -28.   **81.** $+0.032$.   **83.** $+15$.   **85.** -0.083.
87. $-\frac{35}{12}$.   **89.** $-\frac{57}{94}$.   **91.** $-2\frac{1}{9}$.   **93.** -12.   **95.** $+$.   **97.** $+$.

EXERCISE 2-1

1. $+21x$.   **3.** $+7p$.   **5.** $-34s$.   **7.** $-3.8x$.   **9.** $94n - 16m$.   **11.** $+12f$.
13. $-1.6x$.   **15.** $-129a$.   **17.** $-\frac{1}{12}x$.   **19.** $-2.0m$.   **21.** $7r$.   **23.** $-0.2k$.
25. $29c$.   **27.** $34e - 27d$.   **29.** $\frac{5}{4}x$.   **31.** $7xy$.   **33.** 0.   **35.** $4k - 2a$.
37. $-3.1s$.   **39.** 0.   **41.** $3.27x$.   **43.** $\frac{9}{100}ap$.

EXERCISE 2-2

1. a^5.   **3.** d^{13}.   **5.** m^{10}.   **7.** a^7b^7.   **9.** x^9y^4.   **11.** $-10x$.   **13.** $10n^2$.
15. $-8xy$.   **17.** $-15x^3y^3$.   **19.** $-12x^2$.   **21.** $-36ax$.   **23.** $-\frac{2}{5}mn$.
25. $18bx^2$.   **27.** $27s^3t^3$.   **29.** $-24m^2n^2p^2$.   **31.** $+\frac{1}{3}ab^2$.   **33.** $-0.036x^2y$.

EXERCISE 2-3

1. 2^2 or 4.   **3.** $\frac{1}{8}$.   **5.** x^6.   **7.** $\dfrac{1}{(xy)^2}$.   **9.** a^2.   **11.** $-7a$.   **13.** $-4a$.

15. $-5p$.   **17.** -2.   **19.** $\dfrac{128a}{y^2}$.   **21.** $\dfrac{6b^3}{5c}$.   **23.** $\dfrac{24ab}{x}$.   **25.** $-\dfrac{3}{y}$.

27. $\dfrac{b^4}{2a^4}$.   **29.** $-\dfrac{6xy}{5z}$.   **31.** $\dfrac{9}{5}xy^2$.   **33.** $\dfrac{500}{b}$.

EXERCISE 2-4

1. $6xy - 8xy^2$.   **3.** $8xy - 2y + 3z$.   **5.** $3d + 7z$.   **7.** $0.1x - 2.7y$.
9. $-3x + 5y$.   **11.** $3x - 10y - 12$.   **13.** $x + \frac{1}{4}y$.   **15.** $8 + 0.6a - 0.2b$.
17. $x + 6y - 12$ **19.** $3.3a - 3.4b - 2.2c - 6d$.   **21.** $-18b - 10c$.
23. $\frac{29}{30}x - \frac{1}{2}y$.   **25.** $-13x + 10y$.   **27.** $x - 8$.   **29.** $-4x + 3y$.

EXERCISE 2-5

1. $12x^2 + 6xy - 3xz$.   **3.** $-6a^2b + 9ab^2 - 3a^2b^2$.   **5.** $6a^2x - 15ax^2 + 18ax$.
7. $-6a^3c + 8a^2c^2 + 4a^3c^2$.   **9.** $-3p^2qr - 2pq^2r^2 - 7pqr$.   **11.** $10x^2 - 31x - 14$.
13. $2pm - 2pn - m^2 + mn$.   **15.** $9x^3 - 21x^2 + 31x - 35$
17. $-10a^2 + 9a^3 - 8a^4 + 3a^5$.
19. $3a^6b^2 - 18a^5b^3 + 3ab^4 - 4a^4b^3 + 24a^5b^4 - 4ab^5$.
21. $12m^4 + 8m^3n - 49m^2n^3 + 33mn^3 - 6n^4$.
23. $(12x^2 - 13x - 14)$ sq ft.   **25.** $x^3 - y^3$.

EXERCISE 2-6

1. 8.   **3.** -18.   **5.** 5.   **7.** -94.   **9.** $9a - 4x$.   **11.** $4p + 7k + 2x$.
13. $-4a - 2b + c$.   **15.** $13x + 9y$.   **17.** $3m - 10s$.   **19.** $6x - 4$.
21. $2k - 4t$.   **23.** $6f + 12$.   **25.** $2x - (-6y + 3t + 8)$.
27. $ab - (c - gk + h)$.

EXERCISE 2-7

1. $2 - 3a^2$.   **3.** $\dfrac{1}{x} - 6y^2$.   **5.** $3n - 7k$.   **7.** $7a^2 + 4ab - 1$.

9. $-6xy + 11y - \dfrac{9y^3}{x}$.   **11.** $-5a + 4b + 3c^2$.   **13.** $-5a^5 + 15a^2b^2 + 30a^7$.

15. $4m - 2n - 3mn^2$.   **17.** $2a + 3$.   **19.** $-4x - 2$.   **21.** $5a^2 + 21a + 4$.

23. $5x - 2$, $R(7)$ or $5x - 2 + \dfrac{7}{2x + 3}$.   **25.** $x^2 + 2x + 4$.

REVIEW— CHAPTER II

1. $0.7x$ **3.** $31x$.   **5.** $-139y$.   **7.** $-\frac{37}{120}a$.   **9.** $0.6a - 3.1b$.   **11.** $1041a$.

13. $-0.75m$.   **15.** $2\frac{5}{8}x$.   **17.** $-650x$.   **19.** $6.37x$ **21.** a^{15}.   **23.** $-84xy$.
25. $21a^3b^3$.   **27.** $-0.0006a^3b^2$.   **29.** $0.00036m^3$.   **31.** $-\frac{1}{14}x^3y^2z$.   **33.** x^6y^4.
35. $0.00000006x^3y$.   **37.** x^6.   **39.** $(xy)^2$.   **41.** $-3x$.   **43.** $-4ab$.   **45.** $\frac{5}{2}a$.
47. $4b^2$ **49.** $\frac{6}{7}x$.   **51.** $-\dfrac{81m}{80n^2}$.   **53.** $44xy - 3xy^2$.   **55.** $9x - 10y + 7z$.
57. $8x - 16y$.   **59.** $-0.2a + 0.9b$.   **61.** $0.3a - 4.1b - 4.5c + 3.2d$.
63. $\frac{5}{4}x + \frac{3}{8}xy - \frac{13}{15}y$.   **65.** $-1 - 5x + 8y$.   **67.** $5a - 7m + 5k$.
69. $18x^2 + 9xy - 6x$.   **71.** $-27a^5 + 36a^4 - 18a^3m$.   **73.** $6a^2 - 29ax + 28x^2$.
75. $6x^2 - 33xy + 2xz + 15y^2 - 10yz$.   **77.** $a^3 + ab - ac + a^2b + b^2 + a^2c - c^2$.
79. $6a^2 - ac - 3ad + \frac{1}{2}cd$ **81.** -18.   **83.** -24.   **85.** 9.   **87.** $-3y$.
89. $-2a - 4.2b$.   **91.** $2z - (-6m + 4n - 2)$.   **93.** $\dfrac{1}{n} - 12mn^2$.
95. $2x - 3y + 9xy^2$.   **97.** $-6x^2y^2 + 3xz - 5$.   **99.** $-2a^3 + 2a^2 + 6a$.
101. $7x - 5y$.   **103.** $4x - 9 - \dfrac{2}{2x + 7}$.

EXERCISE 3-1

1. 6.   **3.** 2.   **5.** 0.   **7.** $\frac{1}{3}$.   **9.** None of them.   **11.** 3.

EXERCISE 3-2

1. 10.   **3.** 6.   **5.** $10\frac{1}{3}$.   **7.** 0.05.   **9.** 32.   **11.** $\frac{3}{4}$.   **13.** 15.   **15.** -6.
17. -1.99.   **19.** -0.24.   **21.** 48.   **23.** -0.1.

EXERCISE 3-3

1. 2.   **3.** $2\frac{1}{3}$.   **5.** $-\frac{1}{9}$.   **7.** -2.   **9.** 0.   **11.** 2.   **13.** 1.6.   **15.** -24.
17. -70.   **19.** -4.   **21.** 0.   **23.** $1\frac{2}{3}$.   **25.** No; identities.

EXERCISE 3-4

1. $\dfrac{b}{5}$.   **3.** $d - 6$.   **5.** $4a + 3b$.   **7.** $\dfrac{3a + 4b}{2}$.   **9.** $6a$.   **11.** a.   **13.** $2a$.
15. $\frac{1}{3}a - 3b$.   **17.** $b = y - mx$; $x = \dfrac{y - b}{m}$; $m = \dfrac{y - b}{x}$; 4.   **19.** $n = \dfrac{2S}{a + l}$;
$a = \dfrac{2S - nl}{n}$; $l = \dfrac{2S - na}{n}$; 10.

EXERCISE 3-5

1. $\{3, 4\}$.   **3.** $\{3, 4, 5\}$.   **5.** $\{-3, -2\}$.   **7.** $\{-8, -7\}$.   **9.** $\{0, 5, 10, 15\}$.

11. $\{-1, 0, 1, 2, 3\}$. **13.** $\{-10, -5\}$. **15.** ←——○——→ 3 **17.** ←——○——→ -3

19. ←——●——→ 6 **21.** ←——●——→ 3 **23.** ←——●——→ -2 **25.** ←——●——→ 0

EXERCISE 3-6

1. $\{\ldots, 2, 3, 4\}$. **3.** $\{1, 2, 3, \ldots\}$. **5.** $\{\ldots, -5, -4, -3\}$.
7. $\{12, 13, 14, \ldots\}$. **9.** ←——●——→ -2 **11.** ←——○——→ -5 **13.** ←——●——→ -1
15. $\{3, 4, 5, \ldots\}$. **17.** ←——●——→ 7 **19.** ←——●——→ -8 **21.** $\{x \mid x < 5\}$;

$\{x \mid x \le -3\}$; $\{x \mid x > -5\}$; $\{a \mid a \ge -1\}$; $\{k \mid k > 2\}$.

REVIEW—CHAPTER III

1. 4. **3.** $\frac{1}{2}$. **5.** 0.01. **7.** -2. **9.** 24. **11.** $\frac{8}{3}$. **13.** 0. **15.** Identity.
17. -15. **19.** m. **21.** $\frac{3}{7}a$. **23.** $-\frac{4}{3}c$. **25.** Identity. **27.** $6a$.
29. $\{\ldots, 4, 5, 6\}$. **31.** $\{\ldots, -12, -11, -10\}$. **33.** $\{\ldots, -7, -6, -5\}$.
35. $\{19, 20, 21, \ldots\}$. **37.** ←——●——→ -4 **39.** ←——○——→ -2 **41.** $\{x \mid x < 3\}$.
43. $\{x \mid x \le -\frac{2}{3}\}$.

EXERCISE 4-1

1. $(x + 5)$ or $(5 + x)$. **3.** $2x$. **5.** $2x + 3$. **7.** $0.4x$ or $\frac{2}{5}x$ or $\frac{2x}{5}$.
9. $x - 6$. **11.** $x - \frac{1}{6}x$ or $\frac{5}{6}x$. **13.** $(a + b) - \frac{1}{2}ab$ or $(a + b - \frac{1}{2}ab)$.
15. $x + 17$. **17.** $\frac{2x - 20}{5}$ or $\frac{1}{5}(2x - 20)$. **19.** $60x$. **21.** $x - 7 = 15$; 22.
23. $2x - 5 = 16$; $10\frac{1}{2}$. **25.** $\frac{10}{x} = 2$; 5. **27.** $10 + 8 + x = 2x$; 18.
29. $\frac{2}{3}x + 4 = 8$; 6.

EXERCISE 4-2

1. 12 ft; 24 ft. **3.** 90°, 30°, 240°. **5.** 8 in. **7.** $2. **9.** 120 ft $\times$ 160 ft.
11. 16 in.

EXERCISE 4-3

1. 16, 18, 20. **3.** 21, 24, 27, 30. **5.** 417. **7.** $-1, 0, +1, +2$. **9.** 27.
11. 3 and 5.

EXERCISE 4-4

1. 30 lbs of 50¢, 15 lbs of 80¢. **3.** $\frac{1}{4}$ gal. **5.** 40%. **7.** 6 oz.
9. 8 gal cream. **11.** 2 qts.

EXERCISE 4-5

1. 18 mph.   **3.** 24 mi.   **5.** 57 mph.   **7.** 8.7 yd.   **9.** 125 mi.   **11.** 60 mph.

EXERCISE 4-6

1. \$3000 @ 4%; \$4000 @ 7%.   **3.** \$7200   **5.** \$3750.   **7.** $4\frac{1}{2}\%$ and $2\frac{1}{4}\%$.
9. \$11,000.   **11.** \$1800.

EXERCISE 4-7

1. 18.   **3.** 12.   **5.** 9 quarters; 16 dimes.   **7.** 42 pesos.   **9.** 18 ft.
11. 1740 lbs.

REVIEW—CHAPTER IV

1. $x - p$.   **3.** $\dfrac{x - k}{2}$.   **5.** $x + 1.25x$ or $2.25x$.   **7.** $x - p$.   **9.** $\frac{1}{3}x - m$.

11. $x - 22 = 6x$.   **13.** $7 - (x + 12) = -x - 5$.   **15.** $\dfrac{10}{x} = \dfrac{3}{2} + x$.

17. $45°, 60°, 90°, 165°$.   **19.** $41°, 60°, 79°$.   **21.** 9 ft.   **23.** 52.   **25.** -8,
$-1, +6, +13, +20$.   **27.** 96.   **29.** 27 cc/10%; 54 cc/4%.   **31.** 3 gal.
33. $4\frac{2}{7}$ lbs.   **35.** 156 mi.   **37.** $\frac{3}{8}$ mi.   **39.** $12\frac{3}{11}$ min.   **41.** $6\frac{1}{2}\%$.   **43.** \$900.
45. \$5000.   **47.** 39.   **49.** 10.

EXERCISE 5-1

1. $-3a + 18$.   **3.** $4a^2c - 4c^3$.   **5.** $3x^2y - 3xy^2 + 3xyz$.   **7.** $-p^5 - p^4 + 3p^3$.
9. $-0.08x^2y + 0.12xy^2 - 0.2x^2y^2$.   **11.** $-20a^3 + 15a^2 - 10a - 5$.   **13.** $5(x - 2y)$.
15. $3a(a - 4)$.   **17.** $17p(2q - 3)$.   **19.** $2axy(7a - x)$.   **21.** $3a(a - 1)$.
23. $2(mn - 2m + 3n)$.   **25.** $2xy(x^2 + 2xy + 8y^2)$.   **27.** $a^2b^2c^3(ac + b + a^2)$.
29. $0.3xy(30x - 2y + 1)$.   **31.** $x^2(h^2 - 8p + 4y + 6)$.

EXERCISE 5-2

1. $x^2 + 2xy + y^2$.   **3.** $m^2 + 2mn + n^2$.   **5.** $a^2 - 2ab + b^2$.   **7.** $x^2 + 6x + 9$.
9. $9x^2 + 6x + 1$.   **11.** $9a^2 - 12a + 4$.   **13.** $9a^2 - 12ab + 4b^2$.

15. $64x^2 + 112xy + 49y^2$.   **17.** $a^2x^2 - 2abxy + b^2y^2$.   **19.** $\dfrac{x^2}{4} - \dfrac{3xy}{4} + \dfrac{9y^2}{16}$.

21. $100 + 40 + 4 = 144$.   **23.** $400 + 200 + 25 = 625$.   **25.** 2.   **27.** 90.
29. $\frac{1}{3}$.   **31.** $2ac$.   **33.** $(x + y)(x + y)$.   **35.** $(x + 2)(x + 2)$

37. $16a^2 + 8a - 1$.　　**39.** $2(3x + 2y)^2$.　　**41.** $(5x^2 + 8y^2)^2$.　　**43.** $(\frac{1}{5}x + \frac{1}{3}y)^2$.
45. $5(9x + 4y)^2$.　　**47.** $(ax + b)^2$.

EXERCISE 5-3

1. $m^2 - n^2$.　　**3.** $x^2 - 9$.　　**5.** $4x^2 - y^2$.　　**7.** $\dfrac{m^2}{4} - n^2$.　　**9.** $a^2b^2 - c^2$.
11. $1200 + 370 + 21 = 1591$.　　**13.** $(a - b)(a + b)$.　　**15.** $(4a - 1)(4a + 1)$.
17. $(5p - 2q)(5p + 2q)$　　　　　　　　　　**19.** $(0.1x + 0.4y)(0.1x - 0.4y)$.
21. $3(2m + 3n)(2m - 3n)$.　　**23.** $a^3(ab - 1)(ab + 1)$.

EXERCISE 5-4

1. $x^2 + 5x + 6$.　　**3.** $3x^2 + 13x + 12$.　　**5.** $8y^2 - 34y + 35$.　　**7.** $2x^2 + x - 28$.
9. $4m^2 + 16m + 15$.　　　　　　　　**11.** $9x^2 - 42x + 49$.　　　　　　　　**13.** $4a^2 - 9$.
15. $6x^2y^2 - 2xy - 28$.　　**17.** $0.06y^2 - 0.17y + 0.1$.　　**19.** $acx^2 + (ad + bc)x + bd$.
21. $\dfrac{x^2}{4} + \dfrac{5x}{24} - \dfrac{1}{4}$.　　**23.** $2t^2 + 7t + 6 = 276$.

EXERCISE 5-5

1. $(x + 3)(x + 2)$.　　**3.** $(a - 8)(a - 2)$.　　**5.** $(y + 5)(y - 1)$.　　**7.** -8.
9. $(3x + 5)(x + 8)$.　　**11.** $(2x - 3y)(2x - 5y)$.　　**13.** $(2x + 3)(x - 3)$.
15. $(2 - 3b)^2$.　　**17.** $(x - 2)(x - 1)$.　　**19.** $m(m + 6n)(m + 4n)$.
21. $a(18a^3 + 21a - 4)$.　　**23.** $2(5x - y)(3x + y)$.　　**25.** $(0.4c - 0.3)(c - 2)$.
27. $(0.2x - 1)(x + 6)$.　　**29.** -503.　　**31.** $(5x - 3y)(2x + y)$.

EXERCISE 5-6

1. $(a + b)(x + 3)$.　　　　　　**3.** $(d + 1)(x - c)$.　　　　　　**5.** $g(1 + h)(x - 3)$.
7. $(x + 3)(y - 5)$.　　　　**9.** $(a + 1)(a - 1)(x + d)$.　　　　**11.** $(t - s)(b + 2a)$.
13. $(a + 2b)(5c - d)$　　**15.** $(k - m)(x^2 + y^2)$.　　**17.** $[(x + 1) + 1]^2 = (x + 2)^2$.
19. $3(2x + 7)$.　　**21.** $(2x + 2y - 3)(x + y + 1)$.　　**23.** $(2a + 2b + 3)(a + b - 3)$.
25. $5(2x - 1)$.　　　　　　**27.** $(4a - 3b)(2a + b)$.　　　　　　**29.** $(a + 3b)(1 - a + 3b)$
31. $(a^2 - 2)(2a + 5)$.　　**33.** $(0.1a - 0.01b)(0.1a + 0.01b)$.　　**35.** $x^2(\frac{5}{6}y - 1)(\frac{5}{6}y + 1)$.
37. $(\frac{3}{4}s - \frac{2}{3}t)(\frac{3}{4}s + \frac{2}{3}t)$.　　**39.** $(2a + 9)(4a + 1)$.　　**41.** $(4p^2 + 9q^2)(2p + 3q)(2p - 3q)$.
43. $(a + 3b + 2)(a + 3b - 2)$.　　　　　　　　**45.** $2(x)(x + 3)(x - 2)(x + 2)$.
47. $(0.5x - 0.3)^2$.　　**49.** $(2a - 2b - 3)(a - b + 4)$.

EXERCISE 5-7

1. $(x - y)(x^2 + xy + y^2)$.　　**3.** $(x + 2)(x^2 - 2x + 4)$.　　**5.** $(3 - y)(9 + 3y + y^2)$.

7. $(2r + 3t)(4r^2 - 6rt + 9t^2)$.   **9.** $(4ab - c)(16a^2b^2 + 4abc + c^2)$.
11. $(x^2 - y)(x^4 + x^2y + y^2)$.   **13.** $(3x^3 - y^2)(9x^6 + 3x^3y^2 + y^4)$.
15. $8t^3(2t^6 - 1)(4t^{12} + 2t^6 + 1)$.   **17.** $(a + 2)(a^2 - 5a + 13)$.   **19.** $(2x)(x^2 + 27)$.

REVIEW—CHAPTER V

1. $-4m - 28$.   **3.** $-4k^2m + 12km^2 - 8kmn$.
5. $-0.021a^2b - 0.028ab^2 + 0.035a^2b^2$.   **7.** $7(2a - 3b)$.   **9.** $18y(3x - 5)$.
11. $m(7m^2 + 5m - 9)$.   **13.** $8ab(a^2 + 3ab + 2b^2)$.   **15.** $m^3(6a^2 + 8b - 3n^2 - 1)$.
17. $x^2 + 10xy + 25y^2$.   **19.** $64 + 16b + b^2$.   **21.** $49x^2 - 70x + 25$.
23. $16a^2 + 72ab + 81b^2$.   **25.** $\dfrac{c^2}{16} - \dfrac{3}{16}cd + \dfrac{9}{64}d^2$.   **27.** 961.   **29.** 60.   **31.** $6a^2$.
33. $(x + 5)(x + 5)$.   **35.** $(4m + 3n)(2m + n)$.   **37.** $(\frac{2}{5}x - \frac{3}{4}y)^2$.   **39.** $(a^2m + 2y^2)^2$
41. $a^2 - 25$.   **43.** $\dfrac{x^2}{9} - y^2$.   **45.** 396.   **47.** $(6t + 1)(6t - 1)$.
49. $(0.8x + 0.01y)(0.8x - 0.01y)$.   **51.** $m^5(1 + mn)(1 - mn)$.   **53.** $3t^2 + 19t + 28$.
55. $12a^2 + a - 88$.   **57.** $64a^2 - 48a + 9$.   **59.** $20m^2n^2 - 3mn - 9$.
61. $cd + dmx + cnx + mnx^2$.   **63.** 660.   **65.** $(t - 5)(t - 4)$.   **67.** -4.
69. $(3m - 4)(5m - 2)$.   **71.** $3(3 - a)(1 - 2a)$.   **73.** $x(x - 7)(x - 3)$.
75. $4(5t - 4s)(2t + 3s)$.   **77.** $(0.1a - 2)(4a + 3)$.   **79.** 56.   **81.** $(a - 1)(x + y)$.
83. $(m + 7)(n - 5)$.   **85.** $(a - y)(2b + d)$.   **87.** $(3a - b)(s^2 + t^2)$.
89. $5(2t - 1)$.   **91.** $(6x + 2y - 5)(6x + 2y + 1)$.   **93.** $(8m - 5n)(2m + n)$.
95. $(4s + 3t)(1 - 4s + 3t)$.   **97.** $\left(x^2y + \dfrac{z^2}{13}\right)\left(x^2y - \dfrac{z^2}{13}\right)$.   **99.** $8m(m + 5)$.
101. $(3x + 2y + 4)(3x + 2y - 4)$.   **103.** $(0.8a - 0.5)^2$.
105. $(m - n)(m^2 + mn + n^2)$.   **107.** $(5 - x)(25 + 5x + x^2)$.
109. $(5st - u)(25s^2t^2 + 5stu + u^2)$.   **111.** $(m^4 - 5n^2)(m^8 + 5m^4n^2 + 25n^4)$.
113. $(m - 2)(m^2 - 10m + 28)$.

EXERCISE 6-1

1. $\dfrac{3a}{5a - 2}$.   **3.** $\dfrac{b}{5}$.   **5.** $\dfrac{1}{xy - 3}$.   **7.** $\dfrac{x - 3}{x - 2}$.   **9.** $\dfrac{2x - 5}{x - 7}$.   **11.** $\dfrac{x - y}{a - 5}$.
13. $\dfrac{x(2x - 3)}{a - b}$.   **15.** $\dfrac{x + y - 5}{x}$.   **17.** $\dfrac{a - b}{x + y}$.   **19.** $-; +; -$.   **21.** $+;$
$-; -$.   **23.** $-a$.   **25.** $\dfrac{1}{p}$.   **27.** $-\dfrac{x + 2}{x + 1}$.   **29.** -1.

EXERCISE 6-2

1. $\dfrac{2(x^2 - 2)}{x - 2}$.   **3.** $\dfrac{1}{x}$.   **5.** $\dfrac{3(2x - 3)}{xy - 2}$.   **7.** $\dfrac{2(a - 3)}{(a + 2)(2a - 3)}$.   **9.** $\dfrac{3y^2}{5a}$.
11. $\dfrac{x + 3}{x - 1}$.   **13.** $\dfrac{a - 2}{3}$.   **15.** $\dfrac{6x}{x - 3}$.   **17.** 1.   **19.** $\dfrac{1}{3a - b}$.

EXERCISE 6-3

1. $\dfrac{x+3}{x+2}$. **3.** $\dfrac{x+4}{3(x-2)}$. **5.** $\dfrac{(x+2)^2}{(x+4)(x^2-3)}$. **7.** $\dfrac{m+9}{2(m-4)}$. **9.** $\dfrac{x-2}{2x-5}$.

11. $\dfrac{(a+b)^2}{a-3b}$. **13.** $\dfrac{2a+5b}{(a^2+b^2)(a+b)}$. **15.** $\dfrac{x-3}{x-2}$. **17.** $\dfrac{6(a-b-4)}{a-2b}$.

19. $-\frac{1}{3}$.

EXERCISE 6-4

1. $\dfrac{7x+6y}{72}$. **3.** $\dfrac{6a^2-ab-18b^2}{36a^2b^2}$. **5.** $\dfrac{13t-6r}{rt}$. **7.** $\dfrac{4x}{2-3x}$.

9. $\dfrac{x^2-5x+17}{x-7}$. **11.** $\dfrac{5x^2-12x+17}{x^2-x-12}$. **13.** $\dfrac{7}{a-b}$. **15.** $\dfrac{x+4}{x^2-4}$.

17. $\dfrac{16}{(4+a)(5-a)}$. **19.** $\dfrac{d^2-cd-3c^2}{d^2-4c^2}$. **21.** $\dfrac{5}{(x-3)(x+2)}$.

23. $\dfrac{x^2-x-15}{x^2-5x+6}$. **25.** $\dfrac{x^2+5x+2}{(2x+3)(x-3)(3x-2)}$. **27.** $\dfrac{x^2-20x-11}{(2-x)(1+x)(2+x)}$.

EXERCISE 6-5

1. $\dfrac{2x+3}{x+5}$. **3.** $\dfrac{3-x}{x+12}$. **5.** $\dfrac{a^2-2a+1}{2a^2-a}$. **7.** $\dfrac{a^2+a-2}{a^2+a}$.

EXERCISE 6-6

1. $\frac{1}{12}$. **3.** 2. **5.** 1. **7.** $\frac{1}{2}$. **9.** 3. **11.** -10. **13.** -7.

15. $-\frac{16}{7}$. **17.** 3.

REVIEW—CHAPTER VI

1. $+;-;-$. **3.** $\dfrac{2x-3}{x-5}$. **5.** $-\dfrac{1}{a+3}$. **7.** $\dfrac{a-3}{4-a}$. **9.** $\dfrac{x-y}{x+y-1}$.

11. $3m+3n-2$. **13.** $\dfrac{x}{2x-y}$. **15.** $-\dfrac{1}{a+1}$. **17.** 1. **19.** $\dfrac{(a-1)(d-4)}{(c-4)(b+2)}$.

21. $\dfrac{2m-n}{m+5n}$. **23.** $\dfrac{8ax(2x-y)}{7y(a^2-3b)}$. **25.** $\dfrac{p}{2x+3}$. **27.** $\dfrac{(a-6b)(3a+b-2)}{5(2a-3b)}$.

29. $\dfrac{3x-1}{4x+3}$. **31.** $\dfrac{x+1}{2x+1}$. **33.** $\dfrac{7}{5(a-2)}$. **35.** $\dfrac{10}{x^2-4}$. **37.** $\dfrac{y-x-y^2}{xy-2y^2}$.

39. $\dfrac{11a^4-22a^2+6}{(a^4-4)(2a^2-3)}$. **41.** $\dfrac{3a^2+5a}{1-a^2}$. **43.** -1. **45.** $\dfrac{(a-3)(a+1)^2}{(a-2)(a^2+a-1)}$.

47. $\dfrac{3a-19}{7a^2+24a+2}$. **49.** $\frac{1}{2}$. **51.** -12. **53.** $\frac{3}{2}$. **55.** No solution. **57.** 5.

EXERCISE 7-1

1. 8. **3.** 9. **5.** -27. **7.** -64. **9.** $-\frac{8}{27}$. **11.** 0.0081. **13.** a^8.

15. x^9. **17.** x^3. **19.** $\frac{b^2}{a}$. **21.** $\frac{bc}{a}$. **23.** $-b^4$. **25.** 64. **27.** 108.

29. $-a^{15}b^{12}$. **31.** $81a^8b^4c^4$. **33.** -8. **35.** $\frac{a^4}{16}$ **37.** $\frac{x^{12}}{y^{20}}$. **39.** $\frac{81}{v^8}$.

41. $\frac{3^{3n}}{8} = \frac{27^n}{8}$. **43.** x^{3n}. **45.** $\frac{1}{x^n}$. **47.** $9a^2 + 6ax + x^2$. **49.** $3x^{2n}$.

51. $a^{2r}b^{3r}$. **53.** $a^3b^2c^2$. **55.** $\frac{ax}{y^4}$. **57.** $\frac{a^4b^8}{81c^4}$. **59.** $-\frac{a^6y^6}{b^6}$.

EXERCISE 7-2

1. 1. **3.** m^5. **5.** 1. **7.** 1. **9.** $\frac{2xy}{m^2}$. **11.** $\frac{1}{a^{-4}}$ **13.** y^{-2}. **15.** $\frac{1}{y^{-3}}$.

17. $(bx)^{-1}$. **19.** $-\frac{1}{x^{-2}y^{-3}z^{-3}}$. **21.** $\frac{1}{a^{-2}} - \frac{1}{b^{-2}}$. **23.** $\frac{1}{27x^3}$. **25.** x^4. **27.** $\frac{y^2}{x^2}$.

29. $\frac{x^6}{64}$. **31.** a^2b^2. **33.** $a^4x^{12}y^5$. **35.** $\frac{x^4}{y^6}$. **37.** $\frac{3(y^2 + x^2)}{x^2y^2}$ **39.** $\frac{b^3}{a^5}$.

41. $\frac{ab}{a^2 - b^2}$. **43.** $\frac{6 + x^2y}{x^2}$ **45.** $\frac{y}{1 + xy}$.

EXERCISE 7-3

1. $+2$. **3.** $+2$. **5.** -2. **7.** $+\frac{1}{2}$. **9.** None. **11.** $+0.4$. **13.** $-\frac{2}{3}$.

15. -2. **17.** $p^{1/2}$. **19.** $t^{1/10}$. **21.** $p^{2/3}$. **23.** $y^{2/n}$. **25.** $p^{-(s/r)}$.

27. $(c - 6)^{1/2}$. **29.** $\frac{c^{3/4}}{b^{1/4}}$ **31.** $\frac{1}{(c + d^2)^{1/n}}$ **33.** $\sqrt[5]{a}$. **35.** $\sqrt[8]{a^5}$.

37. $\sqrt{3x}$. **39.** 1. **41.** $\sqrt{b - c}$. **43.** $\sqrt[3]{(x + y)^2}$. **45.** $\sqrt[4n]{s^8t}$. **47.** 0.

49. $\sqrt[6]{x}$. **51.** a^3 **53.** $\sqrt[5]{a^3}$. **55.** $\sqrt[12]{m^8n}$. **57.** $\frac{4}{25}$. **59.** $\frac{1}{x + y}$ **61.** $\frac{1}{16}$.

63. $\frac{x^2\sqrt{x}}{8\sqrt[4]{8y}}$. **65.** $\sqrt[10]{s}$. **67.** $\frac{1}{\sqrt[16]{a^3}}$. **69.** $\frac{b^2}{81\sqrt[3]{a}}$. **71.** $\sqrt[m]{y^2}$. **73.** $\sqrt[24]{x}$.

EXERCISE 7-4

1. $\sqrt{3a}$. **3.** $\sqrt[3]{2m}$. **5.** $\sqrt[3]{3m^2}$. **7.** $\sqrt[3]{x^2 + y^2}$. **9.** $\sqrt[4]{9}$. **11.** $\sqrt[5]{16x^2y^2}$.

13. $\sqrt[3]{64}$. **15.** $\sqrt[6]{0.008}$. **17.** $\sqrt{18}$. **19.** $\sqrt[3]{48}$. **21.** $\sqrt{3m^2}$. **23.** $\sqrt[3]{128a^4}$.

25. $\sqrt{32x^5y^3}$. **27.** $\sqrt[6]{(a - b)^7}$. **29.** $5\sqrt{3}$. **31.** $3\sqrt[3]{3}$. **33.** $2\sqrt{2x}$.

35. $3x\sqrt[3]{xy^2}$. **37.** $3xyz^2\sqrt{2y}$. **39.** $6x\sqrt{3x}$. **41.** $\frac{1}{4}\sqrt{x}$. **43.** $2\sqrt[3]{m^2}$.

45. $3(1 + a\sqrt{5})$. **47.** $(a - 5)\sqrt[3]{a - 5}$.

EXERCISE 7-5

1. $\frac{1}{3}\sqrt{3}$.   **3.** $\frac{3x\sqrt{5}}{5}$.   **5.** $\frac{3\sqrt[3]{2}}{2}$.   **7.** $12\sqrt[3]{x^2}$.   **9.** $\frac{\sqrt{7m}}{m}$.   **11.** $\frac{\sqrt[3]{-6abc^2}}{2ba^3}$.

13. $\sqrt{3} + \sqrt{2}$.   **15.** $-(2 + \sqrt{6})$.   **17.** $\frac{x + x\sqrt{x}}{1 - x}$.   **19.** $-5 - 2\sqrt{6}$.

21. $-1 - \frac{1}{2}\sqrt{6}$.   **23.** $1 - \frac{1}{3}\sqrt{6}$.   **25.** $\frac{1}{6}\sqrt{12a - 9b}$.

EXERCISE 7-6

1. $5\sqrt{a}$.   **3.** $3\sqrt{5}$.   **5.** $7\sqrt[3]{3}$.   **7.** $\sqrt{2}$.   **9.** $-\frac{14}{3}\sqrt[3]{6}$.   **11.** $\frac{5}{6}\sqrt{3} - \frac{3}{4}\sqrt{2}$.

13. $6\sqrt[3]{2} - 3\sqrt{2}$.   **15.** $\frac{17\sqrt{3}}{2}$.   **17.** $3 - x\sqrt{6}$.   **19.** $(a - b + 1)\sqrt[3]{(a - b)^2}$.

21. $4\sqrt{0.03}$.   **23.** $6\sqrt{0.1}$.

EXERCISE 7-7

1. $2a$.   **3.** $3a\sqrt{2a}$.   **5.** $2\sqrt[6]{1944}$.   **7.** $72\sqrt{3x} - 18\sqrt{10x} + 252\sqrt{2x}$.

9. $2a^3b\sqrt{72a}$.   **11.** $57 - 12\sqrt{15}$.   **13.** $a - b$.   **15.** $-4 - 3\sqrt{15}$.

17. $a + 19 + 8\sqrt{a + 3}$.   **19.** $2a - 2\sqrt{a^2 - b^2}$.   **21.** $7 - 5\sqrt{2}$.

23. $21 + 14\sqrt{3}$.

EXERCISE 7-8

1. $2\sqrt{2}$.   **3.** $\sqrt{2} - 2\sqrt{5}$.   **5.** $\frac{2}{21}\sqrt{7} - \frac{1}{6}\sqrt{6} + \frac{1}{21}\sqrt{14}$.   **7.** $2\sqrt[6]{3} + \sqrt[3]{18}$.

9. $\frac{1}{3}(\sqrt{10} - 1)$.   **11.** $4\frac{4}{7} + \frac{3}{7}\sqrt{15}$.   **13.** $\frac{8\sqrt{6} + 11\sqrt{3}}{3}$.   **15.** $\frac{7\sqrt{14} - 9}{55}$.

17. -1.   **19.** $\frac{x - \sqrt{ax} - \sqrt{xy} + \sqrt{ay}}{x - y}$.

EXERCISE 7-9

1. $+1$.   **3.** $+1$.   **5.** i.   **7.** $+1$.   **9.** -1.   **11.** $-i$.   **13.** -3.

15. $i - 1$.   **17.** $(\sqrt{3} - 2)i - 3$.   **19.** $4i - 2$.   **21.** $5 - 2i$.   **23.** $6 - 5i$.

25. $12 + i$.   **27.** $\frac{9}{8}$.   **29.** $18 + i$.   **31.** $1 - 21i$.   **33.** $3 - 2i$.   **35.** $9 - 46i$.

37. $91 + 91i$.   **39.** $\frac{10}{17} + \frac{11}{17}i$.   **41.** 1.   **43.** $-\frac{3}{5} + \frac{2}{5}i$.   **45.** $\frac{5 + 11i}{6}$.

REVIEW—CHAPTER VII

1. 81.   **3.** (-0.027).   **5.** $(-\frac{8}{125})$.   **7.** $\frac{1}{m^2}$.   **9.** $-b^{23}$.   **11.** $\frac{27}{8}$.

13. $-\dfrac{243p^5}{q^{20}}.$ **15.** $a^{5n-4}.$ **17.** $a^{2a}b^{am}.$ **19.** $\dfrac{b^8 c^6}{x^4 y^4}.$ **21.** 1.   **23.** 1.

25. $\dfrac{m^2}{xy}.$ **27.** $\dfrac{2^{-3}}{y^{-4}z^{-2}}.$ **29.** $-\dfrac{1}{a^{-2}b^{-3}}.$ **31.** $\dfrac{y^{-3}}{6^{-1}a^{-2}b^{-3}x^{-1}z^{-2}}.$ **33.** $a^{1+m}.$

35. $\dfrac{b^3}{a^{12}c^6}.$ **37.** $\dfrac{yz - az + ay}{ayz}.$ **39.** $\dfrac{1}{s^5 t^5}.$ **41.** $\dfrac{x^4 y^4}{y^4 - 2y^2 x^2 + x^4}.$ **43.** 100.

45. None.   **47.** $-\frac{2}{3}.$ **49.** $-y^{3/5}.$ **51.** $s^{2/r} t^{4/r}.$ **53.** $\dfrac{a^{p/5}}{b^{q/5}}.$ **55.** $3\sqrt[3]{a}.$

57. $\sqrt[2p]{a^4 b}.$ **59.** 0.   **61.** $\dfrac{1}{\sqrt[6]{a}}.$ **63.** $\dfrac{\sqrt[12]{a^5}}{\sqrt{b}}.$ **65.** 9.   **67.** $a^{ba/c^2}.$

69. $\dfrac{ab^{3m}\sqrt[20]{a}}{h^3}.$ **71.** $2a\sqrt{2ab}.$ **73.** $\sqrt[m]{6a^2 b}.$ **75.** $\sqrt[6]{27}.$ **77.** $\sqrt[3]{0.008}.$

79. $\sqrt[3]{40}.$ **81.** $\sqrt[4]{81x^9 y^{10}}.$ **83.** $2\sqrt[3]{-2}$ or $-2\sqrt[3]{2}.$ **85.** $\dfrac{x}{m}\sqrt[3]{amx}.$

87. $\dfrac{3x^2}{4}\sqrt[m]{4}.$ **89.** $a + ab + b^2.$ **91.** $\dfrac{3x\sqrt{p}}{p}.$ **93.** $\dfrac{2\sqrt[3]{b}}{b}.$ **95.** $\dfrac{3\sqrt{2} + 3\sqrt{7}}{-5}.$

97. $\dfrac{\sqrt{x} + x}{1 - x}.$ **99.** $\dfrac{\sqrt[6]{a^5} + \sqrt[6]{a^2 b^3}}{a - b}.$ **101.** $\dfrac{8b + 10\sqrt{ab} + 3a}{4b - a}.$ **103.** $9\sqrt{3}.$

105. $2\sqrt[3]{6}.$ **107.** $\dfrac{7\sqrt{15}}{225}.$ **109.** $(x - 2)\sqrt{3x}.$ **111.** $\dfrac{8x\sqrt{xy}}{15}.$

113. $\dfrac{3\sqrt[5]{16} + 5\sqrt{2}}{2}.$ **115.** $3x\sqrt[4]{3x}.$ **117.** $3m^2 n\sqrt[6]{16m^5 n^3}.$ **119.** $25\sqrt{2} - 40.$

121. $3\sqrt{3} - 2\sqrt{15} - 3\sqrt{5} + 10.$ **123.** $2a + 1 + 2\sqrt{a^2 + a - 6}.$

125. $39 - 16\sqrt{5}.$ **127.** $\frac{2}{3}\sqrt{6} - \frac{2}{3}\sqrt{3} + 1.$ **129.** $\sqrt[3]{18} - 2\sqrt[6]{3}.$

131. $\dfrac{11\sqrt{6} - 24}{25}.$ **133.** $\frac{1}{21}(3\sqrt{2} - 4\sqrt{3} + 6\sqrt{6} - 9).$

135. $\dfrac{x\sqrt{6} + y\sqrt{6} - 5\sqrt{xy}}{2x - 3y}.$ **137.** $-3.$ **139.** $3 + 8i.$ **141.** $2 + i.$

143. $-1 - 8i.$ **145.** $\frac{1}{2} - 2i\sqrt{2}.$ **147.** $-11 - 23i.$ **149.** $-\dfrac{13 + 9i}{20}.$

151. $\dfrac{3 - 5i\sqrt{3}}{21}.$ **153.** $\frac{2}{25} + \frac{29}{25}i.$

EXERCISE 8-1

1. 4, 2.   **3.** $-4, -6.$ **5.** $7, -1.$ **7.** $-2\frac{1}{3}, -5.$ **9.** $\frac{3}{5}, -\frac{2}{3}.$

11. $5a, -3a.$ **13.** $\dfrac{a}{3}, \dfrac{a}{2}.$ **15.** $\dfrac{2}{3a}, -\dfrac{3}{a}.$ **17.** $0, 4a.$ **19.** $-2.$

21. $\frac{5}{3}, -\frac{3}{2}.$ **23.** 4, 6.   **25.** 3 ft.

EXERCISE 8-2

1. $\pm 3.$ **3.** $\pm 3.$ **5.** $\pm\frac{3}{11}\sqrt{11}.$ **7.** $\dfrac{\pm\sqrt{z(a + b)}}{a + b}.$ **9.** $\pm\sqrt{5}.$

11. $\pm\dfrac{3b}{10}\sqrt{10}.$ **13.** $\pm i.$ **15.** $\pm 2i.$ **17.** $\pm i\sqrt{5}.$ **19.** 4.

EXERCISE 8-3

1. $3, 2$.   **3.** $3, 3$.   **5.** $1\frac{1}{2}, 1\frac{1}{3}$.   **7.** $\dfrac{-3+\sqrt{5}}{4}, \dfrac{-3-\sqrt{5}}{4}$.   **9.** $1 + 2i, 1 - 2i$.

11. $-\frac{1}{3} \pm \frac{1}{3}\sqrt{3}$.   **13.** $\dfrac{1 \pm \sqrt{5}}{6}$.   **15.** $1 \pm \frac{1}{4}\sqrt{6}$.   **17.** $0.6, -2$.   **19.** $2 \pm \sqrt{2}$.

21. $a \pm \sqrt{a^2 + c}$.   **23.** $\dfrac{8 \pm i\sqrt{2}}{3}$.   **25.** $\dfrac{(1-a) \pm \sqrt{a^2 - 10a + 17}}{4}$.   **27.** 14.

29. Impossible.

EXERCISE 8-4

1. Rational, equal.   **3.** Imaginary.   **5.** Imaginary.   **7.** Irrational, unequal.
9. Rational, unequal.   **11.** ± 8.   **13.** ± 6.   **15.** $1\frac{1}{3}$.   **17.** -2.
19. $x^2 - 7x + 12 = 0$.   **21.** $2x^2 - 5x + 2 = 0$.   **23.** $49x^2 - 28x + 4 = 0$.
25. $x^2 - 4x + 1 = 0$.   **27.** $x^2 - 6x + 1 = 0$.   **29.** $x^2 + (a - b)x - ab = 0$.

EXERCISE 8-5

1. 33.   **3.** No solution.   **5.** 2.   **7.** 5.   **9.** 2.   **11.** $3, 7$.   **13.** 1.
15. $6, 2$.

EXERCISE 8-6

1. $\{x \mid -2 < x < 2\}$.   **3.** $\{x \mid x < -1 \text{ or } x > 1\}$.   **5.** $\{m \mid -\sqrt{3} < m < \sqrt{3}\}$.
7. $\{x \mid 2 < x < 4\}$.   **9.** $\{n \mid n < 3 \text{ or } n > 5\}$.   **11.** $\{x \mid -3 < x < 2\}$.
13. $\{r \mid r < -4 \text{ or } r > 1\}$.   **15.** $\{m \mid m < -3 \text{ or } m > -2\}$.   **17.** $\{s \mid -6 < s < -2\}$.
19. $\{x \mid -\frac{1}{2} < x < 4\}$.   **21.** $\{x \mid x < 1\frac{1}{2} \text{ or } x > 5\}$.   **23.** $\{y \mid -\frac{2}{3} < y < \frac{1}{2}\}$.

25. $\{n \mid n < -1\frac{2}{3} \text{ or } n > -\frac{1}{2}\}$.   **27.** $\left\{x \mid \dfrac{5-\sqrt{5}}{2} < x < \dfrac{5+\sqrt{5}}{2}\right\}$.

29. $\left\{m \mid m < \dfrac{-3-\sqrt{3}}{2} \text{ or } m > \dfrac{-3+\sqrt{3}}{2}\right\}$.

31. $\left\{t \mid t < \dfrac{-3-\sqrt{5}}{4} \text{ or } t > \dfrac{-3+\sqrt{5}}{4}\right\}$.   **33.** $\{x \mid x < \frac{1}{2} \text{ or } x > \frac{1}{2}\}$.

35. $\{\ \}$.   **37.** $\{x \mid 0 < x < 2\frac{1}{2}\}$.   **39.** $\{k \mid k < -\frac{1}{3} \text{ or } k > 0\}$.

REVIEW—CHAPTER VIII

1. $3, 10$.   **3.** $6, -3$.   **5.** $\frac{3}{4}, -\frac{1}{3}$.   **7.** $\dfrac{a}{5}, \dfrac{2a}{3}$.   **9.** $0, \frac{2}{3}a$.   **11.** 3.

13. 12 ft   **15.** $\pm \frac{3}{2}$.   **17.** $\pm \dfrac{\sqrt{s(p+r)}}{p+r}$.   **19.** $\pm \dfrac{m\sqrt{35}}{10}$.   **21.** $\pm \frac{1}{3}i\sqrt{21}$.

23. -5.   **25.** $36a^2$.   **27.** $\frac{1}{100}$.   **29.** $\frac{1}{4}a^2b^2$.   **31.** $\frac{2}{3}, \frac{1}{4}$.   **33.** $1 \pm \dfrac{i}{2}\sqrt{6}$.

35. $-\frac{2}{5}, -\frac{1}{4}$.   **37.** $\dfrac{-5 \pm \sqrt{19}}{3}$.   **39.** $\dfrac{2 \pm i\sqrt{2}}{3}$.   **41.** $\pm 4i\sqrt{c}$.

43. $1 \pm \sqrt{2}$.   **45.** $0, 3\frac{1}{2}$.   **47.** 20 in.   **49.** Imaginary.   **51.** Rational, equal.   **53.** Imaginary.   **55.** -8.   **57.** $\frac{52}{28}$.   **59.** $25x^2 - 4 = 0$.
61. $x^2 - 8x + 17 = 0$.   **63.** $x^2 + 2x + 5 = 0$.   **65.** 7.   **67.** 4, 1.
69. $-2, -1$.   **71.** $1, -2\frac{1}{4}$.   **73.** $\{x \,|\, x < -5 \text{ or } x > 3\}$.   **75.** $\{x \,|\, 1\frac{1}{2} < x < 4\}$.
77. $\{x \,|\, x < -1 - \sqrt{6} \text{ or } x > -1 + \sqrt{6}\}$.   **79.** $\{\ \ \}$.

EXERCISE 9-1

1. $(1, 2), (2, 0)$.   **3.** $(1, -3), (2, -1), (4, 3)$.   **5.** $(-4, -1)$.   **7.** $(1, -2)$, $(2, 4), (-1, -2)$.   **9.** $8; -2; -4$.   **11.** $-\frac{1}{5}; 2; 1\frac{1}{5}$.

EXERCISE 9-2

1–9.

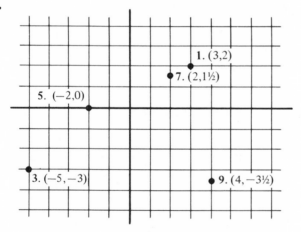

11. A: $(3, 1)$; B: $(-4, 2)$; C: $(-2, -3\frac{1}{2})$; D: $(2\frac{1}{2}, 0)$; E: $(2\frac{1}{2}, -3\frac{1}{2})$.   **13.** Square.
15. $(-3, 4), (2, 4), (2, -1)$.

EXERCISE 9-3

7. $y = 4$.

EXERCISE 9-5

1. $(2, 3)$.   **3.** $(-2, 3)$.   **5.** $(4, -\frac{1}{2})$.   **7.** $(-2\frac{1}{2}, 1)$.   **9.** $(-3\frac{1}{2}, -4\frac{1}{2})$.
11. $(-2, 0)$.   **13.** (a) Yes, $(3, 2)$; (b) Yes; (c) Yes; (d) $(3, 2)$.

EXERCISE 9-6

1. $(2, 3)$.   **3.** $(-\frac{1}{2}, 3\frac{1}{2})$.   **5.** $(2, 1)$.   **7.** $(-2, -3)$.   **9.** $(\frac{5}{7}, -\frac{1}{3})$.
11. Dependent.   **13.** Inconsistent.   **15.** $(6, 2\frac{1}{2})$.   **17.** Inconsistent.
19. $(12, 24)$.   **21.** 23 ft $\times$ 15 ft.   **23.** $1.40 first 3 min; $0.35 each additional
minute.

EXERCISE 9-7

15. $(-4\frac{1}{2}, 3)$, $(1, 3)$, $(1, -\frac{2}{3})$ **17.** $\begin{cases} x < y \\ y < 6 \\ x > 3 \end{cases}$

REVIEW—CHAPTER IX

1. $(3, 4)$, $(1, -2)$.   **3.** $(2, 0)$, $(7, 3)$.   **5.** $6\frac{1}{2}$; -3; $\frac{1}{4}$.   **7.** (a) Parallelogram;
(b) $(1, 1\frac{1}{2})$; (c) $2\sqrt{2}$; 5.   **9.** (a) $(-3, 4)$, $(3, -4)$; (b) $(2, 3)$, $(-2, -3)$;
(c) $(-4, 4)$, $(4, -4)$; (d) $(-3, -1)$, $(3, 1)$.   **11.** $x = -3$; $y = 2$.   **23.** $(3\frac{1}{2}, -2)$.
25. Dependent.   **27.** Inconsistent.   **29.** $(-2, -1)$.   **31.** $(6, 8)$.   **33.** $(-3, 0)$.
35. $(0, -4)$.   **37.** 150 cc/70, 250cc/30%.

EXERCISE 10-1

1. $\log_2 8 = 3$.   **3.** $\log_3 243 = 5$.   **5.** $\log_{16} 2 = \frac{1}{4}$.   **7.** $\log_7 \frac{1}{49} = -2$.
9. $\log_3 0.33 = -1$.   **11.** $4^3 = 64$.   **13.** $0.1^3 = 0.001$.   **15.** $8^{0.33} = 2$.
17. $10^{0.301} = 2$.   **19.** $10^{0.230} = 17$.   **21.** $L = 4$.   **23.** $a = 2$.   **25.** $n = 25$.
27. $a = 36$.   **29.** $l = \frac{2}{3}$.   **31.** 125.   **33.** $a = 0.8$.   **35.** $n = \frac{1}{121}$.

EXERCISE 10-2

1. 4×10^3.   **3.** 1.836×10^3.   **5.** 3×10^6.   **7.** 8.3×10^{-1}.   **9.** 4×10^{-4}.
11. 6.325×10^{-4}.   **13.** (a) $p = 1.4$; (b) $p = 2.4$; (c) $p = -.6$; (d) $p = -1.6$;
(e) $p = 3.4$.   **15.** 0.8657.   **17.** 0.9375.   **19.** 0.9031.   **21.** 1.7803.
23. 3.4771.   **25.** $-1 + 0.8000$ or -0.2.   **27.** $-3 + 0.8$ or -2.2.

EXERCISE 10-3

1. 2.5065.   **3.** 1.7993.   **5.** 1.6232.   **7.** $9.5682 - 10$.   **9.** 3.6990
11. $7.7160 - 10$.   **13.** 2.0000.   **15.** $9.9996 - 10$.   **17.** 3.4829.   **19.** 5.2095.
21. $6.5051 - 10$.   **23.** 0.7042.

EXERCISE 10-4

1. 26.0. **3.** 340. **5.** 0.0399. **7.** 74,300. **9.** 0.674. **11.** 168.
13. 0.0626. **15.** 207,000. **17.** 0.474. **19.** 7970. **21.** 54.9.
23. 0.0000000227 or 2.27×10^{-8}.

EXERCISE 10-5

1. 0.6651. **3.** 1.8714. **5.** 2.7188. **7.** $7.8064 - 10$. **9.** 5.6281.
11. $9.9899 - 10$. **13.** 4.144. **15.** 813.6. **17.** 48,360. **19.** 72.52.
21. 0.001701. **23.** 860,600.

EXERCISE 10-6

1. 504. **3.** 520. **5.** 0.09045. **7.** 0.01762. **9.** 0.172. **11.** 245.9.
13. -26.96. **15.** 1.015×10^{8}. **17.** 24.56. **19.** 252.1.

EXERCISE 10-7

1. 8. **3.** 1.631. **5.** 0.3282. **7.** 1.148. **9.** 0.2613. **11.** 0.9915.
13. -18.87. **15.** 1.159. **17.** 637.3 **19.** 4.27×10^{-5}

EXERCISE 10-8

1. 300,800. **3.** 3.324×10^{6}. **5.** 0.001815. **7.** 2.524×10^{-7}. **9.** 422.7.
11. 9.506×10^{-5}. **13.** 2.331×10^{-9}. **15.** 2.291. **17.** 4.729.
19. 7.06×10^{-25}.

EXERCISE 10-9

1. 27. **3.** 3.7. **5.** 8. **7.** 7.984. **9.** 0.3958. **11.** 1.335. **13.** 0.5947.
15. 0.2153. **17.** 0.8212. **19.** -1.341.

REVIEW—CHAPTER X

1. $\log_{.5} .25 = 2$. **3.** $\log_2 64 = 6$. **5.** $\log_{64} 16 = 0.67$. **7.** $9^2 = 81$.
9. $16^{1/4} = 2$. **11.** $10^{.415} = 2.6$. **13.** $L = 2$. **15.** $n = 4096$. **17.** $n = \frac{1}{512}$.
19. 3.864×10^3. **21.** 4.0×10^7. **23.** 7.89×10^{-4}. **25.** $p = 0.5740$.
27. $p = 0.7067$. **29.** $p = -0.3372$. **31.** 2.9360. **33.** $4.0000 - 10$.

35. 4.5024.   **37.** 2.4524.   **39.** 4.5311 − 10.   **41.** 7.37.   **43.** 130.
45. 1.06×10^{-13}.   **47.** 8.212.   **49.** 74.98.   **51.** 6.314×10^{-14}.   **53.** 297.6.
55. 5.325×10^{-4}.   **57.** 0.06413.   **59.** 3.425×10^{-7}.   **61.** 69.9.   **63.** 0.8628.
65. 0.05721.   **67.** 0.003564.   **69.** 1.040×10^{7}.   **71.** 7.063×10^{-18}.
73. 263.4.   **75.** −10,630.   **77.** 2.892.   **79.** 0.5756.   **81.** 0.5293.
83. 3.941.

EXERCISE A-1

1. 12.   **3.** $4\frac{1}{2}$.   **5.** ± 8.   **7.** $5\frac{1}{7}$.   **9.** 15.   **11.** $\dfrac{7b^2}{2}$.   **13.** 27.

15. 138 mi.   **17.** 81 lbs.   **19.** 24.5 lbs.   **21.** 21, 35.   **23.** $3996.

EXERCISE A-2

1. 1 in 5, or 0.2.   **3.** 6 in 9, or 0.67.   **5.** $\frac{2}{8}$ or 0.25.   **7.** 24 reds/36 whites;
$26r/39w$; $28r/42w$; $30r/45w$.   **9.** 0.23.   **11.** 0.44.   **13.** 0.25.
15. $\frac{10}{51}$ or 0.196.

EXERCISE A-3

1. 6¢.   **3.** 10¢.   **5.** Unfavorable at 10¢, favorable at 3 for 25¢.   **7.** Value
of the prize.   **9.** 16 to 1.

REVIEW—UNIT A

1. 24.   **3.** ± 6.   **5.** $1\frac{1}{3}$.   **7.** 55.   **9.** 171 lbs.   **11.** 27, 63.   **13.** 1 in 8 or 0.125.
15. $\frac{3}{11}$ or 0.27.   **17.** 0.31.   **19.** 0.24.   **21.** 4.5¢.   **23.** Unfavorable at 5¢;
favorable at 6 for 25¢.   **25.** 32 to 1.

EXERCISE B-1

1. 37, 43, 49.   **3.** −50, −102, −154.   **5.** $-\frac{1}{12}, -\frac{1}{3}, -\frac{7}{12}$.   **7.** 119.
9. 54.   **11.** $-1\frac{3}{4}$.   **13.** $-16\sqrt{2}$.   **15.** 13.   **17.** 9.   **19.** $\frac{3}{4}$.   **21.** 4.3.
23. $-\frac{5}{7}$.   **25.** 9, 15, 21.   **27.** −4, −15, −26.   **29.** 2.

EXERCISE B-2

1. $d = 3, s = 282$.   **3.** $a = 7, s = 160$.   **5.** $n = 17, d = 4$.   **7.** $a = 13$,
$l = -92$.   **9.** $a = -7, n = 15$.   **11.** $a = 5, l = 23, d = 3, s = 98$.   **13.** $6750.
15. (a) 624 ft; (b) 6400 ft; (c) $37\frac{1}{2}$ sec.

EXERCISE B-3

1. 54, 162, 486.   **3.** $-27, 81, -243$.   **5.** 0.024, -0.0048, 0.00096.   **7.** 243.
9. 256.   **11.** $32\sqrt{2}$.   **13.** 1.   **15.** 5.   **17.** -2.   **19.** -4.   **21.** 0.1.
23. $-\sqrt{3}$.   **25.** $\pm 12, 36, \pm 108$.   **27.** $1, -1\frac{1}{2}, 2\frac{1}{4}, -3\frac{3}{8}$.   **29.** $\frac{1}{3}$ and 7.

EXERCISE B-4

1. $a = 1, s = 63$.   **3.** $l = 2, s = 86$.   **5.** $a = 16, l = -\frac{1}{2}$.   **7.** $n = 7$,
$r = 2$.   **9.** $729, -243, 81, -27, 9, -3, 1$.   **11.** (a) \$384; (b) \$12,285.
13. $64\sqrt{2n}$ or $91n$ (approx.).   **15.** $117\frac{2}{3}$ ft.   **17.** 16.   **19.** $6\frac{2}{3}$.   **21.** $\frac{7}{9}$
23. $\frac{67}{300}$.

REVIEW—UNIT B

1. 56, 62, 68.   **3.** $-\frac{5}{12}, -\frac{5}{6}, -1\frac{1}{4}$.   **5.** 57.   **7.** $-21\sqrt{3}$.   **9.** $8\frac{1}{2}$.
11. 6.1.   **13.** 11, 17, 23.   **15.** 3.   **17.** $a = 5, s = 549$.   **19.** $a = 60, l = 0$.
21. $a = 1, d = -4, l = -19, s = -54$.   **23.** (a) 496 ft; (b) 4096 ft; (c) 35 sec.
25. $-64, 128, -256$.   **27.** 972.   **29.** $243\sqrt{3}$.   **31.** 5.   **33.** -2.
35. $-\sqrt{2}$.   **37.** $1, -1\frac{2}{3}, 2\frac{7}{9}, -4\frac{17}{27}$.   **39.** $a = 3, s = 363$.   **41.** $a = 256$,
$l = -60\frac{3}{4}$.   **43.** $32, -16, 8, -4, 2, -1, \frac{1}{2}$.   **45.** $32\sqrt{2n}$ or $45n$ (approx.).
47. 20.   **49.** $\frac{8}{9}$.

EXERCISE C-1

1. $\{a, u, n, t\}$.   **3.** $\{1, 4, 9, 16\}$.   **5.** $\{m, a, t, h, e, i, c, s\}$.   **7.** $\{100, 102, 104, \ldots\}$.
9. $\varnothing$.   **11.** $\{1, 2, 3, 4, 5, 6, 7\}$.   **13.** $\{8, 9, 10, \ldots\}$.   **15.** $\varnothing$.   **17.** False.
19. True.   **21.** False.   **23.** False.   **25.** True.   **27.** True.   **29.** False.
31. True.   **33.** True.

EXERCISE C-2

1. True.   **3.** False.   **5.** False.   **7.** False.   **9.** True.   **11.** True.
13. False.   **15.** False.   **17.** True.   **19.** False   **21.** True.   **23.** True.
29.

Four-sided polygons
Rectangles
Squares
U

33.
A B
U
Even counting numbers

35.
c
d
b e f g h
U
$\{i, j, k, \ldots, z\}$

EXERCISE C-3

1. $A \cup B = \{a, b, c, d, f, h\}$; $A \cap B = \{b, d\}$.
3. $A \cup B = \{2, 3, 4, 5, 6, 8\}$; $A \cap B = \{4, 6\}$.
5. $A \cup B = \{\text{Mon., Tues., Wed.}\}$; $A \cap B = \{\text{Mon., Tues., Wed.}\}$.
7. $A \cup B = \{\text{counting numbers}\}$; $A \cap B = \{\text{even counting numbers}\}$.
9. $A \cup B = \{1, 2, 3, \ldots, 12\}$; $A \cap B = \varnothing$.
11. $A' = \{3, 5\}$. 13. $A' = \{\text{odd integers}\}$.
15. $A' = \{a, b, c, d, e\}$. 21. 25.

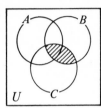

$B \cap C$

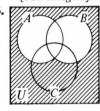

$(A \cup B)'$

REVIEW—UNIT C

1. $\{-2, -1\}$. 3. $\{11, 12, 13, \ldots\}$. 5. $\varnothing$. 7. False. 9. False. 11. False.
13. True. 15. True. 17. False. 19. True. 21. True.
23. $\{a, b, h, i, j, k\}$. 25. $\{a, b, c, d, e, f, h, i\}$. 27. $\{e, f\}$.
29. $\{c, d, e, f, h, i, j, k\}$. 31. $\{3, 5\}$. 33. $\{3, 5, 7, 9, 11, 13\}$. 35. $\{9, 11, 13\}$.
37. All women who have been to Europe, or Canada, or both.
39. All women who have not been to Canada.

EXERCISE D-1

1. $a^8 + 8a^7b + 28a^6b^2 + 56a^5b^3 + 70a^4b^4 + 56a^3b^5 + 28a^2b^6 + 8ab^7 + b^8$.
3. $x^5 - 10ax^4 + 40a^2x^3 - 80a^3x^2 + 80a^4x - 32a^5$.
5. $729x^6 - 2916x^5y + 4860x^4y^2 - 4320x^3y^3 + 2160x^2y^4 - 576xy^5 + 64y^6$.
7. $\frac{1}{32}x^5 - \frac{5}{48}x^4y + \frac{5}{36}x^3y^2 - \frac{5}{54}x^2y^3 + \frac{5}{162}xy^4 - \frac{1}{243}y^5$.
9. $1 + 0.12 + 0.0054 + 0.000108 + 0.00000081$.
11. $a\sqrt{a} - 3a\sqrt{b} + 3b\sqrt{a} - b\sqrt{b}$.
13. $a^8 - 32a^7b + 448a^6b^2 - 3584a^5b^3 + \cdots$.
15. $a^{10} - 40a^9b + 720a^8b^2 - 7680a^7b^3 + \cdots$.
17. $16{,}384x^{14} + 114{,}688x^{13} + 372{,}736x^{12} + 745{,}472x^{11} + \cdots$.
19. $729 - 5832\sqrt{3} + 64{,}152 - 142{,}560\sqrt{3} + \cdots$.
21. $1 + 0.3 + 0.0405 + 0.00324 + \cdots$.

EXERCISE D-2

1. $25{,}344a^7$. 3. $-414{,}720x^7$ 5. 0.0000567. 7. $-220a^3b^9$. 9. $2268a^6b^3$.

REVIEW—UNIT D

1. $a^9 + 9a^8b + 36a^7b^2 + 84a^6b^3 + 126a^5b^4 + 126a^4b^5 + 84a^3b^6 + 36a^2b^7 + 9ab^8 + b^9$.
3. $64x^6 - 576x^5y + 2160x^4y^2 - 4320x^3y^3 + 4860x^2y^4 - 2916xy^5 + 729y^6$.
5. $1 + 0.16 + 0.0096 + 0.000256 + 0.00000256 = 1.16985856$.
7. $a^9 - 27a^8b + 324a^7b^2 - 2268a^6b^3 + \cdots$.
9. $16{,}384x^{14} - 114{,}688x^{13} + 372{,}736x^{12} - 745{,}472x^{11} + \cdots$.
11. $1 + 0.2 + 0.018 + 0.00096 = 1.21896$.   **13.** $414720x^7$.   **15.** $220a^3b^9$.

EXERCISE E-1

1. 3.   **3.** -3.   **5.** 4.   **7.** $\frac{1}{3}$.   **9.** $-\frac{1}{3}$.   **11.** 9.   **13.** $1, \frac{1}{2}$.   **15.** $6, -1$.

EXERCISE E-2

1. 2.   **3.** 0.5.   **5.** 0.5821.   **7.** -6.3.   **9.** 1.650.   **11.** -9.320.   **13.** 6.689
15. 0.8692.   **17.** 1.108.   **19.** -10.32.

REVIEW—UNIT E

1. 5.   **3.** 3.   **5.** $-\frac{1}{5}$.   **7.** $1, -\frac{1}{2}$.   **9.** 2.   **11.** 0.6637.   **13.** -1.650.
15. -0.225.   **17.** 0.3021.

EXERCISE F-1

1. Mean: $9\frac{1}{17}$, Median: 8, Mode: 8.   **3.** Mn: $40\frac{1}{8}$, Mdn: 40, Mode: 37.
5. Mn: -4, Mdn: -3, Mode: -17.   **7.** Mn: 22, Mdn: 22, Mode: 23.
9. Median: 8 yr old.   **11.** Mean: 0.1235 g.

EXERCISE F-2

1. Range $= 22$, $AD = 4.73$, $\sigma = 6.3$.   **3.** Range $= 7$, $AD = 1.4$, $\sigma = 1.8$.
5. Range $= 30$, $AD = 9.2$, $\sigma = 10.4$.   **7.** (a) 21 to 23; (b) 20 to 24; (c) 19 to 25.

REVIEW—UNIT F

1. Mean: $13\frac{6}{17}$, Mdn: 13, Mode: 13.   **3.** Mean: -4.16, Mdn: -2, Mode: -13.
5. Mean: 0.1748 g.   **7.** Range $= 22$, $AD = 4.7$, $\sigma = 6.3$.   **9.** Range $= 7$,
$AD = 1.2$, $\sigma = 1.6$.

TABLES

Table I. Powers—Roots—Reciprocals

n	n^2	n^3	$\sqrt{n}$	$\sqrt[3]{n}$	$1/n$	n	n^2	n^3	$\sqrt{n}$	$\sqrt[3]{n}$	$1/n$
1	1	1	1.000	1.000	1.0000	51	2,601	132,651	7.141	3.708	.0196
2	4	8	1.414	1.260	.5000	52	2,704	140,608	7.211	3.733	.0192
3	9	27	1.732	1.442	.3333	53	2,809	148,877	7.280	3.756	.0189
4	16	64	2.000	1.587	.2500	54	2,916	157,464	7.348	3.780	.0185
5	25	125	2.236	1.710	.2000	55	3,025	166,375	7.416	3.803	.0182
6	36	216	2.449	1.817	.1667	56	3,136	175,616	7.483	3.826	.0179
7	49	343	2.646	1.913	.1429	57	3,249	185,193	7.550	3.849	.0175
8	64	512	2.828	2.000	.1250	58	3,364	195,112	7.616	3.871	.0172
9	81	729	3.000	2.080	.1111	59	3,481	205,379	7.681	3.893	.0169
10	100	1,000	3.162	2.154	.1000	**60**	3,600	216,000	7.746	3.915	.0167
11	121	1,331	3.317	2.224	.0909	61	3,721	226,981	7.810	3.936	.0164
12	144	1,728	3.464	2.289	.0833	62	3,844	238,328	7.874	3.958	.0161
13	169	2,197	3.606	2.351	.0769	63	3,969	250,047	7.937	3.979	.0159
14	196	2,744	3.742	2.410	.0714	64	4,096	262,144	8.000	4.000	.0156
15	225	3,375	3.873	2.466	.0667	65	4,225	274,625	8.062	4.021	.0154
16	256	4,096	4.000	2.520	.0625	66	4,356	287,496	8.124	4.041	.0152
17	289	4,913	4.123	2.571	.0588	67	4,489	300,763	8.185	4.062	.0149
18	324	5,832	4.243	2.621	.0556	68	4,624	314,432	8.246	4.082	.0147
19	361	6,859	4.359	2.668	.0526	69	4,761	328,509	8.307	4.102	.0145
20	400	8,000	4.472	2.714	.0500	**70**	4,900	343,000	8.367	4.121	.0143
21	441	9,261	4.583	2.759	.0476	71	5,041	357,911	8.426	4.141	.0141
22	484	10,648	4.690	2.802	.0455	72	5,184	373,248	8.485	4.160	.0139
23	529	12,167	4.796	2.844	.0435	73	5,329	389,017	8.544	4.179	.0137
24	576	13,824	4.899	2.884	.0417	74	5,476	405,224	8.602	4.198	.0135
25	625	15,625	5.000	2.924	.0400	75	5,625	421,875	8.660	4.217	.0133
26	676	17,576	5.099	2.962	.0385	76	5,776	438,976	8.718	4.236	.0132
27	729	19,683	5.196	3.000	.0370	77	5,929	456,533	8.775	4.254	.0130
28	784	21,952	5.292	3.037	.0357	78	6,084	474,552	8.832	4.273	.0128
29	841	24,389	5.385	3.072	.0345	79	6,241	493,039	8.888	4.291	.0127
30	900	27,000	5.477	3.107	.0333	**80**	6,400	512,000	8.944	4.309	.0125
31	961	29,791	5.568	3.141	.0323	81	6,561	531,441	9.000	4.327	.0123
32	1,024	32,768	5.657	3.175	.0312	82	6,724	551,368	9.055	4.344	.0122
33	1,089	35,937	5.745	3.208	.0303	83	6,889	571,787	9.110	4.362	.0120
34	1,156	39,304	5.831	3.240	.0294	84	7,056	592,704	9.165	4.380	.0119
35	1,225	42,875	5.916	3.271	.0286	85	7,225	614,125	9.220	4.397	.0118
36	1,296	46,656	6.000	3.302	.0278	86	7,396	636,056	9.274	4.414	.0116
37	1,369	50,653	6.083	3.332	.0270	87	7,569	658,503	9.327	4.431	.0115
38	1,444	54,872	6.164	3.362	.0263	88	7,744	681,472	9.381	4.448	.0114
39	1,521	59,319	6.245	3.391	.0256	89	7,921	704,969	9.434	4.465	.0112
40	1,600	64,000	6.325	3.420	.0250	**90**	8,100	729,000	9.487	4.481	.0111
41	1,681	68,921	6.403	3.448	.0244	91	8,281	753,571	9.539	4.498	.0110
42	1,764	74,088	6.481	3.476	.0238	92	8,464	778,688	9.592	4.514	.0109
43	1,849	79,507	6.557	3.503	.0233	93	8,649	804,357	9.644	4.531	.0108
44	1,936	85,184	6.633	3.530	.0227	94	8,836	830,584	9.695	4.547	.0106
45	2,025	91,125	6.708	3.557	.0222	95	9,025	857,375	9.747	4.563	.0105
46	2,116	97,336	6.782	3.583	.0217	96	9,216	884,736	9.798	4.579	.0104
47	2,209	103,823	6.856	3.609	.0213	97	9,409	912,673	9.849	4.595	.0103
48	2,304	110,592	6.928	3.634	.0208	98	9,604	941,192	9.899	4.610	.0102
49	2,401	117,649	7.000	3.659	.0204	99	9,801	970,299	9.950	4.626	.0101
50	2,500	125,000	7.071	3.684	.0200	**100**	10,000	1,000,000	10,000	4.642	.0100

Table II. Mantissas for Common Logarithms

N	0	1	2	3	4	5	6	7	8	9
10	0 000	430	086	128	170	212	253	294	334	374
11	414	453	492	531	569	607	645	682	719	755
12	0 792	828	864	899	934	969	*004	*038	*072	*106
13	1 139	173	206	239	271	303	335	367	399	430
14	461	492	523	553	584	614	644	673	703	732
15	1 761	790	818	847	875	903	931	959	987	*014
16	2 041	068	095	122	148	175	201	227	253	279
17	304	330	355	380	405	430	455	480	504	529
18	553	577	601	625	648	672	695	718	742	765
19	2 788	810	833	856	878	900	923	945	967	989
20	3 010	032	054	075	096	118	139	160	181	201
21	222	243	263	284	304	324	345	365	385	404
22	424	444	464	483	502	522	541	560	579	598
23	617	636	655	674	692	711	729	747	766	784
24	802	820	838	856	874	892	909	927	945	962
25	3 979	997	*014	*031	*048	*065	*082	*099	*116	*133
26	4 150	166	183	200	216	232	249	265	281	298
27	314	330	346	362	378	393	409	425	440	456
28	472	487	502	518	533	548	564	579	594	609
29	624	639	654	669	683	698	713	728	742	757
30	4 771	786	800	814	829	843	857	871	886	900
31	4 914	928	942	955	969	983	997	*011	*024	*038
32	5 051	065	079	092	105	119	132	145	159	172
33	185	198	211	224	237	250	263	276	289	302
34	315	328	340	353	366	378	391	403	416	428
35	441	453	465	478	490	502	514	527	539	551
36	563	575	587	599	611	623	635	647	658	670
37	682	694	705	717	729	740	752	763	775	786
38	798	809	821	832	843	855	866	877	888	899
39	5 911	922	933	944	955	966	977	988	999	*010
40	6 021	031	042	053	064	075	085	096	107	117
41	128	138	149	160	170	180	191	201	212	222
42	232	243	253	263	274	284	294	304	314	325
43	335	345	355	365	375	385	395	405	415	425
44	435	444	454	464	474	484	493	503	513	522
45	532	542	551	561	571	580	590	599	609	618
46	628	637	646	656	665	675	684	693	702	712
47	721	730	739	749	758	767	776	785	794	803
48	812	821	830	839	848	857	866	875	884	893
49	902	911	920	928	937	946	955	964	972	981
50	6 990	998	*007	*016	*024	*033	*042	*050	*059	*067
51	7 076	084	093	101	110	118	126	135	143	152
52	160	168	177	185	193	202	210	218	226	235
53	243	251	259	267	275	284	292	300	308	316
54	7 324	332	340	348	356	364	372	380	388	396
N	0	1	2	3	4	5	6	7	8	9

Proportional Parts

	43	42	41	40
1	4.3	4.2	4.1	4
2	8.6	8.4	8.2	8
3	12.9	12.6	12.3	12
4	17.2	16.8	16.4	16
5	21.5	21.0	20.5	20
6	25.8	25.2	24.6	24
7	30.1	29.4	28.7	28
8	34.4	33.6	32.8	32
9	38.7	37.8	36.9	36

	39	38	37	36
1	3.9	3.8	3.7	3.6
2	7.8	7.6	7.4	7.2
3	11.7	11.4	11.1	10.8
4	15.6	15.2	14.8	14.4
5	19.5	19.0	18.5	18.0
6	23.4	22.8	22.2	21.6
7	27.3	26.6	25.9	25.2
8	31.2	30.4	29.6	28.8
9	35.1	34.2	33.3	32.4

	35	34	33	32
1	3.5	3.4	3.3	3.2
2	7.0	6.8	6.6	6.4
3	10.5	10.2	9.9	9.6
4	14.0	13.6	13.2	12.8
5	17.5	17.0	16.5	16.0
6	21.0	20.4	19.8	19.2
7	24.5	23.8	23.1	22.4
8	28.0	27.2	26.4	25.6
9	31.5	30.6	29.7	28.8

	31	30	29	28
1	3.1	3	2.9	2.8
2	6.2	6	5.8	5.6
3	9.3	9	8.7	8.4
4	12.4	12	11.6	11.2
5	15.5	15	14.5	14.0
6	18.6	18	17.4	16.8
7	21.7	21	20.3	19.6
8	24.8	24	23.2	22.4
9	27.9	27	26.1	25.2

	27	26	25	24
1	2.7	2.6	2.5	2.4
2	5.4	5.2	5.0	4.8
3	8.1	7.8	7.5	7.2
4	10.8	10.4	10.0	9.6
5	13.5	13.0	12.5	12.0
6	16.2	15.6	15.0	14.4
7	18.9	18.2	17.5	16.8
8	21.6	20.8	20.0	19.2
9	24.3	23.4	22.5	21.6

Proportional Parts

Table II. Mantissas for Common Logarithms (continued)

N	0	1	2	3	4	5	6	7	8	9
55	7 404	412	419	427	435	443	451	459	466	474
56	482	490	497	505	513	520	528	536	543	551
57	559	566	574	582	589	597	604	612	619	627
58	634	642	649	657	664	672	679	686	694	701
59	709	716	723	731	738	745	752	760	767	774
60	7 782	789	796	803	810	818	825	832	839	846
61	853	860	868	875	882	889	896	903	910	917
62	924	931	938	945	952	959	966	973	980	987
63	7 993	*000	*007	*014	*021	*028	*035	*041	*048	*055
64	8 062	069	075	082	089	096	102	109	116	122
65	129	136	142	149	156	162	169	176	182	189
66	195	202	209	215	222	228	235	241	248	254
67	261	267	274	280	287	293	299	306	312	319
68	325	331	338	344	351	357	363	370	376	382
69	388	395	401	407	414	420	426	432	439	445
70	8 451	457	463	470	476	482	488	494	500	506
71	513	519	525	531	537	543	549	555	561	567
72	573	579	585	591	597	603	609	615	621	627
73	633	639	645	651	657	663	669	675	681	686
74	692	698	704	710	716	722	727	733	739	745
75	751	756	762	768	774	779	785	791	797	802
76	808	814	820	825	831	837	842	848	854	859
77	865	871	876	882	887	893	899	904	910	915
78	921	927	932	938	943	949	954	960	965	971
79	8 976	982	987	993	998	*004	*009	*015	*020	*025
80	9 031	036	042	047	053	058	063	069	074	079
81	085	090	096	101	106	112	117	122	128	133
82	138	143	149	154	159	165	170	175	180	186
83	191	196	201	206	212	217	222	227	232	238
84	243	248	253	258	263	269	274	279	284	289
85	294	299	304	309	315	320	325	330	335	340
86	345	350	355	360	365	370	375	380	385	390
87	395	400	405	410	415	420	425	430	435	440
88	445	450	455	460	465	469	474	479	484	489
89	494	499	504	509	513	518	523	528	533	538
90	9 542	547	552	557	562	566	571	576	581	586
91	590	595	600	605	609	614	619	624	628	633
92	638	643	647	652	657	661	666	671	675	680
93	685	689	694	699	703	708	713	717	722	727
94	731	736	741	745	750	754	759	763	768	773
95	777	782	786	791	795	800	805	809	814	818
96	823	827	832	836	841	845	850	854	859	863
97	868	872	877	881	886	890	894	899	903	908
98	912	917	921	926	930	934	939	943	948	952
99	9 956	961	965	969	974	978	983	987	991	996
N	0	1	2	3	4	5	6	7	8	9

Proportional Parts

	23	22	21	20
1	2.3	2.2	2.1	2
2	4.6	4.4	4.2	4
3	6.9	6.6	6.3	6
4	9.2	8.8	8.4	8
5	11.5	11.0	10.5	10
6	13.8	13.2	12.6	12
7	16.1	15.4	14.7	14
8	18.4	17.6	16.8	16
9	20.7	19.8	18.9	18

	19	18	17	16
1	1.9	1.8	1.7	1.6
2	3.8	3.6	3.4	3.2
3	5.7	5.4	5.1	4.8
4	7.6	7.2	6.8	6.4
5	9.5	9.0	8.5	8.0
6	11.4	10.8	10.2	9.6
7	13.3	12.6	11.9	11.2
8	15.2	14.4	13.6	12.8
9	17.1	16.2	15.3	14.4

	15	14	13	12
1	1.5	1.4	1.3	1.2
2	3.0	2.8	2.6	2.4
3	4.5	4.2	3.9	3.6
4	6.0	5.6	5.2	4.8
5	7.5	7.0	6.5	6.0
6	9.0	8.4	7.8	7.2
7	10.5	9.8	9.1	8.4
8	12.0	11.2	10.4	9.6
9	13.5	12.6	11.7	10.8

	11	10	9	8
1	1.1	1	0.9	0.8
2	2.2	2	1.8	1.6
3	3.3	3	2.7	2.4
4	4.4	4	3.6	3.2
5	5.5	5	4.5	4.0
6	6.6	6	5.4	4.8
7	7.7	7	6.3	5.6
8	8.8	8	7.2	6.4
9	9.9	9	8.1	7.2

	7	6	5	4
1	0.7	0.6	0.5	0.4
2	1.4	1.2	1.0	0.8
3	2.1	1.8	1.5	1.2
4	2.8	2.4	2.0	1.6
5	3.5	3.0	2.5	2.0
6	4.2	3.6	3.0	2.4
7	4.9	4.2	3.5	2.8
8	5.6	4.8	4.0	3.2
9	6.3	5.4	4.5	3.6

Proportional Parts

INDEX